Springer Series in Statistics

Advisors:
D. Brillinger, S. Fienberg, J. Gani,
J. Hartigan, K. Krickeberg

Springer Series in Statistics

K. Dzhaparidze

Parameter Estimation
and Hypothesis Testing
in Spectral Analysis
of Stationary Time Series

Translated by Samuel Kotz

Springer-Verlag
New York Berlin Heidelberg Tokyo

K. Dzhaparidze
Mathematisch Centrum
Kruislaan 413
Postbus 4079
1098 SJ Amsterdam
The Netherlands

Samuel Kotz *(Translator)*
Department of Management Science
 and Statistics
University of Maryland
College Park, Maryland 20742
U.S.A.

AMS Classification: 62M10, 62F99

Library of Congress Cataloging-in-Publication Data
Dzhaparidze, K. O.
  Parameter estimation and hypothesis testing in
spectral analysis of stationary time series.
  (Springer series in statistics)
  Translation of: Asimptoticheski éffektivnoe
ofsenivanie parametrov spektra gaussovskogo vremennogo
rîada.
  Bibliography: p.
  Includes index.
  1. Time-series analysis. 2. Spectral theory
(Mathematics) 3. Parameter estimation. 4. Statistical
hypothesis testing. I. Title. II. Series.
QA280.D9313   1985   519.5'5   85-22207

The original Russian edition was published by the Publishing House of the University of Tiblissi in 1981 *"Schätzung von Parametern und Prüfung von Hypothesen in der Spektralanalyse von stationären vorläufigen Reihen"*.

9 8 7 6 5 4 3 2 1

ISBN-13:978-1-4612-9325-5   e-ISBN-13:978-1-4612-4842-2
DOI: 10.1007/978-1-4612-4842-2

CONTENTS

INTRODUCTION

1. Traditionally the most important problem of mathematical statistics dealing with random stationary processes X_t, $t = ...,$ $-1,0,1, ...$ is the problem of estimating the second order characteristics, namely the covariance function

$$\beta(\tau) = E\{[X_t - E(X_t)][X_{t+\tau} - E(X_{t+\tau})]\},$$

or its Fourier transform -- the spectral density $f = f(\lambda)$ (under the assumption that the spectral density exists). For this reason, a vast amount of periodical and monographic literature is devoted to the nonparametric statistical problem of estimating the function $\beta(\tau)$ and especially that of $f(\lambda)$ (see, for example, the books [4,21,22,26,56,77,137,139,140,]). However, the empirical value f_n^* of the spectral density f obtained by applying a certain statistical procedure to the observed values of the variables $X_1, ... , X_n$, usually depends in a complicated manner on the cyclic frequency λ. This fact often presents difficulties in applying the obtained estimate f_n^* of the function f to the solution of specific problems related to the process X_t. Therefore, in practice, the obtained values of the estimator f_n^* (or an estimator of the covariance function $\beta_n^*(\tau)$) are almost always "smoothed," i.e., are approximated by values of a certain sufficiently simple function $\tilde{f} = \tilde{f}(\lambda)$ (or $\tilde{\beta}(\tau)$) of argument λ (or of τ) which is then taken as the actual spectral density (or covariance function). Quite often the functions $\tilde{f}$ or $\tilde{\beta}(\tau)$ -- used as approximations -- are chosen in a manner[1] such that they are defined by an analytic formula involving a finite number of unknown parameters; estimation of these parameters is a problem of parametric statistics.   In this approach the nonparametric problem of estimating the unknown function f (or $\beta(\tau)$) plays only an auxiliary role: its solution is utilized solely for choosing a reasonable parametric problem -- under given conditions.   This parametric problem consists of stipulating a hypothesis on the form of function f (or $\beta(\tau)$), testing this hypothesis, and next, estimating the unknown parameters appearing in the expression of the function corresponding to the accepted hypothesis.   This fact is of crucial importance in real practical situations of estimating the parameters of a spectral density (or a covariance function) of a random process[2] and testing the hypothesis about the form of the spectral density. A basic portion of this monograph deals with an investigation of these problems. Chapters II and III are devoted to the problem of determining estimators of a finite number of unknown spectral parameters while Chapters IV and V are concerned with the construction of various criteria for testing hypotheses concerning the form of the spectral density. (Below we shall discuss this problem in somewhat more detail.)

We now note that although the above mentioned problems are perfectly sensible for a fixed (finite) sample size n and in applications one will always be dealing with a finite n, the mathematical results dealing with the case of fixed and not too large n are very few and usually of little interest.   This is

---

[1]The choice of f as a rational function of $\exp(i\lambda)$ is especially handy since solutions of many problems for random processes with rational spectral densities were studied in detail (see, for example [58, 150, 151]).

[2]Recently the practical importance of this problem has become more evident for many investigators.   One now encounters more often in the applied literature that parametric estimators of spectral density are being utilized from the very beginning (instead of nonparametric estimators f_n^*); for example, the so-called "autoregressive estimators" or similar to them, the "maximum entropy estimators" are based on the assumption that $f(\lambda)$ are of a certain special form (cf., e.g., [22, 34, 77, 115, 126]).

due to the fact that a detailed investigation of the properties of statistical procedures for a given finite n are always very cumbersome and hardly ever lead to explicit and visible finite results; these investigations can thus be considered unpromising.

An asymptotic analysis of statistical inference valid in the limit as $n \to \infty$ is a more satisfying venture. The problem of the limiting behavior of statistical procedures as $n \to \infty$ are as a rule mathematically much more tractable than the corresponding finite sample investigations. At the same time, this appears to be of sufficient interest from the point of view of applications as well, since for the values of n which are not too large, the desired statistical inference (in particular, the inference concerning the spectral density f) will usually be quite inaccurate and is thus often considered practically useless, while for a fixed large n the results obtained are usually close to those obtained for $n \to \infty$. This is the reason that in the work below, we shall be mainly concerned only with the asymptotic results corresponding to $n \to \infty$.

For simplicity, we shall always assume that the process X_t has a zero expected value $E(X_t) = 0$. This assumption usually does not result in a loss of generality since even if $E(X_t)$ is unknown, in applications of asymptotic results valid for $n \to \infty$, it is usually sufficient to replace X_t by $X_t - (1/n)\sum_{j=1}^{n} X_j$ in order to assume that the process under consideration possesses a zero mean value.

We now proceed to a brief description of the content of this monograph.

Properties of the Likelihood Function for Gaussian Processes

2. Assume that X_t is a real-valued stationary random process with discrete time $t = ..., -1, 0, 1, ...$, zero expected value $E(X_t) = 0$, and covariance function $\beta(\tau)$ sufficiently rapidly decreasing at infinity and represented as

$$E(X_t X_{t+\tau}) = \beta(\tau) = \int_{-\pi}^{\pi} e^{i\lambda\tau} f(\lambda) d\lambda.$$

When investigating mathematically the problems of statistical inference concerning a spectral density f of a stationary process X_t one should clearly first impose certain assumptions on the corresponding probability distributions which will specify the problem and permit a comparison of various possible recommendations. Here it is natural to start with the case of Gaussian processes which represent the simplest and most important class of random functions, those most widely studied, and often occurring in applications.[3]

Assuming that X_t is a Gaussian process, we can in principle obtain an explicit expression for the logarithm of the likelihood function

(1)
$$L_n = \log p_n(X_1, ..., X_n)$$
$$= -\frac{1}{2}\{n \log 2\pi + \log \det(B_f) + \mathbf{X}'B_f^{-1}\mathbf{X}\}$$

of random variables $X_1, ... , X_n$ (which are components of the random column-vector $\mathbf{X}$), where p_n is an n-dimensional probability density and $B_f = [\beta(\tau-s)]$, τ, $s = 1, ..., n$, is a Toeplitz matrix associated with the function f. For this purpose one is required, however, to solve the complicated problem of determining the explicit expressions for $\det(B_f)$ and B_f^{-1} (particular cases of this are dealt with in [3,87,93,113, 133, and 149]). As it will be seen in Example 1 in Section 1 of Chapter I, even in the simplest case, when X_t is an autoregressive process, an explicit expression for L_n turns out to be quite involved even for autoregressive processes of the first order. Moreover, as the order of autoregression increases, the formula for L_n becomes more complicated (cf. [101,111,118]). Formulas contained in papers [3,93,113,133, and 149] allow us, in principle, to obtain an explicit expression for L_n also in the case when X_t is a moving average process or even a mixed autoregressive-moving average process; however, the formulas involved are unavoidably, very cumbersome. This is easily seen by considering the formula (1.1.16) -- (and the results of the Examples 2-4 in Section 1 of Chapter I following from it). This formula is an expression (actually simpler than it was

------

[3]We also note that the experience accumulated in the course of study of many other statistical problems (including random processes problems) gives us the confidence to suppose that methods which are of high accuracy in the Gaussian case will also be useful when applied to numerous non-Gaussian probability distributions.

envisioned previously) for L_n applicable to a class of processes which include all the mixed autoregressive-moving average processes.

Taking into account the complexity of obtaining a suitable statistical inference concerning the Gaussian process X_t based solely on utilization of the expression for L_n, Mann and Wald [86] in 1943 (as applied to autoregressive processes) and Whittle [121] in the early fifties (as applied to general processes with a positive spectral density $f > 0$) suggested to utilize, instead of the exact expression for L_n, its "principle part" $\tilde{L}_n$ which satisfies the condition

(2) $\qquad n^{-1/2}(L_n - \tilde{L}_n) \to 0, \quad \text{as} \quad n \to \infty$

(in the sense of convergence in probability). As it was expected, it turns out that in the most important case, when the sample size n substantially exceeds the typical time of damping the correlations between the variables X_t, the statistical inference obtained in this manner possesses the same asymptotic properties as the inference based on the utilization of the exact value of L_n. Moreover, and this is most important, it turns out that under very general conditions $\tilde{L}_n$ can be chosen to be of much simpler form than L_n.

Indeed, if the Fourier coefficients $\beta(\tau)$ of a positive spectral density $f > 0$ of a Gaussian random process X_t satisfy the condition

(3) $\qquad \sum_{\tau=1}^{\infty} \tau |\beta(\tau)|^2 < \infty$

(as it is demonstrated in Chapter I, cf. Theorem 2.1), L_n can be expressed[4] as

(4) $\qquad \tilde{L}_n = -\dfrac{n}{2} \left\{ \log 2\pi + \dfrac{1}{2\pi}\int_{-\pi}^{\pi} \log[2\pi f(\lambda)]d\lambda \right.$
$\qquad\qquad \left. + \dfrac{1}{2\pi}\int_{-\pi}^{\pi} \dfrac{I_n(\lambda)}{f(\lambda)}d\lambda \right\},$

---

[4]Formula (4) for $\tilde{L}_n$ was first derived by Whittle [121] who obtained it based on purely heuristic arguments and did not indicate any exact conditions for the validity of (2); different (and narrower) exact conditions for the validity of fomulas (2) and (4) may be also found in [103].

where $I_n(\lambda)$ is the periodigram of the process X_t.

Assume now that for some distinct fixed q values of λ_1, ..., λ_q of a circular frequency λ the spectral density f vanishes. Namely, let

(5) $$f(\lambda) = f_q(\lambda) = f_0(\lambda)|(z-z_1)\cdots(z-z_q)|^2,$$

where $z = e^{i\lambda}$, $z_j = e^{i\lambda_j}$, and f_0 is a bounded positive integrable function satisfying Assumption 1 of Section 2 in Chapter I. In this case, Whittle's formula becomes meaningless since the last integral in the right-hand-side of the equation diverges. The inapplicability of Whittle's formula in the case of vanishing spectral density cannot be considered puzzling: it is known that in such a case many statistical procedures which are optimal for random processes with an everywhere positive spectral density become nonoptimal (and often even possess zero efficiency); in this case so often quite different and much more complex procedures turn out to be optimal (cf., e.g., the papers [1,107], and [161, pp. 238-244], devoted to linear statistical problems of estimating an unknown mean value or regression coefficients of Gaussian processes). The situation in the case under consideration of an essentially nonlinear problem of determining the principal part of the likelihood function is analogous to the above. As it will be shown in Chapter I (cf. Corollary I.2.1) if the spectral density is of the form (5) then $\tilde{L}_n$ can be chosen as

$$\tilde{L}_n = -\frac{n}{2}\left\{\log 2\pi + \frac{1}{2\pi}\int_{-\pi}^{\pi}\log[2\pi f_0(\lambda)]d\lambda\right.$$

(6)

$$\left. + \frac{1}{2\pi}\int_{-\pi}^{\pi}\frac{I_n(\lambda,\tilde{Y})}{f_0(\lambda)}\,d\lambda\right\},$$

where $I_n(\lambda,\tilde{Y})$ is the periodogram of the process constructed from quantities $(\tilde{Y}_1, ... , \tilde{Y}_n) = \tilde{Y}$ related to vector Y by the relation $\tilde{Y} = (I_n - V(V^*V)^{-1}V^*)Y$. The components of vector Y are:

$$Y_t = \sum_{j_1=1}^{t}\sum_{j_2=1}^{j_1}\cdots\sum_{j_q=1}^{j_{q-1}}\exp\{i[(t-j_1)\lambda_1 + (j_1-j_2)\lambda_2$$

(7)

$$+... + (j_{q-1}-j_q)\lambda_q]\}\,X_{j_q}, \quad t = 1, ..., n$$

(cf. formula (I.2.31)). Here V is a matrix whose columns are: $col(z_j, ..., z_j^n)$, $j = 1, ..., q$. Formula (6) appears to be more complicated than Whittle's formula (4). Actually, as it will be seen from the above, it is also quite suitable for solving specific statistical problems concerning the process X_t (cf. the examples on pages 58 and 59, for instance). In particular, in the simplest case,[5] when $q = 1$ and $\lambda_1 = 0$, $Y_k = X_1 + X_2 + \cdots + X_k$, $\tilde{Y}_k = Y_k - (1/n)\sum_{t=1}^{n} Y_t$, and

$$(8) \qquad \tilde{L}_n = -\frac{n}{4\pi} \left\{ \int_{-\pi}^{\pi} \left[\log(4\pi^2 f_0(\lambda)) + \frac{I_n(\lambda, \tilde{Y})}{f_0(\lambda)} \right] d\lambda \right\}.$$

3. In Section 3 of Chapter I we shall again be considering the case when X_t is a Gaussian process with the spectral density f satisfying the condition $m \leqslant f \leqslant M$, where m and M are positive numbers.

Let

$$(9) \qquad g_n(\lambda) = f(\lambda)(1 + n^{-1/2}a_n(\lambda)), \quad n = 1, 2, ... ,$$

be a sequence of spectral densities where the sequence of square integrable functions $a_n(\lambda)$, $n = 1, 2, ...$, converges in the mean square as $n \to \infty$ to a square integrable function a on $[-\pi, \pi]$. The class of such functions g_n (or more often the subclass of functions of the form $g = f(1 + n^{-1/2}a)$, where $a(\lambda)$ is a fixed function) is usually used in the theory of statistical hypothesis testing as an alternative to the hypothesis that the spectral density equals f. A study of asymptotic power (as $n \to \infty$) of various tests for testing this hypothesis can be carried out based on the contiguity of sequences of Gaussian measures $P_n(f)$, $n = 1, 2, ...$, and $P_n(g)$, $n = 1, 2, ...$, corresponding to the spectral densities f and g respectively (cf. [110]). The validity of contiguity in this case is the subject of Theorem I.3.1.

---

[5]Note that if Y_t is a Gaussian process with stationary increments and a strictly positive spectral density f_0, then the spectral density of a stationary process $X_t = Y_{t+1} - Y_t$ which is equal to $f_0|z-1|^2$ vanishes for $\lambda = 0$; therefore, the expression for the "principal part" of the logarithm of the likelihood ratio $\tilde{L}_n$ of the process Y_t coincides with the expression (8) (cf. [55]). Note that dealing with the simplest case of $q = 1$ Pham Dinh [202] comes to a different type of approximation.

Next we prove in Section 3 that under the general conditions stated above, the logarithm of the likelihood ratio $\Lambda(f_1,f_2) = \log(dP_n(f_2)/dP_n(f_1))$ satisfies the following asymptotic relations

(10) $\Lambda(f,g_n) - \Lambda(f,g) \to 0$

and

$$\Lambda(f,g) - \frac{1}{2}\left\{n^{-1/2}[\mathbf{X}'B_f^{-1}B_{a/2\pi}\mathbf{X} - \text{tr}(B_{a/2\pi})]\right.$$

(11)

$$\left. - \frac{1}{4\pi}\int_{-\pi}^{\pi}a^2(\lambda)d\lambda\right\} \to 0$$

in $P_n(f)$ probability as $n \to \infty$ (cf. Theorems I.3.2 and I.3.3). These relations, and especially the second one, play an important role in obtaining asymptotic statistical inference about a Gaussian process X_t. Therefore, from an applications aspect, it is quite essential that under the general conditions indicated in Lemma I.3.1 on page 63, formula (11) can be substantially simplified utilizing the fact that in this case

$$\left|n^{-1/2}\mathbf{X}'B_f^{-1}B_a\mathbf{X} - \frac{n^{1/2}}{2\pi}\int_{-\pi}^{\pi}\frac{I_n(\lambda)}{f(\lambda)}a(\lambda)d\lambda\right| \to 0$$

in $P_n(f)$ as $n \to \infty$ probability (cf. Lemma I.3.1).

An especially important particular case considered in this subsection of the introduction is the case when the spectral density f depends on the vector-valued parameter θ belonging to an open set Θ of the R_p space, i.e., $f = f_\theta$, $\theta \in \Theta$, and the derivatives (viewed as limits in $L_2[-\pi,\pi]$) $\dot{\phi}_{k,\theta} = \dot{\phi}_{k,\theta}(\lambda)$ of the function $\log f_\theta$ with respect to the k-th entry of the vector θ exist (cf. (I.3.17)). Introducing the notation $a_n = n^{1/2}[f_{\theta+n^{-1/2}\mathbf{h}_n} - f_\theta]/f_\theta$, where $\mathbf{h}_n$, $n = 1,2, ...$, is a sequence of

p-dimensional vectors converging to a p-dimensional vector $\mathbf{h}$ (and here $\theta+n^{-1/2}\mathbf{h}_n \in \Theta$), we obtain that $f_n = f_{\theta+n^{-1/2}\mathbf{h}_n}$ if f

is the density f_θ and a is defined by the equality $a = \mathbf{h}'\dot{\phi}_\theta$ where $\dot{\phi}_\theta = (\dot{\phi}_{1,\theta}, ..., \dot{\phi}_{p,\theta})$ (cf. (I.3.19)).

In this particular case the results stated in this subsection can be formulated as the following three assertions (cf. Corollary I.3.1).

1) Sequences of Gaussian measures $P_n(f_\Theta)$, $n = 1,2,...$, and $P_n(f_{\Theta+n^{-1/2}h})$, $n = 1,2, ...$ are contiguous.

2) $\Lambda(f_\Theta, f_{\Theta+n^{-1/2}h}) - \Lambda(f_\Theta, f_{\Theta+n^{-1/2}h_n}) \to 0$

    in $P_n(f_\Theta)$ probability as $n \to \infty$.

3) $\Lambda(f_\Theta, f_{\Theta+n^{-1/2}h}) - \dfrac{n^{1/2}}{4\pi} \displaystyle\int_{-\pi}^{\pi} \dfrac{I_n(\lambda)-f_\Theta(\lambda)}{f_\Theta(\lambda)} \, \mathbf{h}' \dot{\phi}_\Theta(\lambda)d\lambda$

    $+ \dfrac{1}{8\pi} \displaystyle\int_{-\pi}^{\pi} [\mathbf{h}' \dot{\phi}_\Theta(\lambda)]^2 d\lambda \to 0$

    in $P_n(f_\Theta)$ probability as $n \to \infty$.

The last assertion is evidently valid under the additional conditions of Lemma I.3.1. If, moreover, the spectral density f_Θ is a continuous function of $\Theta \in \Theta$ then as it is shown at the end of Section 3 of Chapter I, the properties 1)-3) easily imply that the family of distributions $P_n(f_\Theta)$, $\Theta \in \Theta$ is asymptotically differentiable in LeCam's sense [80-82] (cf. conditions (D1)-(D4) below).

Assertions of the type 1)-3) under different (more specialized) conditions were proved by Davies [61] who also discussed the problem of applying these results to the problems of statistical inference concerning a Gaussian random process X_t. We shall yet return to these discussions.

Now we proceed to a substantially less studied case when the spectral density f vanishes for some values of $\lambda = \lambda_j$, $j = 1, ..., q$, i.e., it is of the form (5).

It turns out that in this case the assertion about the contiguity of the sequences of Gaussian measures $P_n(f)$, $n = 1,2, ...$ and $P_n(g)$, $n = 1,2, ...$ is valid; so are the relations (10) and (11) above (cf. Assertions 1)-3) of Theorem I.4.1).

Under certain additional conditions also included in the statement of Theorem I.4.1 the following relation is valid:

$$\left| n^{-1/2} \mathbf{X}' B_f^{-1} B_{a/2\pi} \mathbf{X} - \frac{n^{1/2}}{2\pi} \int_{-\pi}^{\pi} [\, I_n(\lambda, \tilde{Y}) - f_0(\lambda)] \, r(\lambda) d\lambda \right| \to 0$$

in $P_n(f)$ probability as $n \to \infty$, where $r = a/f_0$ and $I_n(\lambda, Y)$ is a periodogram of the process Y_t (cf. formulae (6) and (7) above).

Assume now that the function f_0 in formula (5) depends on a vector-valued parameter $\theta \in \Theta$ and once more denote by $\dot{\phi}_\theta$ the vector of derivatives of the function $\log f_\theta$ in the L_2 sense. It is proved in Section 4 of Chapter I (cf. Corollary 1) that in this case the assertions 1) and 2) of this subsection are valid and that assertion 3) is replaced here by

3')
$$\Lambda(f_\theta, f_{\theta+n^{-1/2}h}) - \frac{n^{1/2}}{4\pi} \int_{-\pi}^{\pi} \frac{I_n(\lambda,Y)-f_{0,\theta}(\lambda)}{f_{0,\theta}(\lambda)} \, h'\dot{\phi}_\theta(\lambda)d\lambda$$

$$+ \frac{1}{8\pi} \int_{-\pi}^{\pi} [h'\dot{\phi}_\theta(\lambda)]^2 d\lambda \to 0$$

in $P_n(f)$ probability as $n \to \infty$.

Finally, we note that since in the course of the proofs of the assertions presented in Chapter I, it is often necessary to utilize cumbersome algebraic calculations, part of these calculations is relegated to Appendices 1 and 2 at the end of the chapter.

Estimation of Parameters of Spectral Density

4. In Chapters II-IV the above stated results are utilized to obtain statistical inference about a stationary process X_t. In particular, in Chapters II and III, the problem of estimating an unknown vector-valued parameter $\theta \in \Theta$ appearing in the expression for the spectral density $f = f_\theta$ of the process X_t is considered. The main attention in these chapters is devoted to the study of statistical properties of the estimators obtained for the parameter θ (these are particular measurable functions of the random variables $X_1, ..., X_n$).

The maximum likelihood method is traditionally considered the most important general method -- from the theoretical point of view -- for the derivation of statistical estimators of unknown parameters. In the case of a Gaussian process X_t a maximum likelihood estimator $\bar{\theta}$ is the value of parameter θ which maximizes the expression (1) of the logarithm of the likelihood function $L_n = L_{n,\theta}$. In a regular case this estimator possesses all those asymptotic optimal properties which are enjoyed by maximum likelihood estimators in the case of independent observations -- namely, $\bar{\theta}$ is a consistent, asymptotically normal, and asymptotically efficient estimator of a vector-valued parameter θ (cf. [70]).

Unfortunately, as it was already mentioned above, an explicit derivation of for a maximum likelihood estimator $\bar{\theta}$ is most often a practically insoluble (or, at least, highly cumbersome and complex) problem.

However, since the main virtue of the estimator $\bar{\theta}$ is its optimality in the limit as $n \to \infty$ it seems plausible that this valuable property will remain intact if $\bar{\theta}$ is altered but in such a manner that the alteration will be insignificant for very large values of n. Therefore a natural possibility to overcome the above-mentioned difficulty seems to be to use instead of the whole expression $L_{n,\theta}$ its "principal part" $\tilde{L}_n = \tilde{L}_{n,\theta}$, which is substantially simpler, i.e., the estimator $\bar{\theta}$ is replaced by the estimator $\tilde{\theta}$ which maximizes the quantity $\tilde{L}_{n,\theta}$.

It was Whittle who, in 1953, introduced the idea of singling out the principal part $\tilde{L}_{n,\theta}$ and its utilization for estimating unknown parameters θ. Whittle applied it to the case of general strictly positive spectral densities $f_\theta > 0$ for more general processes than the Gaussian, namely linear processes X_t. Thus the estimator $\tilde{\theta}$ of the parameters of the spectral density of the linear process X_t defined above is often called "Whittle's estimator." We shall be using this terminology occasionally, however, in the particular case of the Gaussian process X_t, we shall more often utilize a more precise term for this procedure "asymptotic maximum likelihood (m.l.) estimator," and in the case of a linear process X_t the term "least squares estimator."

The statistical properties of the estimator $\tilde{\theta}$ are studied in Chapter II. In most of this chapter, it is assumed that the process X_t is Gaussian, however, in Section 6 a number of results obtained under this condition are carried over to a more general class of linear processes, while in the concluding Section 7 a process satisfies the "strong mixing" condition in Rosenblatt's sense [109].

The discussion above allows us to expect that in the case of the Gaussian process X_t the asymptotic m.l. estimator $\tilde{\theta}$ of the parameter θ will possess under very general regularity conditions the same properties of consistency, asymptotic normality, and asymptotic efficiency as the exact m.l. estimator $\bar{\theta}$. Sections 2 and 3 of Chapter II are mainly devoted to the proof of this assertion.

In Section 2 the simplest case is discussed when the spectral density $f = f_\theta$ is everywhere positive and such that $\tilde{L}_n$ can be

represented by (4) so that the estimator $\tilde{\theta}$ becomes the value of θ
maximizing the r.h.s. of (4). It is this very case that was
considered by Whittle [121] whose results were later proved
again in a more rigorous fashion by Walker [131]. Unlike the
case of our Section 2, it was assumed in papers [121] and [131]
that one of the entries of the vector θ is the parameter σ^2 which
is the mean square error of the linear forecast of the process X_t
for one step forward (cf. formula (II.1.3)) while only the ratio
$g(\lambda) = f(\lambda)/\sigma^2$ depends on the other entries. This assumption is
suitable in the case of a general linear process and it is often but
not always assumed to be valid in applied problems. Since,
however, we are going to discuss an important problem in
Section 5 of Chapter II, for which the assumption of the papers
[121] and [131] is not satisfied, in general, we shall devote
Section 2 to a generalization of known results by Whittle and
Walker concerning the properties of the estimators $\tilde{\theta}$ for the case
of an arbitrary vector-valued parameter θ. We shall show that
in the general case, the proof of consistency of the estimator $\tilde{\theta}$
requires only a minor modification of arguments presented in
[131] while for the proof of the asymptotic normality and
asymptotic efficiency of $\tilde{\theta}$ we shall utilize a different method
based substantially on the well-known results by Ibragimov [66].
The final result turns out to be very similar in its form to the
one which was obtained in [131] for the special class of
vector-valued parameters described above. Under general
conditions (stipulated in the statement of Theorem II.2.2 which
include, in particular, the requirement that the limit as $n \to \infty$ of
Fisher's information matrix be equal to

$$(12) \qquad \Gamma_\theta = \left[\frac{1}{4\pi} \int_{-\pi}^{\pi} \frac{\partial}{\partial \theta_k} \log f_\theta(\lambda) \frac{\partial}{\partial \theta_l} \log f_\theta(\lambda) d\lambda \right]_{k,\, l=1,\,...,p}$$

is nonsingular; θ_k being the k-th component of the vector θ)
one can show that the distribution of the random vector
$n^{1/2}(\tilde{\theta}-\theta)$ approaches, as $n \to \infty$, the distribution $N(0, \Gamma_\theta^{-1})$ (i.e.,
the normal distribution with zero mean and covariance
matrix Γ_θ^{-1}).

Section 3 of Chapter II is devoted to the problem of
estimating the unknown parameter θ of the spectral density
in the case which was not discussed previously, that is, when
the spectral density $f = f_\theta$ is of the form (5) where $f_0 = f_{0,\theta}$,
and is such that relation (6) is fulfilled. Defining the
asymptotic m.l. estimator $\tilde{\theta}$ of a parameter θ as the value of θ

maximizing the r.h.s. of (6) -- as in the case of the strictly positive spectral density -- we prove the consistency, asymptotic normality, and asymptotic efficiency of the estimator $\tilde{\theta}$, and also observe that the limit of Fisher's information matrix, in this case as well, coincides with the matrix Γ_θ.

In Section 4 of Chapter II several examples of models of Gaussian processes often encountered in practice are discussed. These processes are determined by a finite number of parameters for which explicit expressions of asymptotic likelihood equations can be written out (roots of these equations represent the components of the vector $\tilde{\theta}$); also an explicit form of the asymptotic covariance matrix of the estimator $\tilde{\theta}$ (i.e., the form of the matrix Γ_θ determined by formula (12)) can be given in this case. Here the common models of the autoregressive process, moving average process, and the mixed autoregressive-moving average process are considered; also the model of a stationary process with an exponential spectral density recently introduced by Bloomfield [23] is discussed. Some of the examples studied in Section 4 result in spectral densities $f_\theta(\lambda)$ with fixed zeroes -- their examples are collected in Subsection 4.5 of this section. Examples presented in Section 4 show that although asymptotic likelihood equations are substantially simpler than the strict likelihood equations and thus can be written up explicitly in a relatively short amount of space, nevertheless, these equations are still quite complicated from a practical point of view, being of the form of rather cumbersome nonlinear equations which can only be solved numerically with a substantial amount of effort. At the same time the asymptotic covariance matrix Γ_θ^{-1} of the estimators $\tilde{\theta}$ can often be efficiently calculated relatively easily; this allows us to utilize it for the estimation of the "degree of efficiency of various simplified estimators," i.e., to determine whether it makes sense to further improve on these simplified estimators or not.

It is observed in the beginning of Section 5 of Chapter II that all the examples collected in Section 4 do not take into account the important fact that under real-world conditions, the observations of the values of a random process X_t are never absolutely accurate, but always contain certain "observation errors" (or "noise") which are often quite substantial. In those cases, when the effect of "observation errors" cannot be neglected, one has to assume that the observation data is a realization of the sum of two random processes "signal" and "noise," which in practice may very often be considered to be independent of each other; furthermore usually the "noise" is considered to be "white," i.e., representing a sequence of independent identically-distributed random variables. In this case, the basic problem consists in estimating the parameters appearing in the expression of the spectral density of the "signal", along with a single parameter of the "noise" -- its intensity -- by means of a finite number of observations on the sum of the "signal" and the "noise." This problem is briefly discussed in Section 5 of Chapter II, based on the joint paper [51] by the author and G.I. Marr. Here we are assuming that the "signal" and the "noise" are both Gaussian random processes and that the main attention is devoted to an analysis of several simple examples of the spectral density of the signal, for which there exist asymptotic m.l. estimators, and that it is possible to write down less cumbersome explicit expressions of the corresponding asymptotic equations and the matrix Γ_Θ which defines the covariance matrix of estimator $\tilde{\Theta}$.

5. It is shown in a number of papers devoted to mathematical statistics that many results dealing with the case of Gaussian independent observations are actually valid for a number of non-Gaussian probability distributions. The experience obtained in the investigation of statistical problems for Gaussian processes also shows that quite often the results obtained for Gaussian random processes remain valid also for a more general class of processes, of which first and foremost, is the class of general linear processes X_t represented in the form $X_t = \sum_{T=0}^{\infty} g_T \epsilon_{t-T}$, where $g_0 = 1$, ϵ_t, $t = 0, \pm 1, \pm 2, ...$, is a sequence of independent, identically distributed random variables with $E(\epsilon_t) = 0$, $E(\epsilon_t^2) = \sigma^2 > 0$, $E(\epsilon_t^4) = \kappa_4 + 3\sigma^4 < \infty$, and the coefficients $g_1, g_2, ...$, are such that the series $g_1^2 + g_2^2 + \cdots$ is convergent (cf., e.g., [35]). In the same vein, the results concerning the consistency and asymptotic normality of the estimators $\tilde{\Theta}$ maximizing the r.h.s. of (4) were originally proved by Whittle [124] and Walker [131] at once for the general linear case (but under the assumption that $f_\Theta > 0$ and that $\Theta = (\Theta_1, ... , \Theta_{p-1}, \sigma^2)$, where $\sigma^2 = E(\epsilon_t^2)$ is given by formula (II.1.3) on page 103, and only the ratio $g(\lambda) = f(\lambda)/\sigma^2$ depends on $\Theta_1, ..., \Theta_{p-1}$). The last

assumption may be to some extent justified because it yields a very simple form of the limiting covariance matrix of the random vector $n^{1/2}(\tilde{\Theta}-\Theta)$ which will be discussed somewhat below; however, this assumption is not always justified in practice.

In view of the latter state of affairs, we shall begin in Section 6 of Chapter II with a discussion of the properties of Whittle-type estimators $\tilde{\Theta}$ (i.e., estimators maximizing the r.h.s. of (4)) in the case of a linear process X_t with everywhere positive spectral density $f_\Theta(\lambda)$ depending on the vector-valued parameter $\Theta = (\Theta_1, ..., \Theta_p)$ in an arbitrary manner and which uniquely determines the function $f_\Theta(\lambda)$. In this case one succeeds to show that under the usual conditions stipulated in the statement of Theorem 2 of page 109 the estimator $\tilde{\Theta}$ turns out to be consistent and asymptotically normal, while the distribution of the vector $n^{1/2}(\tilde{\Theta}-\Theta)$ as $n \to \infty$ tends to the normal distribution $N(0, \Gamma_\Theta^{-1} + \Gamma_\Theta^{-1}C_{\kappa_4,\Theta}\Gamma_\Theta^{-1})$ where Γ_Θ is again

defined by the formula (12) above, while

(13)
$$C_{\kappa_4,\Theta} = \left[\kappa_4 \int_{-\pi}^{\pi} \frac{\partial}{\partial\Theta_k}\log f_\Theta(\lambda)d\lambda\right.$$
$$\left. \times \int_{-\pi}^{\pi} \frac{\partial}{\partial\Theta_\ell}\log f_\Theta(\lambda)d\lambda\right]_{k,\ell=1,\ldots,p} .$$

In the particular case when $\Theta = (\Theta_1, ..., \Theta_{p_3 1}, \sigma^2)$, where $\sigma^2 = E(\varepsilon_t^2)$ and only the ratio $g_\Theta(\lambda) = f_\Theta(\lambda)/\sigma^2$ depends on Θ_1, ..., Θ_{p-1}, this result naturally reduces to the known result of the papers [121,131] according to which the distribution $n^{1/2}(\tilde{\Theta}-\Theta)$ for $n \to \infty$ tends to the normal distribution with zero mean and covariance matrix

(14)
$$\begin{bmatrix} \Gamma_\Theta^{(p-1)} & 0 \\ 0 & \sigma^4(2+\kappa_4) \end{bmatrix},$$

where $\Gamma_\Theta^{(p-1)}$ is a matrix of the $(p-1)$-th order defined by the formula (12) with indices k and ℓ running only between and including the values 1 up to $p-1$.

Formula (14) indeed explains the meaning of the assumption about $\Theta = (\Theta_1, ..., \Theta_{p-1}, \sigma^2)$ which was introduced in the papers [131,124] -- we discover that under this assumption the estimators $\tilde{\theta}_1$, ..., $\tilde{\theta}_{p-1}$ for the parameters θ_1, ..., θ_{p-1} only on which the normalized spectral density $g(\lambda) = f(\lambda)/\sigma^2$ depends are "robust" in the sense that their limiting distribution (as $n \to \infty$) does not depend at all on the form of distribution of the random variables ε_t (i.e., it is exactly the same as in the Gaussian case). At the same time, the limiting distribution of the estimator $\tilde{\sigma}^2 = \tilde{\theta}_p$ of parameter $\sigma^2 = \theta_p$ turns out to be normal with the variance $\sigma^4(2+\kappa_4)/n$ depending on κ_4, so that it is not a "robust" estimator in the sense in which the estimators $\tilde{\theta}_1$, ..., $\tilde{\theta}_{p-1}$ are.

The "robustness" of the estimators $\tilde{\theta}_1$, ..., $\tilde{\theta}_{p-1}$ of parameters θ_1, ..., θ_{p-1} means that for any linear process the limiting dispersion of the estimators $\tilde{\theta}_1$, ..., $\tilde{\theta}_{p-1}$ coincides with the limiting dispersion of analogous estimators of the same parameters in the case of a Gaussian process X_t. Since for a Gaussian process X_t the estimators $\tilde{\theta}_1$, ..., $\tilde{\theta}_{p-1}$ are asymptotically efficient, while in the general linear case this assertion is not valid, it follows that (from the aspect of achievable asymptotic accuracy of estimators of parameters θ_1, ..., θ_{p-1}) the Gaussian random processes are "the worst" among all linear processes. Actually, however, estimators of parameters θ_1, ..., θ_{p-1}, which for some special linear process X_t turn out to be more accurate than Whittle-type estimators, are usually such that their asymptotic covariance matrix depends substantially on the particulars of the distribution of variables ε_t while for some other distribution they may be substantially inferior to estimators $\tilde{\theta}_1$, ..., $\tilde{\theta}_{p-1}$. It is also important that estimators $\tilde{\theta}_1$, ..., $\tilde{\theta}_{p-1}$ and $\tilde{\theta}_p = \sigma^2$ of parameters θ_1, ..., θ_{p-1}, $\theta_p = \sigma^2$ for all linear processes possess the smallest limiting dispersion in a wide class of estimators of these parameters (which in fact coincides with the class of possible estimators dependent only on "statistics of the second order"), i.e., within this class of estimators they are optimal also for linear processes. A precise proof of this last assertion is presented in Subsection 6.4 of Section 6 in Chapter II.

The class of linear processes X_t is rather a wide class; nevertheless, it does not contain a number of important cases of stationary random processes. In connection to this we shall consider in Section 7, the properties of the estimators $\tilde{\theta}_1$, ..., $\tilde{\theta}_p$ for another but also very wide class of stationary random processes, namely, the processes X_t satisfying the condition of strong mixing with a specified rate of decreasing mixing coefficient $\alpha(\tau)$. Establishing the rate of decrease which is sufficient to assure the validity of the corresponding central limit theorem, we shall prove in Section 7 that in the case under consideration the estimators $\tilde{\theta}_1$, ..., $\tilde{\theta}_p$ will also be consistent and asymptotically normal while the distribution of the vector $n^{1/2}(\tilde{\theta}-\theta)$ approaches here, as $n \to \infty$, the distribution $N(0, \Gamma_\theta^{-1}(\Gamma_\theta + C_{f_4, \theta})\Gamma_\theta^{-1})$, where

(15)
$$C_{f_4, \theta} = \left[\frac{1}{8\pi} \int_{-\pi}^{\pi} \int f_4(\lambda_1, -\lambda_1, \lambda_2, -\lambda_2) \right.$$
$$\left. \times \frac{\partial}{\partial\theta_k} \frac{1}{f_\theta(\lambda_1)} \frac{\partial}{\partial\theta_l} \frac{1}{f_\theta(\lambda_2)} d\lambda_1 d\lambda_2\right]_{k, l=1, \ldots, p},$$

and $f_4(\lambda_1, \lambda_2, \lambda_3, \lambda_4)$ is a spectral density of the fourth order for the process X_t. In the particular case of a Gaussian process this result, of course, reduces to the result of Theorem 2 in Section 2.

6. Above it was noted that the Whittle-type estimator $\tilde{\theta}$ of parameter θ of the spectrum of a stationary process X_t possesses under general conditions a number of "nice" statistical properties which render this estimator very attractive. Nevertheless, the utilization of "Whittle-type estimators" $\tilde{\theta}$ in many practically important cases is hindered by the fact that it is required to solve a complicated system of nonlinear equations (cf. in particular, the examples investigated in Subsections 4.2-4.5 and 5.2 of Chapter II). For this reason, we shall devote Chapter III of this monograph to the problem of constructing simplified estimators of parameter θ, whose determination does not require carrying out cumbersome calculations, while the estimators obtained are as "nice" asymptotically as the estimators $\tilde{\theta}$.

In the beginning of Section 1 we shall consider the general problem of determining estimators of unknown vector-valued parameters $\theta \in \Theta$ appearing in the expression for finite-dimensional probability densities of a random process X_t based on data of observed values of n random variables X_1, ..., X_n. We introduce a bounded in probability p-dimensional random column-vector $\Phi_{n, \theta} = \Phi_{n, \theta}(X_1, \ldots, X_n)$ satisfying the condition

(16) $\Phi_{n,\Theta*} - \Phi_{n,\Theta} + J_\Theta \tau_n(\Theta*-\Theta) \to 0$

in probability as $n \to \infty$. Here $\Theta*$ is a τ_n^*-consistent estimator of the value of the parameter Θ, J_Θ is a nondegenerate $(p \times p)$-matrix generally dependent on Θ with nonrandom entries, and τ_n^* and τ_n, $n = 1,2, \dots$ are infinitely increasing sequences of positive numbers such that τ_n increases at least as fast as τ_n^* but $\tau_n^{1/2}/\tau_n^* \to 0$ as $n \to \infty$. Assume that the vector $\Phi_{n,\Theta}$ as $n \to \infty$ possesses a limiting p-dimensional normal distribution $N(0,J)$ with zero mean and fixed covariance matrix J, and let $J_* = J_*(X_1, \dots, X_n)$ be a τ_n^*-consistent estimator of matrix J_Θ. Then it is possible to show that the p-dimensional random column-vector

(17) $\vec{\Theta}_n(X_1, \dots, X_n) = \vec{\Theta} = \Theta* + \dfrac{1}{\tau_n} J_*^{-1}\Phi_{n,\Theta*}$

is a τ_n-consistent and asymptotically normal estimator of the value of the parameter Θ, such that the distribution of the vector $\tau_n(\vec{\Theta}-\Theta)$ tends to the p-dimensional normal distribution $N(0,J_\Theta^{-1}JJ_\Theta^{-1})$ as $n \to \infty$.

In Subsection 1.3 of Chapter III applications of the above stated general assertion to the problem of mathematical statistics dealing with estimation of the unknown parameter Θ are clarified. Numerous methods of constructing estimators result in estimators $\vec{\Theta}$ which are roots of a system of equations (with respect to unknown Θ) of the form

(18) $\Phi_{n,\Theta}(X_1, \dots, X_n) = 0,$

where $\Phi_{n,\Theta}$ is a random p-dimensional column-vector dependent on Θ. Often it is also possible to prove that the root $\vec{\Theta}$ of equation (18) is a τ_n^*-consistent estimator of Θ (where τ_n^* is a certain rapidly increasing numerical sequence). Finally, it is often possible to prove (usually by applying the mean value theorem) that $\vec{\Theta}$ satisfies the condition of form (16), where $\Theta*$ is replaced by $\vec{\Theta}$ and the distribution of the vector $\Phi_{n,\Theta}$ tends to the normal distribution $N(0,J)$ as $n \to \infty$. In such a case, in view of (16) with $\Theta* = \vec{\Theta}$ and $\Phi_{n,\Theta} = 0$ by the very definition of $\vec{\Theta}$, the distribution of the random vector $\tau_n(\vec{\Theta}-\Theta)$ will tend to a p-dimensional normal distribution $N(0,J_\Theta^{-1}JJ_\Theta^{-1})$ as $n \to \infty$. Thus the estimator $\vec{\Theta}$ automatically turns out to be τ_n-consistent and asymptotically normal with an easily calculated limiting covariance matrix.

However, quite often the determination of the root $\tilde{\theta}$ of a system of equations (18) turns out to be a complicated and cumbersome problem which requires substantial effort and is time-consuming even for modern computers. In such cases, some other more easily constructed estimators become of substantial interest provided they have the same asymptotic distribution as $\tilde{\theta}$. In view of that stated above, if certain τ_n^*-consistent estimators θ_* and J_* of parameter θ and the matrix J_θ respectively are known, then the estimator $\tilde{\theta}$ of the form (17) can be used as such a simplified estimator (here $\Phi_{n,\theta}$ coincides with the r.h.s. of relation (18)). This is indeed the basic route for utilizing the general assertion concerning the quantity $\vec{\theta}_n$.

Remark 1. Assume that the elements of the vector $\Phi_{n,\theta}$ are sufficiently smooth functions in θ. In this case, utilizing the mean value theorem (cf. formula (III.1.9) on page 203), formula (16), and the fact that $\tau_n^{1/2}/\tau_n^* \to 0$ as $n \to \infty$, it is easy to verify that the random matrix J_{θ_*}, where J_θ is the related Jacobian with the multiplicative factor τ_n^{-1}, is a τ_n^*-consistent estimator of the value of matrix J_θ.

   Therefore the estimator

(19)
$$\vec{\theta}^{(1)} = \theta_* - \frac{1}{\tau_n} J_{\theta_*}^{-1} \, \Phi_{n,\theta_*} \, ,$$

of the value of parameter θ, is in this case asymptotically equivalent to the root $\tilde{\theta}$ of the system of equations (18). Now let the roots of the system (18) be determined using the well-known Newton-Raphson approximating iteration method (cf., e.g., [94]). Then under the condition that a certain τ_n^*-consistent estimator θ_* is chosen for the initial value of the estimator the first iteration cycle already results in an estimator of the form (19).

Remark 2. If all the elements of the matrix J_θ are continuous in θ, and θ_* is as above a τ_n^*-consistent estimator of parameter θ, then the matrix J_{θ_*} is a τ_n^*-consistent estimator of matrix J_θ. Therefore in such a case the estimator

(20)
$$\vec{\theta}^{(2)} = \theta_* + \frac{1}{\tau_n} J_{\theta_*}^{-1} \Phi_{n,\theta_*}$$

of parameter Θ is also asymptotically equivalent to the root $\tilde{\Theta}$
of the system of equations (18).

The results presented above are well-known when applied
to the classical problem of determining the m.l. estimators of
parameter Θ in a probability density $p(x,\Theta)$ based on
independent observations of variables $X_1, ..., X_n$ (possessing
this density). In this case (18) are interpreted as the usual
likelihood equation,

$$\frac{1}{n^{1/2}} \sum_{j=1}^{n} \frac{\partial}{\partial \Theta_k} \log p(X_j,\Theta) = 0, \quad k = 1,...,p, \quad \tau_n = n^{1/2},$$

and

(21)
$$J_\Theta = \left[\int_{-\infty}^{\infty} \frac{\partial}{\partial \Theta_k} \log p(x,\Theta) \frac{\partial}{\partial \Theta_j} \log p(x,\Theta) \right.$$
$$\left. \times p(x,\Theta)dx \right]_{k,\, j=1,\, ...,\, p}$$

is the Fisher's information matrix. The estimators $\tilde{\Theta}^{(1)}$ and
$\tilde{\Theta}^{(2)}$ in this particular case are discussed in many basic
textbooks on mathematical statistics (cf., e.g., [106], Section 5d,
[64] Section 5.2, [71] item 18.21). Asymptotic properties of the
corresponding estimator $\tilde{\Theta}^{(2)}$ are studied in detail in LeCam's
well-known paper [79]. In the general case, the class of
estimators $\tilde{\Theta}$ which includes both $\tilde{\Theta}^{(1)}$ and $\tilde{\Theta}^{(2)}$ was
introduced by the author in [42]. It is worthwhile to
emphasize that this general case also incorporates many of
those cases when Θ is a parameter of a spectrum of a
Gaussian or a general linear process X_t. We shall discuss the
latter case below.
    In what follows some results related to the particular case
studied by LeCam in the papers [80-82] will be required.
Assume that the distribution $P_n = P_{n,\Theta}$ of random variables
$X_1, ..., X_n$ is uniquely determined by the value of the
vector-valued parameter Θ which must be estimated (in
particular, this may be the case of estimating the parameter Θ
which uniquely determines the spectral density of the
observed Gaussian process X_t). Following LeCam [81] we
shall assume that the family of distributions $P_{n,\Theta}, \Theta \in \Theta$,
where Θ is the set of all possible values of Θ (which is an
open subset of R_p) is "asymptotically differentiable" for some

increasing sequence of positive numbers τ_n, $n = 1,2, \ldots$ in the sense that it satisfies the following conditions:

(D1)   The sequences $P_{n,\theta}$, $n = 1,2, \ldots$ and $P_{n,\theta+h/\tau_n}$, $n = 1$,

2, ... are contiguous for any $\theta \in \Theta$ and h, such that $\theta+h/\tau_n \in \Theta$ for all $n = 1,2, \ldots$.

(D2)   For any $\theta \in \Theta$ there exists a sequence of p-dimensional random vectors $\Delta_{n,\theta}$, $n = 1,2, \ldots$, and a $(p{\times}p)$-matrix Γ_θ such that

$$\Lambda(\theta,\theta + h/\tau_n) - h'\Delta_{n,\theta} + \frac{1}{2}\, h'\Gamma_\theta h \to 0$$

in $P_{n,\theta}$ probability as $n \to \infty$, for any h such that $\theta + h/\tau_n \in \Theta$, where $\Lambda(\theta,\theta+h/\tau_n) = \log\{dP_{n,\theta+h/\tau_n}/dP_{n,\theta}\}$.

(D3)   If $h_n \to h$ in R_p and $\theta \in \Theta$ then

$$\Lambda(\theta,\theta + h/\tau_n) - \Lambda(\theta,\theta + h_n/\tau_n) \to 0$$

in $P_{n,\theta}$ probability as $n \to \infty$.

(D4)   If $\mathfrak{U}_n$ is a σ-algebra defined on the sample space and $A \in \mathfrak{U}_n$ then the function $\theta \rightsquigarrow P_{n,\theta}(A)$ is Lebesgue measurable.

In Subsection 1.4 of Chapter III additional conditions on $P_{n,\theta}$, $\theta \in \Theta$ are imposed such that: (a) the random vector $\Delta_{n,\theta}$ and the $(p{\times}p)$-matrix Γ_θ appearing in condition (D2) satisfies the relation (16) for $\Delta_{n,\theta} = \Phi_{n,\theta}$, $J_\theta = \Gamma_\theta$ and some τ_n-consistent estimator $\theta*$ and (b) the distribution of the random vector $\Delta_{n,\theta}$ for $n \to \infty$ tends to the normal distribution $N(0,\Gamma_\theta)$, where Γ_θ is a nondegenerate matrix.[6] In view of the arguments presented above we obtain that under the conditions indicated above the estimator

(22) $$\vec{\theta} = \theta* + \frac{1}{\tau_n}\, \Gamma_*^{-1} \Delta_{n,\theta*}$$

of parameter θ (here Γ_* is a τ_n-consistent estimator of the matrix Γ_θ) is τ_n-consistent and asymptotically normal, while

---

[6]Under these conditions the family of distributions $P_{n,\theta}$, $\theta \in \Theta$ turns out to be locally asymptotically normal in the LeCam sense; see Definition A2.1 in Appendix 2 to Chapter II.

the random vector $\tau_n(\tilde{\Theta}-\Theta)$ possesses the normal distribution $N(0,\Gamma_\Theta^{-1})$ as $n \to \infty$.

In Subsection 1.5 of Chapter III the important particular case, when X_t is a Gaussian process with a spectral density $f = f_\Theta > 0$ satisfying certain regularity conditions (cf., Theorem II.2.2), is discussed separately. In this case the k-th component of the vector $\Delta_{n,\Theta}$ is of the form

$$(23) \qquad \frac{n^{1/2}}{4\pi} \int_{-\pi}^{\pi} \frac{I_n(\lambda)-f_\Theta(\lambda)}{f_\Theta(\lambda)} \frac{\partial}{\partial\Theta_k} \log f_\Theta(\lambda)d\lambda,$$

the matrix Γ_Θ coincides with the limit of Fisher's information matrix (cf. formula (12) above) and $\tau_n = \sqrt{n}$. Next, another particular case is discussed when the spectral density $f = f_\Theta$ of a Gaussian process X_t is of the form (5), where $f_0 = f_{0,\Theta}$ satisfies the regularity conditions of the Theorem II.3.2. Here the k-th component of the vector $\Delta_{n,\Theta}$ is of the form

$$(24) \qquad \frac{n^{1/2}}{4\pi} \int_{-\pi}^{\pi} \frac{I_n(\lambda,\tilde{Y}) - f_{0,\Theta}(\lambda)}{f_{0,\Theta}(\lambda)} \frac{\partial}{\partial\Theta_k}\log f_{0,\Theta}(\lambda)d\lambda,$$

where $I_n(\lambda,\tilde{Y})$ is as above the periodogram of the process $\tilde{Y}_t$ (cf. formula (7)) and Γ_Θ as above coincides with matrix (12).

In Subsection 1.6 of Chapter III a more general case is studied, where X_t is an arbitrary linear process with spectral density $f = f_\Theta$ satisfying the conditions of Theorem II.6.2. It is shown here that if $\Theta*$ is a $\sqrt{n}$-consistent estimator of the parameter Θ then the random vector $\Delta_{n,\Theta} = \Phi_{n,\Theta}$ whose k-th component is given by (23) satisfies the condition (16) for $\tau_n = \sqrt{n}$ and $J_\Theta = \Gamma_\Theta$, while, as $n \to \infty$, this vector possesses the normal distribution $N(0,\Gamma_\Theta + C_{K_{4,\Theta}})$ (cf. formula (13) above).

It follows from there that in this case the estimator (17) is asymptotically normal and the limiting distribution of the vector $n^{1/2}(\tilde{\Theta}-\Theta)$ is the normal distribution $N(0,\Gamma_\Theta^{-1} + \Gamma_\Theta^{-1}C_{K_{4,\Theta}}\Gamma_\Theta^{-1})$. This distribution is also the limiting distribution of the vector $n^{1/2}(\tilde{\Theta}-\Theta)$, where $\tilde{\Theta}$ is the least squares estimator.

Finally, we observe that in the general case of the processes X_t satisfying the strong mixing condition in the sense of Section 7 of Chapter II the estimator $\vec{\Theta}$ of the parameter Θ is asymptotically equivalent to the Whittle estimator $\tilde{\Theta}$.

7. As it was stated in the preceding subsection, in order

to construct simplified estimators $\vec{\theta}$ with "nice" asymptotic properties based on formula (22) it is necessary to start with some consistent estimators θ_*. Since θ_* is required only to be consistent, it is reasonable to choose the simplest possible computable consistent estimators. In connection to this, in Section 2 of Chapter III, for all the examples considered in Sections 4 and 5 of Chapter II, methods of determining relatively simple consistent estimators of unknown parameters of the spectral density are presented. In all the examples in Section 2 of Chapter III -- except for the model of a process with exponential spectral density discussed in subsection 5 -- the spectral density f of the observed process is a rational function of $z = e^{i\lambda}$. Under these conditions the root of a system of equations in θ obtained by equating the values of the covariance function $\beta_\theta(\tau)$ for $\tau = 0,1, ..., p\text{-}1$, to their consistent estimators $\beta_n^*(\tau)$, $\tau = 0,1, ..., p\text{-}1$, may serve as a very simple $\sqrt{n}$-consistent estimator θ_* of some unknown p-dimensional vector-valued parameter θ.

In Section 3 of Chapter III we offer a number of specific procedures for constructing simplified estimators $\vec{\theta}$ of the parameter θ of a spectral density. In Subsection 3.2 the case is considered, where the spectral density depends linearly on a finite number, say p, of unknown parameters $(\theta_1, ..., \theta_p) = \theta$ (cf. formula (II.4.20)). A special example of such a case is the example considered in [74] of a moving average process X_t of the r-th order with spectral density of the form (II.4.19), where the covariances $\beta(0)$, $\beta(1)$, ..., $\beta(\tau)$ are unknown parameters and $p = r+1$. In this case, starting with equation (22) it is easy to show that the root $\vec{\theta}$ of a system of linear equations (III.3.1) is an estimator of the parameter θ asymptotically equivalent to the Whittle estimator $\tilde{\theta}$ (cf. Subsection 4.2 in Chapter II). The papers [43,74] present an alternative proof of this fact.

Next, in Subsection 3.2 of Chapter III we consider the case, where X_t is a moving average process of the r-th order with spectral density f of the form (II.4.13) (cf. page 118), where $(\alpha_1, ..., \alpha_r)'$ are unknown parameters. It is shown here that if one utilizes formula (22) for determining the estimators $\vec{\theta} = \vec{\alpha} = (\vec{\alpha}_1, ..., \vec{\alpha}_r)'$ of the parameters $\alpha = (\alpha_1, ..., \alpha_r)'$ where $\tau_n = \sqrt{n}$, $\Gamma_* = \Gamma_{\alpha_*}^{(r)}$

(cf. (II.4.17)) and $\alpha_* = (\alpha_{1*}, ..., \alpha_{r*})'$ are certain $\sqrt{n}$-consistent estimators of $\alpha = (\alpha_1, ... , \alpha_r)'$ (for example, those defined in Subsection 2.2 of Chapter III, then $\vec{\alpha}$ becomes a root of a simple system of r linear equations (III.3.2). We observe at once that for some more specialized choice of estimator Γ_* of the matrix $\Gamma_\alpha^{(r)}$ utilizing formulas (22) for the determination of estimators results in a system of linear equations which actually coincide with the equations suggested by Hannan in his paper [141] and his book [140]. This fact simply explains the asymptotic equivalence of Hannan's and Whittle's estimators $\tilde{\theta}$ while a large amount of space is devoted in [141] and [140] to the proof of this fact.

In Subsection 3.3 of Chapter III we consider the case when the spectral density f is a general rational function in $z = e^{i\lambda}$ of the form (II.4.28) with unknown coefficients of polynomials in the numerator and denominator. Here it is shown that applying the general formula (22) with $\tau_n = \sqrt{n}$ and choosing in an appropriate manner the initial consistent estimators $\iota_{1*}, ..., \iota_{q*}, \alpha_{1*}, ..., \alpha_{r*}$ of the parameters $\iota_1, ..., \iota_q$, $\alpha_1, ..., \alpha_r$ we may reduce the determination of the estimators $\vec{\iota}_1, ..., \vec{\iota}_q, \vec{\alpha}_1, ..., \vec{\alpha}_r$ which are asymptotically equivalent to Whittle's estimators $\tilde{\iota}_1, ..., \tilde{\iota}_q, \tilde{\alpha}_1, ..., \tilde{\alpha}_r$ to the solution of several systems of q and r linear equations. It is also shown that under a specific, quite specialized choice of estimators of the matrices $\Gamma_\iota^{(q)}$, $\Gamma_\alpha^{(r)}$, and Ω (cf. formula (II.4.33)) the

estimators $\vec{\iota}_1, ..., \vec{\iota}_q, \vec{\alpha}_1, ..., \vec{\alpha}_r$ coincide with the estimators proposed by Hannan in [141] and [140].

In Subsection 3.3 the case is considered when the spectral density (25) is written up in the form (II.4.37) where $\iota_1, ..., \iota_q$, $\beta_y(0), \beta_y(1), ..., \beta_y(r)$ are unknown parameters. The procedure for determining the estimators $\vec{\iota}_1, ..., \vec{\iota}_q, \vec{\beta}_y(0), \vec{\beta}_y(1), ..., \vec{\beta}_y(r)$ of these parameters suggested herein is an appreciable simplification of Parzen's procedure [100].

In Subsection 3.4 of Chapter III we shall consider the problem of determining simplified estimators of parameters possessing the same "nice" asymptotic properties as Whittle's estimators in the case when the "signal" is observed on a "white noise" background. Based on the results of the examples in Section 5 of Chapter II and Subsection 2.4 of Chapter III we propose for all these examples specific "recommendations" for determining each estimator. In the concluding subsections 3.5 and 3.6 of Section 3 in Chapter III, simplified estimators are proposed which possess "nice" asymptotic properties of parameters in models of a process X_t with an exponential spectral density of the form (II.4.47) (cf.

page 125), and respectively, for a process X_t with a spectral density of the form (II.4.53) (cf. page 128) vanishing for $\lambda = 0$.

Hypothesis Testing About Spectral Density

8. Following LeCam's general ideas [80-82] (cf. also [110]) in Chapter IV we consider a sequence of experiments $\bar{E}_n = \{X_n,$ $\mathfrak{U}_n, P_{n, \Theta}, \Theta \in \Theta\}$ $n = 1,2, ...,$ where X_n is a set of possible outcomes of the n-th experiment (i.e., for example, the set of possible values of the vector $\mathbf{X} = (X_1, ..., X_n)')$. $\mathfrak{U}_n$ is a σ-algebra defined on X_n, and $P_{n, \Theta}, \Theta \in \Theta$ is a family of distributions on $\mathfrak{U}_n$, such that for a certain choice of a random asymptotically normal vector $\Delta_{n, \Theta} = \Delta_{n, \Theta}(\mathbf{x}), \mathbf{x} \in {}_n$ and nonrandom matrix Γ_Θ, it satisfies conditions (D1)-(D4) with $\tau_n = n^{1/2}$. Assume for definiteness that the set $\Theta \in R_p$ of possible values of a vector-valued parameter Θ contains the origin. Under this condition, in Section 1 the problem of testing hypothesis H_0 that parameter Θ takes on zero value is considered.

We shall assume as it is done in [80-82] that the alternative hypothesis H_1 is that the parameter Θ takes on the value $n^{-1/2}\mathbf{h} \in \Theta, \mathbf{h} \neq 0$.

As it is known (see, e.g., the book [100] whose results can be easily carried over to the more general situation considered herein) the test-statistic $\hat{\Phi}_n$ defined by the critical region

(25) $\hat{\Phi}_n = \{\mathbf{x}: \Delta'_{n, \Theta}(\mathbf{x}) \Gamma_\Theta^{-1} \Delta_{n, \Theta}(\mathbf{x}) < d_\alpha\},$

where d_α is the quantile of a χ^2-distribution with p degrees of freedom corresponding to the fixed level of significance α, $0 < \alpha < 1$, possesses several properties of asymptotic optimality (as $n \to \infty$). Theorems describing these properties are assembled in Chapter 6 of the book [110] and are actually generalizations to a large class of experiments (not only to the case of a stationary Markov chain considered in [110]) of results contained in the basic paper by Wald [29] dealing with the particular case when the random variables $X_1, ..., X_n$ are mutually independent and identically distributed. As early as 1960, LeCam pointed out that this kind of generalization is possible (cf. also [81,82]). Theorem 1 of Section 1 in Chapter IV represents exactly an extension of one of the "optimality properties" stated in [110] to the general situation considered in our monograph.

In Section 1 of Chapter IV it is proved that the test-statistic $\hat{\phi}_n$ is asymptotically equivalent (in the sense of Definition 1 on page 241) to test-statistics $\phi_n^{(1)}$ and $\phi_n^{(2)}$ (the second is the likelihood ratio criterion) defined by the critical regions

$$\phi_n^{(1)} = \{x: \tilde{\delta}_n'(x)\Gamma_\Theta\tilde{\delta}_n(x) > d_\alpha\},$$

$$\phi_n^{(2)} = \{x: 2\Lambda(0,\tilde{\delta}_n(x)) > d_\alpha\},$$

correspondingly, where $\tilde{\delta} = \tilde{\delta}_n$ is an asymptotically efficient estimator of the parameter Θ, $\Lambda(\Theta_1,\Theta_2) = \log(dP_{n,\Theta_2}/dP_{n,\Theta_1})$ and d_α is as above.

The general results which are discussed here can be applied to the case which is of special interest for our purposes (but which apparently was not considered from this aspect), when $P_{n,\Theta}$, $\Theta \in \Theta$ is a family of Gaussian distributions corresponding to a stationary random process. Indeed, it is easy to show that if the spectral density f_Θ satisfies certain (sufficiently general) regularity conditions then the corresponding family of distributions $P_{n,\Theta}$, $\Theta \in \Theta$ satisfies the conditions described above. Moreover, if $f_\Theta > 0$, then the k-th component of the vector $\Delta_{n,\Theta}$ is defined by formula (23) and, if f_Θ is of the form (5), it is defined by formula (24). The matrix Γ_Θ is defined here by formula (12). This immediately implies that in order to test the hypothesis H_0: f_Θ, $\Theta = 0$, one can utilize asymptotically equivalent test-statistics either $\hat{\phi}_n$ or $\phi_n^{(1)}$ or $\phi_n^{(2)}$ which are optimal in a certain sense. In the concluding part of Section 1 of Chapter IV a number of specific examples of constructing such statistics is considered; they are related to naturally stated problems of testing hypotheses about parameters of spectral densities.

We now proceed to the problem of testing composite hypotheses. For convenience we shall denote $\gamma = (\Theta_1, ..., \Theta_s)'$ and $\delta = (\Theta_{s+1}, ..., \Theta_{s+k})'$, where $k = p-s > 0$ so that $\Theta = (\gamma,\delta)'$ and $P_{n,\Theta} = P_{n,(\gamma,\delta)}$. Let Γ and D be sets of vectors γ and δ respectively corresponding to all $\Theta \in \Theta$; assume that D contains the value $\delta = 0$. Let us suppose that the sequence of experiments $E_n = \{X_n, \mathfrak{U}_n, P_{n,\Theta}, \Theta \in \Theta\}$, $n = 1, 2, ...$ satisfies the above stated conditions imposed on these sequences. In

Section 2 of Chapter IV under this assumption the problem of testing a composite hypothesis H_0 that the distribution on the space X_n belongs to the family of distributions $P_{n, (\gamma, 0)}$, $\gamma \in \Gamma$, versus the alternative H_1 that it belongs to the family of distributions $P_{n, (\gamma, \delta)}$, $\gamma \in \Gamma$, $\delta \in D$, $\delta \neq 0$ is considered; the dependence of δ on n is defined here as usual by the condition $\delta = n^{-1/2}d$.

Three different test statistics are introduced for the solution of this problem in Section 2 of Chapter IV; two of them are similar to statistics $\hat{\phi}_n$ and $\phi_n^{(1)}$, while the third one is new. These three statistics are, in essence, generalizations to the general case considered herein, of the well-known Wald's [29], Rao's [105] (cf. also [106], page 418), and Neyman's test statistics [90], which were produced for testing a composite hypothesis about a parameter of probability density of a random variable based on independent observations. As it is known in the case of independent observation, these three statistics are asymptotically equivalent in the sense of the definition in Section 1 of Chapter V to the likelihood ratio criteria and their power, as $n \to \infty$, possesses several "asymptotic properties" indicated in paper [29].[7]   In Section 2 it is proved that all these assertions about these criteria remain valid in the general case under consideration.

In the framework of this monograph it is especially important that the general results of Section 2 can be used in the particular case when $P_{n, \Theta}$, $\Theta \in \Theta$ is a family of Gaussian distributions with spectral density $f_\Theta = f_{(\gamma, \delta)}$ where $\Theta = (\gamma, \delta)' \in \Theta$ satisfying the conditions of Section 2 or Section 3 in Chapter II. The problem of constructing "asymptotically optimal tests" for testing the composite hypothesis $H_0: f_\Theta = f_{(\gamma, 0)}$, $\gamma \in \Gamma$ versus the alternative $H_1: f_\Theta = f_{(\gamma, \delta)}$, $\gamma \in \Gamma$, $\delta \in D$, $\delta = n^{-1/2}d$ where $d \neq 0$ in this particular case is investigated in Section 3 of Chapter IV. Here also are presented a number of specific examples of constructing such criteria related to Gaussian autoregressive processes, mixed autoregressive-moving average processes, and processes with exponential spectral density.

---

[7]See [106], pages 418-420 concerning the equivalence of Wald's, Rao's, and likelihood ratio criteria.   The equivalence of Neyman's and likelihood ratio criteria was observed by LeCam in [79] (see also the discussion of this problem in [88]).

9. In the final Chapter V the problem of testing the hypothesis H_0 concerning the form of spectral density f of a stationary random process X_t is considered. In the first two sections of this chapter, it is assumed that when H_0 is valid, the spectral density f is of a fixed form and, in the last section, that it belongs to a certain parametric family f_Θ, $\Theta \in \Theta$. Unlike in the preceding chapter, the assumptions on the process X_t are more general, i.e., it is assumed that X_t is a general linear process representable in the form $X_t = \sum_{\tau=0}^{\infty} g_\tau \, \varepsilon_{t-\tau}$ where the coefficients $g_0, g_1, \dots$ and the sequence of independent, identically distributed random variables ε_t, $t = 0, \pm1, \dots$, satisfy the conditions described in item 5 of this Introduction.

The construction of goodness-of-fit tests for testing the hypothesis H_0 in Chapter V is based on the remarkable fact that, independently of the form of spectral density f and the distribution of random variables ε_t, the sequence of random functions

$$\zeta_n(\tau) \equiv \sqrt{2} \, [\xi_n(\pi\tau) - \tau\xi_n(\pi)], \quad 0 \leqslant \tau \leqslant 1, \quad n = 1,2,\dots$$

where
$$\xi_n(x) = \frac{\sqrt{n}}{2\pi} \int_0^x \frac{I_n(\lambda)}{f(\lambda)} \, d\lambda,$$

converges to to a Brownian bridge $\zeta(\tau)$, $0 \leqslant \tau \leqslant 1$ in distribution in the space $C[0,1]$ of continuous functions $c(\tau)$, $0 \leqslant \tau \leqslant 1$ (cf. Proposition 1 on page 274). Clearly, from the aspect of the asymptotic theory the random function $\zeta_n(\tau)$, $0 \leqslant \tau \leqslant 1$, plays the same role here as the random function $\sqrt{n} \, |F_n(\tau)-\tau|$, $0 \leqslant \tau \leqslant 1$, in the problems of nonparametric statistics (here F_n is the empirical distribution function corresponding to independent random variables $X_1, \dots, X_n$ uniformly distributed on $[0,1]$).

In Section 1 of Chapter V a class of goodness-of-fit tests is introduced for testing the simple hypothesis H_0 about the form of the spectral density f determined by the critical regions of the form $\{x\colon V(\zeta_n) > d_\alpha\}$, where V is a continuous functional in $C[0,1]$, α is a predetermined significance level, and d_α is a quantile of the distribution $L\,[V(\zeta)]$. The following continuous functionals on $C[0,1]$: either

$$\max_{0 \leqslant \tau \leqslant 1} |c(\tau)| \quad \text{or} \quad \max_{a \leqslant \tau \leqslant 1} |c(t)|/\tau, \; 0 < a < 1, \quad \text{or} \quad \int_0^1 c^2(\tau) d\tau$$

can be chosen for $V(c)$. Such a choice of $V(c)$ results in goodness-of-fit tests related to Kolmogorov-Smirnov, Renyi, and ω^2 tests respectively, which were proposed for testing the hypothesis about the distribution of the random variable X based on an independent sample (cf., e.g., [25] or [31], Chapter V, Section 3).

In Subsection 1.2 of Chapter V, a finite class of "close" alternatives H_1 is introduced. Namely, it is assumed here that under the (alternative) hypothesis H_1 the linear process X_t can be represented in the form $X_t = \sum_{\tau=0}^{\infty} g_{\tau n} \varepsilon_{t-\tau}$, $n = 1, 2,$ ... where the coefficients are such that the finite limit

$$(26) \qquad \lim_{n \to \infty} n^{1/2} \frac{\tilde{g}_n(z) - \tilde{g}(z)}{\tilde{g}(z)} = h(z),$$

where

$$\tilde{g}(z) = \sum_{\tau=0}^{\infty} g_\tau z^\tau, \qquad \tilde{g}_n(z) = \sum_{\tau=0}^{\infty} g_{\tau n} z^\tau,$$

exists. In that case

$$f_n(\lambda) = \frac{\sigma^2}{2\pi} |\tilde{g}_n(z)|^2$$

under the alternative H_1 while

$$\lim_{n \to \infty} n^{1/2} [f_n - f]/f = a$$

uniformly in λ, where $a = a(\lambda) = 2\mathrm{Re}\, h(z)$. Under these conditions it is proved (cf. Proposition 2 on page 276) that when H_1 is valid the random function $\zeta_n(\tau)$, $0 \leqslant \tau \leqslant 1$, as $n \to \infty$, converges (in the sense indicated above) to a random function $\zeta(\tau) + A(\tau)$, $0 \leqslant \tau \leqslant 1$, where ζ is a Brownian bridge and

$$A(\tau) = \sqrt{2}[\tilde{a}(\pi\tau) - \tau\tilde{a}(\pi)], \qquad \tilde{a}(x) = \frac{1}{2\pi} \int_0^x a(\lambda) d\lambda.$$

In view of this, the asymptotic power (as $n \to \infty$) of the test with the critical region $\{x : V(\zeta_n) > d_\alpha\}$ equals

$$\int_{d_\alpha}^{\infty} l(x, A) dx$$

where $l(x, A)$ is the density of distribution $L[V(\zeta + A)]$.

In Section 2 of Chapter V the very important practical particular case when

$$V(c) = 2 \sum_{j=1}^{m} \left\{ \int_{0}^{1} \tilde{\Phi}_j(\tau) dc(\tau) \right\}^2,$$

where $\tilde{\Phi}_j$, $j = 1, ..., m$ are certain continuous and bounded functions satisfying the conditions

$$\tilde{\Phi}_j(\tau) = \tilde{\Phi}_j(-\tau), \quad \int_{-1}^{1} \tilde{\Phi}_j(\tau) d\tau = 0,$$

$$\int_{-1}^{1} \tilde{\Phi}_j(\tau) \tilde{\Phi}_k(\tau) d\tau = \delta_{jk}$$

is discussed in detail. It is easy to verify that in this case the critical region becomes

(27) $\{x: \Phi_n'(x)\Phi_n(x) > d_\alpha\},$

where $\Phi_n = \Phi_n(x)$ is an m-dimensional random column-vector the k-th entry of which equals

$$\frac{\sqrt{n}}{4\pi} \int_{-\pi}^{\pi} \Phi_k(\lambda) I_n(\lambda) f^{-1}(\lambda) d\lambda, \quad \Phi_k(\lambda) = 2\tilde{\Phi}_k(\lambda/\pi),$$

and d_α is a quantile of a χ^2-distribution with m degrees of freedom. It is also easy to verify that $L[V(\xi+A)]$ coincides here with the noncentral χ^2-distribution with m degrees of freedom and the noncentrality parameter $\mu'\mu$, where μ is a column-vector whose k-th entry equals

$$\mu_k = \frac{1}{4\pi} \int_{-\pi}^{\pi} \Phi_k(\lambda) a(\lambda) d\lambda.$$

The asymptotic value (as $n \to \infty$) of the power of the test equal to

$$\int_{d_\alpha}^{\infty} l_m(x, \mu'\mu) dx,$$

where $l_m(x, \mu'\mu)$ is the density of this distribution, and it depends to a large extent on the form of orthogonal functions $\Phi_1, ..., \Phi_m$. It is advisable to attempt to choose these functions in a manner such that the power of the test is maximal. Observe that for a fixed m the value of the quantity

$$\int_{d_\alpha}^{\infty} l_m(x, d) dx$$

increases with the value of the noncentrality parameter d, but it decreases with the number of degrees of freedom m. Therefore the best choice of functions Φ_1, ..., Φ_m will be in general not a simple problem even in the relatively simple case when the alternative H_1 is fixed and thus the function a is completely determined. Nevertheless, in the important particular case when the class of alternatives can be described by the function a, representable as the finite combinations of orthogonal functions Φ_1, ..., Φ_p, i.e., $a(\lambda) = h_1\Phi_1(\lambda) + \cdots + h_p\Phi_p(\lambda)$, it is recommended (in Subsection 2.2 of Chapter V) to utilize the same orthogonal functions when constructing the critical region (27). For such a choice of the critical region, the degrees of freedom m will coincide with p and the noncentrality parameter $\mu'\mu$ in the asymptotic expression of the power of the test will attain its maximal value equal to

$$\frac{1}{4\pi} \int_{-\pi}^{\pi} a^2(\lambda)d\lambda.$$

In Subsection 2.3 of Chapter V a number of specific examples for constructing goodness-of-fit tests determined by the critical region of the form (27) are presented. The first of these examples deals with testing the hypothesis H_0 that X_t is a linear autoregressive process of order q, against the "close" (approaching) alternative H_1 that the order of autoregression equals q', where $q' \geqslant q$. The test statistic corresponding to this problem -- which is determined by the critical region (27) -- includes, in particular, the well known Quenouille test [72] (cf. also [138] or [139] page 95).

The next example deals with the more general problem -- which was apparently not discussed previously -- of testing the hypothesis H_0 that X_t is a mixture autoregressive process of order q and moving-average process of order r, versus the "close" alternative H_1 that these orders are q' and r' respectively, where $q' \geqslant q$ and $r' \geqslant r$. Finally the last example concerns testing hypotheses for processes with an exponential spectral density.

10. As it was stated above, the final section of Chapter V is devoted to the important practical problem of testing the composite hypothesis H_0 that the spectral density f of a linear process X_t belongs to a parametric family of spectral densities $f = f_\Theta$, $\Theta \in \Theta$ where Θ is an open set of a p-dimensional Euclidean space R_p.

Subsections 2 and 3 of this section are devoted to the construction of goodness-of-fit tests for testing this hypothesis H_0 and the investigation of asymptotic properties of their powers under a general class of alternatives. First, however, we shall consider a more general case when the n-dimensional distribution $P^{(n)}$ of the observed random vector $\mathbf{X} = (X_1, ..., X_n)$ depends on the unknown p-dimensional parameter Θ (which, however, does not completely determine this distribution). Assume that there exists an m-dimensional (with $m > p$) random column-vector $\clubsuit_{n,\Theta} = \clubsuit_{n,\Theta}(\mathbf{X})$; such that as $n \to \infty$ its distribution converges to the m-dimensional distribution $N(\mu, I_m)$ with mathematical expectation μ and the unit covariance matrix I_m. Assume also that if Θ_* is a $\sqrt{n}$-consistent estimator of Θ then

$$\clubsuit_{n,\Theta_*} - \clubsuit_{n,\Theta} + B \sqrt{n} (\Theta_* - \Theta) \to 0$$

in $P^{(n)}$ probability as $n \to \infty$, where B is a nonrandom $(m \times p)$-matrix of rank p. Next, let there exist a consistent estimator B_* of the matrix B and an estimator $\hat{\Theta}$ of the parameter Θ such that $\sqrt{n}(\hat{\Theta}-\Theta) - (B'B)^{-1}B'\clubsuit_{n,\Theta} \to 0$ as $n \to \infty$ in $P^{(n)}$ probability. (Remarks 1 and 2 in Subsection 3.1 of Chapter V are devoted to the justification of such a construction.) Then as $n \to \infty$ the distribution of random variables $\clubsuit'_{n,\Theta}\clubsuit_{n,\Theta}$ and $\clubsuit'_{n,\Theta_*}A_*\clubsuit_{n,\Theta_*}$, where $A_* = I_m - B_*(B_*B_*)^{-1}B_*'$, converge to a noncentral χ^2-distribution with $m-p$ degrees of freedom and the noncentrality parameter $\mu'A\mu$, where $A = I_m - B(B'B)^{-1}B'$ (cf. Lemmas 1 and 2 on pages 285 and 286).

Along with $\hat{\Theta}$ in Subsection 3.1 the estimator $\tilde{\Theta}$ of the parameter Θ is introduced, such that $\sqrt{n}(\tilde{\Theta}-\Theta) - W^{-1}L_{n,\Theta} \to \infty$ as $n \to \infty$ in $P^{(n)}$-probability, where W is a nondegenerate $(p \times p)$-matrix and the p-dimensional random vector $L_{n,\Theta} = L_{n,\Theta}(\mathbf{X})$ possesses the property that the distribution of $(m+p)$-dimensional vector $(\clubsuit_{n,\Theta}, L_{n,\Theta})$ converges as $n \to \infty$ to the normal distribution

$$N\left(\begin{bmatrix} I_m & B \\ B' & W \end{bmatrix} \begin{bmatrix} \mu \\ \kappa \end{bmatrix}, \begin{bmatrix} I_m & B \\ B' & W \end{bmatrix} \right),$$

where κ is a p-dimensional column-vector. It is proved that if there exists a consistent estimator W_* of the matrix W and $L_{n, \Theta}$ -- in addition to the condition stated above -- satisfies also the relation

$$L_{n, \Theta_*} - L_{n, \Theta} + W\sqrt{n}(\Theta_* - \Theta) \to 0$$

in $P^{(n)}$ probability as $n \to \infty$, then the distribution of the random variable

$$[\Phi_{n, \Theta_*} - B_* W_*^{-1} L_{n, \Theta_*}]' C_*^{-1} [\Phi_{n, \Theta_*} - B_* W_*^{-1} L_{n, \Theta_*}]$$

where $C_* = I_m - B_* W_*^{-1} B_*'$ as $n \to \infty$ tends to a noncentral χ^2-distribution with m degrees of freedom and noncentrality parameter $\mu' C \mu$ (cf. Lemma 4 on page 288).

The general results presented here are widely applicable (cf. a discussion of this in the Appendix to Chapter V). For us it is especially important that these results can be applied to the problem of constructing goodness-of-fit tests for testing a composite hypothesis H_0 that the spectral density f of a linear process $X_t = \varepsilon_t + g_1 \varepsilon_{t-1} + g_2 \varepsilon_{t-2} + \cdots$ belongs to the family of functions f_Θ, $\Theta \in \Theta$ (where Θ is an open set in the Euclidean space R_p). In Subsection 3.2 this problem is solved under the assumption that the distribution of random variables ε_t is unknown (it is only known that $E(\varepsilon_t) = 0$ and $E(\varepsilon_t^4) < \infty$). Denote by $\Phi_{n, \Theta}$ and $L_{n, \Theta}$ the m-dimensional and p-dimensional vectors whose k-th components are equal to

$$\frac{\sqrt{n}}{4\pi} \int_{-\pi}^{\pi} \Phi_{k, \Theta}(\lambda) \frac{I_n(\lambda)}{f_\Theta(\lambda)} d\lambda$$

and

$$\frac{\sqrt{n}}{4\pi} \int_{-\pi}^{\pi} \frac{\partial}{\partial \Theta_k} \log f_\Theta(\lambda) \frac{I_n(\lambda)}{f_\Theta(\lambda)} d\lambda,$$

respectively, where the functions $\Phi_{k, \Theta}$, $k = 1, ..., m$ are orthogonal for all $\Theta \in \Theta$. In such a case if the hypothesis H_0 is valid, then as it can be shown under general conditions the vectors $\Phi_{n, \Theta}$ and $L_{n, \Theta}$ will satisfy the above stated conditions in which μ and κ are assumed to be zero vectors and $B = B_0$ and $W = W_0$ be matrices whose (k, l)-th entries are defined by the formulas (V.3.16) and (V.3.17) respectively. This assertion permits us to construct -- based on given estimators Θ_* and $\hat{\Theta}$ of parameter Θ and B_* and W_* of matrices B and W

respectively -- three different goodness-of-fit tests[8] for testing the hypothesis H_0 determined by the critical regions (V.3.20) - (V.3.22).

Assume now that according to the "close alternative" H_1, the process X_t is a linear process satisfying condition (26). Under this condition in Subsection 3.3 of Chapter V the asymptotic (as $n \to \infty$) value of the power of the proposed tests is determined (cf. formula (V.3.23)).

Finally at the end of Subsection 3.3 in Chapter V a number of specific examples of constructing goodness-of-fit tests of testing the composite hypothesis H_0 concerning the spectral density f of a linear proces X_t are considered.

In many papers cited herein, the statistical inference concerning the spectral density f of a random process X_t was considered simultaneously for both discrete time processes with $t = ..., -1, 0, 1, ...$ and processes with continuous time t, $-\infty < t < \infty$. However, the limited scope of this monograph does not allow us to include results given in these papers (as well as in the adjoining papers [41,48-50] by the author) dealing with processes with continuous time (which in many cases are related to the results discussed herein dealing with processes with discrete time). In order to fill this gap at least partially, we mention briefly analogous results dealing with the continuous case also in the appendices to each one of the five chapters of this monograph and refer the reader to the literature which contains these results. In addition, we also discuss very briefly the feasibility of carrying over some of the results of this work to more general cases of multidimensional (vector-valued) processes and random fields (i.e., process depending on a multivariate parameter t).

---

[8]As it is stated in the Appendix to Chapter V, the first of these tests is related to the well-known Pearson's χ^2-test for testing the composite hypothesis about the form of the distribution of independent identically distributed random variables based on grouped observations (cf., e.g., [71,76,147]) while the second and third are modifications of Pearson's test proposed in [52] and [91] respectively.

Chapter I
PROPERTIES OF MAXIMUM LIKELIHOOD FUNCTION FOR A GAUSSIAN TIME SERIES

1. General Expression for the log Likelihood

1. Let X_t, $t = ..., -1, 0, 1, ...$ be a Gaussian stationary process with zero expected value $E(X_t) = 0$, finite variance $D(X_t) = E(X_t^2) < \infty$, and absolutely continuous spectral function

$$F(\lambda) = \int_{-\pi}^{\lambda} f(\lambda) d\lambda, \quad -\pi \leqslant \lambda \leqslant \pi,$$

where $f = f(\lambda)$ is the spectral density of the process X_t.[1]
Denote by

$$B_f = [\beta(\tau-s)]_{\tau,s=1,...,n} = \begin{bmatrix} \beta(0) & \beta(1) & ... & \beta(n-1) \\ \beta(1) & \beta(0) & ... & \beta(n-2) \\ \vdots & \vdots & & \vdots \\ \beta(n-1) & \beta(n-2) & ... & \beta(0) \end{bmatrix}$$

the Toeplitz matrix associated with the function f, where

$$\beta(\tau) = \int_{-\pi}^{\pi} f(\lambda) e^{i\lambda\tau} d\lambda$$

---

[1] It is assumed, in addition that $\int_{-\pi}^{\pi} \log f(\lambda) d\lambda > -\infty$, that is, X_t is regular in Kolomogorov's sense.

is the covariance function, i.e.,

$$\beta(\tau) = E(X_t X_{t+\tau}).$$

As is well-known, the n-dimensional probability density $p_n(x_1, ..., x_n)$ of random variables $X_1, ..., X_n$ is of the form

(1) $$p_n(x_1, ..., x_n) = \frac{1}{(2\pi)^{n/2}[\det(B_f)]^{1/2}} e^{-\frac{1}{2} x' B_f^{-1} x}.$$

where $\mathbf{x} = (x_1, ..., x_n)' \in R_n$. Denote by L_n the logaritm of the likelihood function

(2) $$L_n = \log p_n(X_1, ..., X_n).$$

It then follows from (1) and (2) that

(3) $$L_n = -\frac{1}{2} \{n \log 2\pi + \log \det(B_f) + X' B_f^{-1} X\}$$

where $\mathbf{X}$ is an n-dimensional random vector-column, whose k-th element equals X_k.

When solving various problems dealing with statistical inference about the spectral density f (or the covariance function $\beta(\tau)$) in the case when f is known only up to a certain number of unknown parameters, it is often required to obtain an explicit expression for $\det(B_f)$ and B_f^{-1} in formula (3). The latter problem almost always turns out to be very complicated. Even in the relatively simple case when X_t is an autoregressive process of low order, the formulas for B_f^{-1} become cumbersome as long as the order of autoregression exceeds $q = 1$ (cf. [111,118]). The simplies case when $q = 1$ is considered below.

Example 1. Let X_t satisfy the difference equation

(4) $$X_t - \theta X_{t-1} = \varepsilon_t,$$

where ε_t is a sequence of independent Gaussian random variables with zero expected value $E(\varepsilon_t) = 0$ and variance $E(\varepsilon_t^2) = \sigma^2 > 0$ and $|\theta| < 1$. As it is known (cf., e.g., [24]) the covariance function $\beta(\tau)$ and the spectral density f are given in this case by the formulas

(5) $\qquad \beta(\tau) = \dfrac{\sigma^2}{1-\Theta^2}\, \Theta^{|\tau|}$

and

(6) $\qquad f(\lambda) = \dfrac{\sigma^2}{2\pi}\, |1 - \Theta e^{i\lambda}|^{-2}$

respectively.   Hence the (k,ℓ)-th entry of the matrix B_f is

equal to $\sigma^2 \Theta^{|k-\ell|}/(1-\Theta^2)$ and it is easy to verify that $\sigma^2 B_f^{-1}$ is an $(n \times n)$-matrix whose (k,ℓ)-th entry is different from zero if $k = \ell$, in which case it equals $1 + \Theta^2(1 - \delta_{1k}-\delta_{nk})$ or if $|k-\ell| = 1$, in which case it equals Θ.

Since

$$\det(B_f) = \frac{\sigma^2}{1-\Theta^2}$$

and

$$\mathbf{X'}B_f^{-1}\mathbf{X} = \sigma^{-2}\left[(1-\Theta^2)X_1^2 + \sum_{j=2}^{n} (X_j-\Theta X_{j-1})^2 \right],$$

it follows that

$$L_n = -\frac{1}{2}\left\{ n \log 2\pi\sigma^2 - \log(1 - \Theta^2) \right.$$

(7)

$$\left. + \sigma^{-2}\left[(1-\Theta^2)X_1^2 + \sum_{j=2}^{n} (X_j-\Theta X_{j-1})^2 \right]\right\}.$$

For $q > 1$ the formulas for L_n are far more cumbersome (cf. [111]).   Even more involved is the case when X_t is a moving average process or moreso when X_t is a mixed autoregressive and moving average process (formulas from which explicit expressions for L_n can be determined are given for example in [3,87,93,113,149]).   This can be substantiated from the examples presented below pertaining to the cases of low orders of autoregression and moving averages.   Before proceeding to consider these examples we shall first present in the next section an alternative expression for L_n for a class of processes which includes the case of mixed autoregressive and moving average processes.

1.2.   Let $f_j(\lambda)$, $j = 0,1, ..., q$, $-\pi \leqslant \lambda \leqslant \pi$ be the nonnegative summable functions related by the recursive formulas

(8) $f_j(\lambda) = f_{j-1}(\lambda)|z-z_j|^2, \quad z = e^{i\lambda}, \quad |z_j| \leqslant 1, \quad j = 1, ..., q,$

and let $\beta_j(\tau)$ be their Fourier coefficients

(9) $\beta_j(\tau) = \int_{-\pi}^{\pi} f_j(\lambda)e^{i\lambda\tau}d\lambda, \quad j = 0, 1, ..., q.$

In view of (8) the functions f_q and f_0 are related as follows:

(10) $f_q(\lambda) = f_0(\lambda)|Q_q(z)|^2, \quad Q_q(z) = (z-z_1)\cdots(z-z_q),$

where $z = e^{i\lambda}$ and $z_1, ..., z_q$ are the roots of the polynomial $Q_q(z)$ whose absolute value are less than one.

In the case when $Q_q(z)$ possesses r distinct roots ($r \leqslant q$), say $\zeta_1, ..., \zeta_r$ of multiplicity $q_1, ..., q_r$ with $q_1 + \cdots + q_r = q$ we shall assume for simplicity that $z_1 = \cdots = z_{q_1} = \zeta_1, z_{q_1+1} =$

$\cdots = z_{q_1+q_2} = \zeta_2$, and so on.   Obviously

$$Q(z) = Q_q(z; \zeta_1, ..., \zeta_r) = \prod_{1 \leqslant j \leqslant r} (z-\zeta_j)^{q_j}.$$

The Toeplitz matrices

$$B_{f_0}^{(n)} = [\beta_0(\tau-s)]_{\tau, s=1, ..., n}, \quad B_{f_q}^{(n)} = [\beta_q(\tau-s)]_{\tau, s=1, ..., n}$$

corresponding to functions f_0 and f_q respectively are related by the following simple equation

(11) $CB_{f_0}^{(n+q)}C^* = B_{f_q}^{(n)},$

where C is an $(n \times n+q)$-matrix with the (k,ℓ)-th entry different from zero only if $0 \leqslant \ell-k \leqslant q$, in which case it is equal to the coefficient at $z^{\ell-k}$ of the polynomial $Q_q(z)$. The $(n+q \times n)$-matrix C^* is the transpose of the conjugate entries of C.

The following properties of the matrix C are easily verified:

(1) The matrix of dimensionality n consisting of the last n columns of C can be written in the form of the product $C_{z_1}\cdots C_{z_q}$, where C_z is a lower triangular matrix of dimensionality n, the (k,ℓ)-th entry of which equals 1 or

-z for $k = \ell$ and $k-\ell = 1$ respectively, while it equals zero otherwise.

(2) Consider the $(t-s+1)\times n$-matrix $V_{s,t} = [V_{s,t}^{(1)}, ..., V_{s,t}^{(r)}]$, $s \leqslant t$, where

$$V_{s,t}^{(j)} = \begin{bmatrix} \zeta_j^s & s\zeta_j^s & \cdots & \dfrac{s^{q_j-1}}{(q_j-1)!}\,\zeta_j^s \\ \vdots & & & \vdots \\ \zeta_j^t & t\zeta_j^s & \cdots & \dfrac{t^{q_j-1}}{(q_j-1)!}\,\zeta_j^t \end{bmatrix}.$$

If $t-s = n+q-1$, then $CV_{s,t}$ will be a zero $(n\times q)$-matrix[2](to be denoted $CV_{s,t} = 0$). In particular, $CV_{1-q,n} = 0$.

Remark 1. Clearly C_z^{-1} is also a lower triangular matrix of dimensionality n with (k,ℓ)-th entry equal to $z^{k-\ell}$ for $k \geqslant \ell$. Denote $S = (C_{z_1} ... C_{z_q})^{-1}$. Evidently det $S = 1$.

Remark 2. In the particular case of simple roots when $q_1 = \cdots = q_r = 1$ and $r = q > 1$ the matrix $V_{1-q,0}$ of dimensionality q is a Vandermonde matrix(obviously $V_{0,0} = 1$ for $q = 1$) whose determinant is known to be $\Pi_{k<j}(\zeta_k^{-1} - \zeta_j^{-1})$.

In the other particular case of a single root ζ_1 of multiplicity $q_1 = q > 1$, the determinant of the matrix $V_{1-q,0}^{(1)} = V_{1-q,0}$ equals $\zeta_1^{-q(q-1)/2}$. To verify this we multiply sucessively $V_{1-q,0}$ on the left by the following two matrices of dimensionality q: first by a matrix whose (k,ℓ)-th entry differs from zero and equals one, only for $k+\ell = q+1$ (its determinant equals $(-1)^{q(q-1)/2}$) and next by a lower triangular matrix with coefficient at $z^{k-\ell}$ of the polynomial $(z-\zeta_1)^{k-1}$ at the intersection of the k-th row and ℓ-th column $(k \geqslant \ell)$. This will result in an upper triangular matrix with ones on the main diagonal. Thus the required assertion follows from the fact that the determinant of the second factor equals $(-\zeta_1)^{q(q-1)/2}$.

---

[2]This follows from the fact that for every $j = 1, ..., r$, the matrix $CV_{s,t}^{(j)}$ possesses the (k, ℓ)-th entry $a_{k\ell}(z_j)$ where $a_{k1}(z) = z^{s+k-1}Q_q(z)$ and $a_{k\ell}(z) = z(\partial/\partial z)a_{k,\ell-1}(z)/\ell$ for $\ell = 2, ..., q$.

In the general case we proceed analogously by multiplying $V_{1-q, 0}$ on the left first by the same matrix as above (with the determinant $(-1)^{q(q-1)/2}$), and then again by a lower triangular matrix but this time with the coefficient at $z^{k-\ell}$ of the polynomial $(z-z_1)...(z-z_{k-1})$ at the intersection of the k-th row and ℓ-th column ($k \geqslant \ell$). The determinant of the latter matrix is the product of diagonal elements 1, $(-z_1)$, ..., $\Pi_{1 \leqslant k \leqslant q}(-z_k)$, and thus it equals

$$(-1)^{q(q-1)/2} \prod_{1 \leqslant k \leqslant q} z_1 \cdots z_k.$$

As a result of the operations on $V_{1-q, 0}$ described in the preceding paragraph we obtain an upper triangular matrix of dimensionality q.[3] The first q_1 diagonal entries of the matrix all equal 1, the next q_2 diagonal entries all equal $(1-\zeta_1/\zeta_2)^{q_1}$, and so on, the last q_r diagonal entries being equal to $\Pi_{1 \leqslant k \leqslant r}(1-\zeta_k/\zeta_q)^{q_k}$. To summarize, the determinant here equals one in the case of a single root of multiplicity $q_1 = q$ (as it was established above), while in the case of distinct roots it equals $\Pi_{k<j}(1-\zeta_k/\zeta_j)^{q_k q_j}$.

Thus we have

$$\prod_{1 \leqslant k < q} z_1 ... z_k \det V_{1-q, 0} = \begin{cases} 1, & \text{for } r = 1 \\ \prod_{k<j}(\zeta_k^{-1}-\zeta_j^{-1})^{q_k q_j} & \text{for } r > 1 \end{cases}$$

Finally expressing the product $\Pi z_1 \cdots z_k$ in terms of the quantities $\zeta_1, ..., \zeta_r$ we state the result obtained as:

Lemma 1. *The matrix* $V_{1-q, 0} = [V_{1-q, 0}^{(1)}, ..., V_{1-q, 0}^{(r)}]$ *of dimensionality* q *is nondegenerate and its determinant satisfies the relation*

---

[3]To verify this we carry out the same operation on the matrix $v_{1-q, 0}^{(j)}$. This will result in a $(q \times q_j)$-matrix $[a_{k\ell}^{(j)}]$, where $a_{k1}^{(j)} = \zeta_j^{1-k} Q_k(\zeta_j)$ with $Q_k(z) = (z-z_1) \cdots (z-z_{k-1})$, then $a_{k2}^{(j)} = (\partial/\partial \zeta_j) a_{k1}^{(j)}/\zeta_j$, and so on.

$$\prod_{1 \leqslant j \leqslant r} \zeta_j^{q_j(q_j-1)/2} \det V_{1-q,0}$$

$$= \begin{cases} 1, & \text{for } r = 1, \\ \prod_{k<j} (\zeta_k^{-1}-\zeta_j^{-1})^{q_k q_j}, & \text{for } r > 1. \end{cases}$$

In view of Lemma 1 and the properties (1) and (2) of the matrix C we have

$$SC = [-V_{1,n}V_{1-q,0}^{-1}, I_n].$$

Therefore subdividing the matrix of dimensionality $2q+n$

$$(12) \qquad \mathbb{B}\{_{f_0}^{(2q+n)}]^{-1} = \begin{bmatrix} 0 & V_{1-q,n}^* \\ V_{1-q,n} & B_{f_0}^{(n+q)} \end{bmatrix}^{-1}$$

into blocks after the $2q$-th row and $2q$-column[4] in the lower right-hand side we obtain a matrix which is the inverse of

$$(13) \qquad \begin{aligned} &[-V_{1,n}V_{1-q,0}^{-1}, I_n]B_{f_0}^{(n+q)}[-V_{1,n}V_{1-q,0}^{-1}, I_n]^* \\ &= SCB_{f_0}^{(n+q)} C^*S^* = SB_{f_q}^{(n)} S^* \end{aligned}$$

(the last equality follows from (11)).

This last result serves as the basis for the proof of the following assertions

Lemma 2. *Let* $\mathbf{X}_n = \text{col}(X_1, ..., X_n)$ *and* $\mathbf{Y}_{n+m} = \text{col}(0, ..., 0, Y_1, ..., Y_n) = (0, S\mathbf{X}_n)$, *i.e., in view of Remark 1*

$$(14) \qquad Y_{k_0} = \sum_{k_0 \geqslant ... \geqslant k_q \geqslant 1} z_1^{k_0-k_1} ... z_q^{k_{q-1}-k_q} X_{k_q},$$

$$k_0 = 1, ..., n.$$

---

[4] Here we utilize the well-known formula:

$$\begin{bmatrix} A & C \\ C^* & B \end{bmatrix}^{-1} = \begin{bmatrix} A^{-1}+A^{-1}CD^{-1}C^*A^{-1*} & -A^{-1}CD^{-1} \\ -D^{-1}C^*A^{-1*} & D^{-1} \end{bmatrix}$$

where $D = B - C^*A^{-1}C$.

Then the quadratic form $\mathbf{X}_n^*[B_{f_q}^{(n)}]^{-1}\mathbf{X}_n$ *can be written out in the following three alternative ways:*

(15)
$$
\begin{aligned}
\mathbf{X}_n^*[B_{f_q}^{(n)}]^{-1}\mathbf{X}_n &= \mathbf{Y}_{n+2q}^*[B_{f_0}^{(2q+n)}]^{-1}\mathbf{Y}_{n+2q} \\
&= \mathbf{Y}_{n+q}^* D_{f_0}^{(n+q)}\mathbf{Y}_{n+q} \\
&= \overline{\mathbf{Y}}_{n+q}^*[B_{f_q}^{(n+q)}]^{-1}\overline{\mathbf{Y}}_{n+q},
\end{aligned}
$$

where

$$
\begin{aligned}
D_{f_0}^{(n+q)} &= [B_{f_0}^{(n+q)}]^{-1} \\
&\quad - [B_{f_0}^{(n+q)}]^{-1}V_{1-q,n}(V_{1-q,n}^*[B_{f_0}^{(n+q)}]^{-1}V_{1-q,n})^{-1} \\
&\quad \times V_{1-q,n}^*[B_{f_0}^{(n+q)}]^{-1}
\end{aligned}
$$

and

$$
\overline{\mathbf{Y}}_{n+q} = B_{f_0}^{(n+q)}D_{f_0}^{(n+q)}\mathbf{Y}_{n+q}.
$$

Proof.   The first of the equations (15) is a direct corollary of the relation between the matrix and the corresponding submatrices in the subdivision (12) into blocks after the $2q$-th row and the $2q$-th column.   Indeed,

$$
\begin{aligned}
\mathbf{X}_n^*[B_{f_q}^{(n)}]^{-1}\mathbf{X}_n &= (S\mathbf{X}_n)^*(SB_{f_q}^{(n)}S^*)^{-1}(S\mathbf{X}_n) \\
&= \mathbf{Y}_{n+2q}^*[B_{f_0}^{(n+2q)}]^{-1}\mathbf{Y}_{n+2q}.
\end{aligned}
$$

If, however, the matrix (12) is subdivided into blocks after q rows and q columns then in the right-hand side lower corner the matrix $D_{f_0}^{(n+q)}$ will appear.   From here and the definition of the vector $\mathbf{Y}_{n+q}$ follows the second equality in (15).   The last equality in (15) is a simple consequence of idempotency of the matrix

$$
[B_{f_0}^{(n+q)}]^{1/2}D_{f_0}^{(n+q)}[B_{f_0}^{(n+q)}]^{1/2}. \qquad\qquad \square
$$

Remark 3. In view of property (1) of matrix C and Remark 1, the components of the vector $\mathbf{Y}_{n+q}$ are particular solutions to the system of difference equations:

$$[z^{-q}Q_q(z)]_{z^{-1}=B}Y_t = X_t, \quad t = 1, ..., n$$

under the boundary conditions $Y_{1-q} = ... = Y_0 = 0$, where B is a backward shift operator: $BY_t = Y_{t-1}$ (this can easily be verified from the matrix notation of this system: $C\mathbf{Y}_{n+q} = \mathbf{X}_n$).

Thus, in view of property (2) of matrix C any solution of the system presented above may be written in the form $\mathbf{Y}_{n+q} + V_{1-q,0}\mathbf{\eta}_q$ with an arbitrary q-dimensional vector $\mathbf{\eta}_q$.

Remark 4. It is of interest to connect the arguments presented herein with a linear regression problem in a system of observations consisting of vector $\mathbf{Y}_{n+q}$ with the mean value $V_{1-q,n}\mathbf{\eta}_q$, where $\mathbf{\eta}_q$ is a vector of unknown regression coefficients while $V_{1-q,n}$ is a known regresor of the form given above. Moreover, let $B_{f_0}^{(n+q)}$ be a covariance matrix of the vector $\mathbf{Y}_{1-q,n}$. Then, as it is well-known (see, e.g., [35]) the best linear unbiased estimator (BLUE) (or the Gauss-Markov estimator) of the parameter $\mathbf{\eta}_q$ is given by

$$\mathbf{\eta}_0 = (V^*B^{-1}V)^{-1}V^*B^{-1}\mathbf{Y}_{n+q}, \quad V = V_{1-q,n}, \quad B = B_{f_0}^{(n+q)}.$$

This estimator is determined from the condition

$$\min[\mathbf{Y}_{n+q}-V\eta]^*B^{-1}[\mathbf{Y}_{n+q}-V\eta] = \mathbf{Y}_{n+q}^* D_{f_0}^{(n+q)} \mathbf{Y}_{n+q},$$

$$V = V_{1-q,n}, \quad B = B_{f_0}^{(n+q)}.$$

Now we can state the basic result of this subsection:

Theorem 1. Let X_t be a (complex-valued) Gaussian stationary process with zero expectation and spectral density f_q represented in the form (10).

Then the logarithm of the likelihood function[5]

$$\mathbf{L}_n = -\frac{1}{2}\{n\log 2\pi + \log \det[B_{f_q}^{(n)}] + \mathbf{X}_n^*[B_{f_q}^{(n)}]^{-1}\mathbf{X}_n\}$$

---

[5]See the next page.

of the vector $\mathbf{X}_n$ *with components* $X_1, \ldots, X_n$ *can be written out in the form*

(16)
$$L_n = -\frac{1}{2}\{n \log 2\pi + \log \det[B_{f_0}^{(n+q)}] + \mathbf{Y}_{n+q}^* D_{f_0}^{(n+q)} \mathbf{Y}_{n+q}$$
$$+ \log \det(V_{1-q,n}^*[B_{f_0}^{(n+q)}]^{-1}V_{1-q,n}) - \log|\det V_{1-q,0}|^2\},$$

where the last summand is determined from Lemma 1 and $\mathbf{Y}_{n+q}$ *is related to* $\mathbf{X}_n$ *in the manner indicated in Lemma 2.*

The quadratic form in (16) can also be represented in two other alternative forms as given by relations (15).

Proof. In view of Lemma 2 it is only required to prove the relation:

$$\det[B_{f_q}^{(n)}] = \det B \ \det(V^*B^{-1}V)|\det V_{1-q,0}|^2,$$

where we use the notations $B = B_{f_0}^{(n+q)}$ and $V = V_{1-q,n}$.

This follows from the following formulas:

(17) $\det[\mathcal{B}_{f_0}^{(n+2q)}] = (-1)^q \det(V^*B^{-1}V)\det B,$

$V = V_{1-q,n},\quad B = B_{f_0}^{(n+q)}$

(18) $\det[\mathcal{B}_{f_0}^{(n+2q)}] = \det[B_{f_0}^{(2q)}]\cdot\det(WBW^*),$

$W = [-V_{1,n}V_{1-q,0}^{-1}, \ I_n],$

since

$$\det[\mathcal{B}_{f_0}^{(2q)}] = \det \begin{bmatrix} 0 & V_{1-q,0}^* \\ V_{1-q,0} & B_{f_0}^{(q)} \end{bmatrix} = (-1)^q|\det V_{1-q,0}|^2$$

and in view of (13)

---

[5] As it is known (cf., e.g., [26]) the probability density of complex-valued observations $X_1, \ldots, X_n$ indeed satisfies the relation $p_n(X_1, \ldots, X_n) = 2^n \exp(2L_n)$.

(19) $\det WBW^* = \det[B_{f_q}^{(n)}]$

(recall that $\det S = 1$).

Relations (17) and (18) are obtained by applying the well-known formulas for computing determinants of block matrices[6] by subdividing $B_{f_0}^{(2q+n)}$ into blocks after the q-th

row and qth-column and then after the $2q$-th row and $2q$-th column. The theorem is proved. $\square$

1.3. As an application of the formula (16) we shall now, in particular, deduce from this formula an explicit expression for the logarithm of likelihood L_n in the case of processes of moving averages of the first and the second orders as well as of a mixed autoregressive-moving average process of the first orders.

Example 2. Let X_t be a moving average process of the first order of the form

$$X_t = z_1 \varepsilon_t - \varepsilon_{t-1}, \qquad |z_1| \leqslant 1,$$

where ε_t here and in all the examples below is the same sequence as in Example 1.

Then, as it is known, the spectral density of process X_t is of the form

$$f(\lambda) = f_1(\lambda) = \frac{\sigma^2}{2\pi} |z - z_1|^2, \qquad z = e^{i\lambda},$$

i.e., here $q = 1$ and

$$f_0(\lambda) = \sigma^2/2\pi, \qquad \beta_0(\tau) = \sigma^2 \delta_{0\tau}.$$

Since here $B_{f_0}^{(n)} = \sigma^2 I_n$ and $V_{0,n}$ is an $n+1$-dimensional

vector whose k-th component equals z_1^{k-1} it follows that

---

[6] Here we have in mind the formulas

$$\det \begin{bmatrix} A & C \\ C^* & B \end{bmatrix} = \det B \cdot \det(A - CB^{-1}C^*)$$

$$= \det A \cdot \det(B - C^*A^{-1}C).$$

$$\sigma^2 \mathbf{Y}_{n+1}^* D_{f_0}^{(n+1)} \mathbf{Y}_{n+1} = \mathbf{Y}_{n+1}^* \mathbf{Y}_{n+1} - \frac{1}{N_n(z_1)} |V_{0,n}^* \mathbf{Y}_{n+1}|^2$$

$$= \sum_{t=1}^{n} |Y_t|^2 - \frac{1}{N_n(z_1)} \left| \sum_{t=1}^{n} \overline{z}_1^t Y_t \right|^2,$$

where

$$N_n(z) = \sum_{j=0}^{n} |z|^{2j} = \frac{1-|z|^{2(n+1)}}{1-|z|^2},$$

and in view of formula (14)

$$Y_k = \sum_{j=1}^{k} z_1^{k-j} X_j.$$

The validity of the nonrandom summands on the r.h.s. of (16) can also be easily verified. This yields the final result

$$L_n = -\frac{1}{2}\left\{ n \log 2\pi\sigma^2 + \log N_n(z_1) \right.$$

$$\left. + \frac{1}{\sigma^2}\left[\sum_{t=1}^{n} |Y_t|^2 - \frac{1}{N_n(z_1)} \left| \sum_{t=1}^{n} \overline{z}_1^t Y_t \right|^2 \right] \right\}.$$

Thus, for example, for $z_1 = 1$ we have

$$L_n = -\frac{1}{2}\left\{ n \log 2\pi\sigma^2 + \log(n+1) \right.$$

(20)

$$\left. + \frac{1}{\sigma^2}\left[\sum_{j=1}^{n} Y_j^2 - \frac{1}{n+1} \left(\sum_{j=1}^{n} Y_j \right)^2 \right] \right\},$$

where

$$Y_j = X_1 + X_2 + \cdots + X_j, \qquad j = 1, ..., n.$$

Example 3. Let X_t be a moving average process of the second order represented in the form

$$X_t = \varepsilon_t - 2\varepsilon_{t-1} + \varepsilon_{t-2}.$$

The spectral density

$$f(\lambda) = f_2(\lambda) = \frac{\sigma^2}{2\pi}|1-z|^4, \quad z = e^{i\lambda}$$

is determined by the equations (8) for $q = 2$, $z_1 = z_2 = 1$, $f_0(\lambda) = \sigma^2/2\pi$,

$$f_1(\lambda) = \frac{\sigma^2}{2\pi}|1-z|^2, \quad z = e^{i\lambda}.$$

Since

$$\beta_0(\tau) = \sigma^2 \delta_{0\tau},$$

it follows that $B_{f_0}^{(n)} = \sigma^2 I_n$. The matrix $V_{-1,n}$ consists of two columns of length $n+2$: $\text{col}\{1, ..., 1\}$ and $\text{col}\{-1,0,1, ..., n\}$.
Therefore

$$\det[V_{-1,0}] = \det\begin{bmatrix} 1 & -1 \\ 1 & 0 \end{bmatrix} = 1.$$

and

$$\det(V'_{-1,n}V_{-1,n}) = \det\begin{bmatrix} n+2 & \dfrac{n(n+1)}{2}-1 \\ \dfrac{n(n+1)}{2}-1 & \dfrac{n(n+1)(n+2)}{6}+1 \end{bmatrix}$$

$$= \frac{1}{12}(n+1)(n+1)(n^2+5n+6).$$

Denote the last quantity by d_n.
Since

$$\sigma^2 Y^*_{n+2} D_{f_0}^{(n+2)} Y_{n+2} = Y^*_{n+2} Y_{n+2}$$

$$-\begin{bmatrix} \sum\limits_1^n Y_t \\ \sum\limits_1^n tY_t \end{bmatrix}' \begin{bmatrix} n+2 & \dfrac{n(n+1)}{2}-1 \\ \dfrac{n(n+1)}{2}-1 & \dfrac{n(n+1)(n+2)}{6}+1 \end{bmatrix}^{-1} \begin{bmatrix} \sum\limits_1^n Y_t \\ \sum\limits_1^n tY_t \end{bmatrix},$$

we thus finally obtain that

$$L_n = -\frac{1}{2}\left\{ n \log 2\pi\sigma^2 + \log d_n \right.$$

$$+ \frac{1}{\sigma^2}\left[\sum_{j=1}^{n} Y_j^2 - \frac{1}{d_n}\left(\left(\sum_{j=1}^{n} Y_j\right)^2 \left[1 + \frac{n(n+1)(2n+1)}{6}\right]\right)\right.$$

$$\left.\left. + (2+n)\left(\sum_{j=1}^{n} jY_j\right)^2 + (2-n-n^2)\sum_{j=1}^{n} Y_j \sum_{j=1}^{n} jY_j\right]\right\},$$

where $Y_j = X_1 + 2X_2 + \cdots + jX_j$, $j = 1, ..., n$.

Example 4. Let X_t be a mixed autoregressive-moving average process of the first orders satisfying the difference equation

$$X_t - \Theta X_{t-1} = z_1\varepsilon_t - \varepsilon_{t-1},$$

where $|z_1| \leqslant 1$, $-1 < \Theta < 1$.
It is known that the spectral density f of the process X_t is of the form

$$f(\lambda) = f_1(\lambda) = \frac{\sigma^2}{2\pi}|z_1 - z|^2|1 - \Theta z|^{-2}, \quad z = e^{i\lambda}.$$

Here $q = 1$

$$f_0(\lambda) = \frac{\sigma^2}{2\pi}|1 - \Theta z|^{-2}, \quad z = e^{i\lambda},$$

and

$$B_0(\tau) = \sigma^2 \frac{\Theta^{|\tau|}}{1 - \Theta^2},$$

so that $B_{f_0}^{(n)}$ coincides with the matrix B_f defined in Example 1. Therefore

(21) $\qquad . \sigma^2 x'[B_{f_0}^{(n)}]^{-1}y = (1-\Theta^2)x_1y_1 + \sum_{j=2}^{n}(x_j - \Theta x_{j-1})(y_j - \Theta y_{j-1}),$

where x and y are arbitrary column vectors whose k-th entry equals x_k and y_k respectively.
Since $V_{0,n}$ is here as in Example 2, we have in view of (21)

$$\sigma^2 V_{0,n}^*[B_{f_0}^{(n+1)}]^{-1}V_{0,n} = (1-\Theta^2) + |z_1-\Theta|^2 N_{n-1}(z_1),$$

where $N_n(z)$ is determined as in Example 1. Moreover,

$$\det[B_{f_0}^{(n)}] = \frac{\sigma^{2n}}{1-\Theta^2}$$

(cf. Example 1) so that the nonrandom summand in the braces on the r.h.s. of (16) is equal to

$$n \log 2\pi\sigma^2 + \log\left[1 + N_{n-1}(z_1)\frac{|z_1-\Theta|^2}{1-\Theta^2}\right].$$

Applying formula (21) we also determine the random summand

$$Y_{n+1}^* D_{f_0}^{(n+1)} Y_{n+1} = Y_{n+1}^*[B_{f_0}^{(n+1)}]^{-1}Y_{n+1}$$

$$- \frac{1}{1-\Theta^2+|z_1-\Theta|^2N_{n-1}(z_1)} |V_{0,n}^*[B_{f_0}^{(n+1)}]^{-1}Y_{n+1}|^2,$$

where the $n+1$-dimensional vector Y_{n+1} possesses the zero first component while the remaining components are equal to:

$$Y_j = \sum_{k=1}^{j} X_k z_1^{j-k}, \quad j = 1, ..., n.$$

We thus obtain

$$L_n = -\frac{1}{2}\left\{n \log 2\pi\sigma^2 + \log\left[1 + \frac{|z_1-\Theta|^2 N_{n-1}(z_1)}{1-\Theta^2}\right]\right.$$

$$+ \frac{1}{\sigma^2}\left[\sum_{j=1}^{n} |Y_{j,\Theta}|^2 - |z_1-\Theta|^2\left|\sum_{j=1}^{n} \bar{z}_1^{j-1}Y_{j,\Theta}\right|^2 (1-\Theta^2\right.$$

$$\left.\left. + |z_1-\Theta|^2 N_{n-1}(z_1))^{-1}\right]\right\},$$

where $Y_{1,\Theta} = X_1$, $Y_{j,\Theta} = Y_j - \Theta Y_{j-1}$, and $j = 2, ..., n$. In particular, we have for $z_1 = 1$

$$L_n = -\frac{1}{2}\left\{ n \log 2\pi\sigma^2 + \log\left[1 + n\frac{1-\Theta}{1+\Theta}\right]\right.$$

$$\left. + \frac{1}{\sigma^2}\left[\sum_{j=1}^{n} Y_{j,\Theta}^2 - \left(n + \frac{1+\Theta}{1-\Theta}\right)^{-1}\left(\sum_{j=1}^{n} Y_{j,\Theta}\right)^2\right]\right\},$$

where

$$Y_{1,\Theta} = X_1, \quad Y_{j,\Theta} = X_j + (1-\Theta)\sum_{k=1}^{j-1} X_k, \quad j = 1, ..., n.$$

2. Asymptotic Expression for the "Principal Part" of the log Likelihood

2.1. As we have seen in the preceding section, an explicit expression for the log likelihood L_n is, as a rule, very cumbersome (in those cases when it can be written down at all explicitly). This fact substantially complicates the possibility of deriving statistical inference about the process X_t based on a study of the expression for L_n. However, most often, the interest centers around the case when n exceeds manifold of a typical damping time of correlations between values of X_t. (We note in connection to this that in general statistical inference based on the utilization of L_n is usually optimal in some sensible manner only asymptotically as $n \to \infty$.) If n is very large, then one could utilize not the whole exact expression for L_n but just its "principal part" $\tilde{L}_n$ with hardly any changes in the properties of the resulting statistical inference. This principal part satisfies

(1) $\qquad n^{-1/2}(L_n - \tilde{L}_n) \to 0.$

as $n \to \infty$ (the convergence is in probability). As we shall see in this Section, under very general conditions, one can choose $\tilde{L}_n$ to be of a much simpler form than L_n.

Lemma 1. *Let the spectral density f and the covariance function $\beta(\tau)$ of a random process X_t satisfy the following conditions*

1) *There exists a positive number m such that $m \le f(\lambda)$, $-\pi \le \lambda \le \pi$.*

2) $\qquad \sum_{\tau=1}^{\infty} \tau|\beta(\tau)|^2 < \infty.$

Then the expected value and the variance of the quadratic form

(2) $$\mathbf{X}'(B_f^{-1} - B_{1/(2\pi)^2 f})\mathbf{X}$$

are bounded.

Remark 1. As it is shown at the end of Appendix 1 to this Chapter, it follows from conditions 1) and 2) that

$$\sum_{\tau=1}^{\infty} \tau |\rho(\tau)|^2 < \infty;$$

where

$$\rho(\tau) = \frac{1}{(2\pi)^2} \int_{-\pi}^{\pi} \frac{e^{i\lambda\tau}}{f(\lambda)} \, d\lambda$$

are the Fourier coefficients of the function $1/(2\pi)^2 f$. Recall that since B_f denotes the Toeplitz matrix of dimension n associated with the function f, it follows that

$$B_{1/(2\pi)^2 f} = [\rho(k - \ell)]_{k, \ell=1, \ldots, n}.$$

Proof. We shall use the well-known formulas for the expected value and variance of a quadratic form $X'AX$ corresponding to an arbitrary symmetric matrix A:

(3) $$E(\mathbf{X}'A\mathbf{X}) = \text{tr}(B_f A).$$

and

(4) $$D(\mathbf{X}'A\mathbf{X}) = 2\,\text{tr}(B_f A B_f A).$$

From (2)-(4) we have

(5) $$E[\mathbf{X}'(B_f^{-1} - B_{1/(2\pi)^2 f})\mathbf{X}] = \text{tr}[I_n - B_f B_{1/(2\pi)^2 f}]$$

and

(6) $$D[\mathbf{X}'(B_f^{-1} - B_{1/(2\pi)^2 f})\mathbf{X}] = 2\,\text{tr}[(I_n - B_f B_{1/(2\pi)^2 f})]^2,$$

hence the proof of the lemma follows from the assertion of Lemma A1.4 in Appendix 1 taking $a = 1/2\pi$.   □

Theorem 1. *Under the conditions of Lemma 1 the relation* (1) *is valid,*[7] *where*

(7)
$$\tilde{L}_n = -\frac{n}{2} \left\{ \log 2\pi + \frac{1}{2\pi} \int_{-\pi}^{\pi} \log[2\pi f(\lambda)]d\lambda \right.$$

$$\left. + \frac{1}{2\pi} \int_{-\pi}^{\pi} \frac{I_n(\lambda,X)}{f(\lambda)} d\lambda \right\},$$

and

(8)
$$I_n(\lambda,X) = \frac{1}{2\pi n} \left| \sum_{j=1}^{n} X_j e^{-i\lambda j} \right|^2$$

is the periodogram of the process X_t.

Proof. Clearly

(9)
$$X'B_{1/(2\pi)^2 f} X = \sum_{k,j=1}^{n} X_k X_j \frac{1}{(2\pi)^2} \int_{-\pi}^{\pi} \frac{e^{i\lambda(k-j)}}{f(\lambda)} d\lambda$$

$$= \frac{n}{2\pi} \int_{-\pi}^{\pi} \frac{I_n(\lambda,X)}{f(\lambda)} d\lambda.$$

In view of [67] under the conditions of Lemma 1 we have

(10)
$$\left| \log \det(B_f) - \frac{n}{2\pi} \int_{-\pi}^{\pi} \log[2\pi f(\lambda)]d\lambda \right| < \infty.$$

The proof of the theorem now follows from (1.3), (7)-(10) and the assertion of Lemma 1. □

2.2. In this section an expression for $\tilde{L}_n$ satisfying relation (1) under the conditions on f, which are different from the conditions of Lemma 1, will be obtained. Specifically, let the following assumption be valid.

Assumption 1. *The spectral density f can be represented in the form*

---

[7]Actually, a stronger assertion is proved: the expected value and the variance of the quadratic form L_n - $\tilde{L}_n$ are bounded and thus this form is bounded in probability (cf. also [38,160]).

$$f(\lambda) = f_q(\lambda) = f_0(\lambda)|(z-z_1)\cdots(z-z_q)|^2, \qquad z = e^{i\lambda},$$

(11)

$$z_j = e^{i\lambda_j}, \qquad j = 1, ..., q,$$

where $\lambda_1, ..., \lambda_q$ are all unequal to each other and the positive summable function f_0 and its Fourier coefficients $\beta_0(\tau)$ satisfy the conditions 1) and 2) of Lemma 1.

Furthermore, let $\lambda_1, ..., \lambda_p$ be points of continuity of f_0 so that in view of Fejér's theorem [65, p. 89]

(12) $\qquad |\sigma_n(\lambda_j, f_0) - f_0(\lambda_j)| \to 0$

for all λ_j, $j = 1, ..., q$, where

(13) $\qquad \sigma_n(\lambda, f_0) = \dfrac{1}{2\pi} \sum_{k=-n}^{n} \left[1 - \dfrac{|k|}{n}\right] \beta_0(k) e^{-i\lambda k}$

are Fejér's sums of the function f_0.

Remark 2. From these conditions in particular, the relation

(14) $\qquad |s_n(\lambda_j, f_0) - f_0(\lambda_j)| \to 0, \qquad j = 1, ..., q,$

follows as $n \to \infty$, where

(15) $\qquad s_n(\lambda, f_0) = \dfrac{1}{2\pi} \sum_{k=-n}^{n} e^{-i\lambda k} \beta_0(k)$

are partial sums of the Fourier series for the function f_0. Indeed,

(16)
$$|s_n(\lambda_j, f_0) - f_0(\lambda_j)| \leqslant |\sigma_n(\lambda_j, f_0) - f_0(\lambda_j)|$$
$$+ \dfrac{1}{2\pi n} \sum_{k=-n}^{n} |k| \, |\beta_0(k)| \to 0$$

(cf. [65], p. 79).

Remark 3. Under the above-stated conditions in addition to (12) the relation

(17) $\qquad \dfrac{1}{n} V^* B_{f_0} V \to 2\pi \, \text{diag}\{f_0(\lambda_1), ..., f_0(\lambda_q)\}$

is valid, where $V = V_{1,n}$ is an $(n \times q)$-matrix consisting of columns $(z_j^1, ..., z_j^n)$, $j = 1, ..., q$. Indeed, the assertion about the diagonal entries is identical to (12); as far as the off-diagonal entries are concerned, they converge to zero, since we have (as in the case in (16)) for $k \neq j$

$$\frac{1}{n} \left| \sum_{\ell,r=1}^{n} \bar{z}_j^\ell z_k^r \, \beta_0(\ell-r) \right| = \frac{1}{n} \left| \beta_0(0) \sum_{\ell=1}^{n} (z_k\bar{z}_j)^\ell \right.$$

$$\left. + \sum_{m=1}^{n-1} [\beta_0(m)\bar{z}_j^m + \overline{\beta_0(m)}\, z_k^m] \sum_{\ell=1}^{n-m} (z_k\bar{z}_j)^\ell \right|$$

$$(18) \qquad = \frac{1}{n} \left| z_k\bar{z}_j (1-z_k\bar{z}_j)^{-1} \left[\beta_0(0) + \sum_{m=1}^{n-1} (\beta_0(m)\bar{z}_j^m + \overline{\beta_0(m)}z_k^m) \right. \right.$$

$$\left. \left. - (z_k\bar{z}_j)^n \left[\beta_0(0) + \sum_{m=1}^{n-1} (\beta_0(m)\bar{z}_k^m + \overline{\beta_0(m)}z_j^m) \right] \right] \right|$$

$$\leqslant \frac{C}{n} \sum_{k=-n}^{n} |\beta_0(k)| \to 0,$$

where C is a constant (which depends on the distance $\lambda_k - \lambda_j$ but not on n).

In fact, a much more general assertion is valid: if f_1, ..., f_m are functions satisfying the same conditions as f_0 then

$$(19) \qquad \frac{1}{n} V^* B_{f_1} \cdots B_{f_m} V \to (2\pi)^m \mathrm{diag}\{f_1(\lambda_j) \cdots f_m(\lambda_j),$$

$$j = 1, ..., q\}.$$

(The proof is presented in Appendix 2, Proposition A2.1.)

Remark 4. Clearly, λ_1, ..., λ_q are points of continuity of the function $h_0 = 1/(2\pi)^2 f_0$ so that the corresponding Fejér's sums satisfy

$$(20) \qquad |\sigma_n(\lambda_j, h_0) - h_0(\lambda_j)| \to 0$$

for all λ_j, $j = 1, ..., q$. In view of Remark 1 the assertion

$$(21) \qquad |s_n(\lambda_j, h_0) - h_0(\lambda_j)| \to 0, \qquad j = 1, ..., q,$$

is valid. Also

$$(22) \qquad \frac{1}{n} V^* B_{h_0} V \to \mathrm{diag}\left\{ \frac{1}{2\pi f_0(\lambda_1)}, ..., \frac{1}{2\pi f_0(\lambda_q)} \right\}.$$

The last relation implies that in particular

(23)
$$\frac{1}{n} V^* B_{f_0}^{-1} V \to \text{diag}\left\{ \frac{1}{2\pi f_0(\lambda_1)}, ..., \frac{1}{2\pi f_0(\lambda_q)} \right\}$$

(cf. assertion 1) of Corollary A2.1 presented in Appendix 2.)

Lemma 2. *Let the spectral density* $f = f_q$ *of a Gaussian random process* X_t, $t = ...,-1,0,1, ...$ *satisfy Assumption 1. Then the mathematical expectation and the variance of the quadratic form*

(24)
$$X^*[B_{f_q}^{-1} - S^*(I_n - P_n)B_{h_0}(I_n - P_n)S]X$$

are bounded where as usual $X = \text{col}(X_1, ..., X_n)$ *with covariance matrix* $B_{f_q} = B_{f_q}^{(n)}$, S *is the matrix defined in Remark 1 of Subsection 1.2 and* $P_n = V(V^*V)^{-1}V^*$ *is an orthogonal projection into the subspace generated by the columns* $v_j = \text{col}(z_j^1, ..., z_j^n)$, $j = 1, ..., q$, *of the matrix* $V = V_{1,n}$.

Proof. We shall utilize the representation of the covariance matrix of the vector $Y_n = SX_n$ (equal to $EY_nY_n^* = SB_{f_q}S^*$) in the form (1.13), where $V_{1-q,n}$ now consists of the columns $\text{col}(z_j^t, -q < t \leqslant n)$. We have for $\tilde{Y}_n = (I_n - P_n)Y_n$

(25)
$$E\tilde{Y}_n\tilde{Y}_n^* = (I_n - P_n)B_{f_0}(I_n - P_n).$$

Hence it follows from the formulas[8] (3) and (4) and the Assertion of Lemma A1.4 of Appendix 1 (for $a = 1/2\pi$, $f = f_0$) that both the expectation and the variance of quadratic form (24) are bounded,

(26)
$$\text{tr}[I_n - B_{h_0}(I_n - P_n)B_{f_0}(I_n - P_n)] - \text{tr}(I_n - B_{h_0}B_{f_0})$$
$$= \text{tr}[B_{h_0}(P_nB_{f_0} + B_{f_0}P_n - P_nB_{f_0}P_n)]$$
$$= \text{tr}[(V^*V)^{-1}V^*(B_{f_0}B_{h_0} + B_{h_0}B_{f_0})V$$
$$- (V^*V)^{-1}V^*B_{f_0}V(V^*V)^{-1}V^*B_{f_0}V] \to q$$

---

[8]The following should be taken into account here: in view of the usual convention that in the case of a complex X_t, $EX_nX_n^* = B_{f_q}$ is valid

and

(27)
$$\text{tr}[I_n - B_{h_0}(I_n - P_n)B_{f_0}(I_n - P_n)]^2 - \text{tr}[(I_n - B_{h_0}B_{f_0})^2]$$
$$= 2\text{tr}(I_n - B_{h_0}B_{f_0})[B_{h_0}(P_n B_{f_0} + B_{f_0}P_n - P_n B_{f_0}P_n)]$$
$$+ \text{tr}[B_{h_0}(P_n B_{f_0} + B_{f_0}P_n - P_n B_{f_0}P_n)]^2 \to q$$

(the passage to the limit as $n \to \infty$ is carried out taking (19) into account; cf. also its partial cases, formulas (17) and (22)). Lemma 2 is thus proved. □

Remark 5. One can arrive at the approximation of the quadratic form $X_n^* B_{f_q}^{-1} X_n$ by the quantity $\tilde{Y}_n^* B_{h_0} \tilde{Y}_n$ carried out herein by using the following informal arguments.

In view of the last equalities (1.15) and Remark 3 in the preceding section, $X_n^* B_{f_q}^{-1} X_n$ can be written as

$$(Y_{n+q} - V_{1-q,n}\eta_0)^* [B_{f_0}^{(n+q)}]^{-1}(Y_{n+q} - V_{1-q,n}\eta_0),$$

where $V_{1-q,n}$ is a trigonometric "regressor" and η_0 is a BLUE for the imaginary "regression parameter" η_q. Moreover, it is well-known (cf. [35] Section 7.5) that in the case of such a regressor, a simpler least squares estimator

$$\eta_{LS} = (V_{1-q,n}^* V_{1-q,n})^{-1} V_{1-q,n} Y_{n+q}$$

possesses the same asymptotic properties as η_0. Thus the quantity

$$\tilde{Y}_{n+q}^* [B_{f_0}^{(n+q)}]^{-1} \tilde{Y}_{n+q},$$

where

$$\tilde{Y}_{n+q} = (I_{n+q} - P_{n+q})Y_{n+q} = Y_{n+q} - V_{1-q,n}\eta_{LS}$$

should in principle serve as a nice approximation to $X_n^* B_{f_q}^{-1} X_n$.

---

together with $EX_n X_n' = 0$, the variance of a quadratic form $X_n^* A X_n$ generated by a Hermitian matrix A is the trace of the square of matrix AB_{f_q}.

Evidently, it is desirable to introduce further simplifications by reducing the dimensionality $n+q$ to n. Now it is easy to arrive at the desired approximation by substituting B_{h_0} in place of $B_{f_0}^{-1}$ and hoping that (as Lemma 1 indicates) the required precision of approximation will be retained.

Analogously to (9),

(28)
$$X^*S^*(I_n-P_n)B_{h_0}(I_n-P_n)SX$$
$$= \tilde{Y}^* B_{h_0} \tilde{Y} = \frac{n}{2\pi} \int_{-\pi}^{\pi} \frac{I_n(\lambda, \tilde{Y})}{f_0(\lambda)} \, d\lambda,$$

where

(29)
$$I_n(\lambda, \tilde{Y}) = \frac{1}{2\pi n} \left| \sum_{j=1}^{n} \tilde{Y}_j e^{-i\lambda j} \right|^2$$

is a periodogram constructed by using the components $\tilde{Y}_1$, ..., $\tilde{Y}_n$ of the vector $\tilde{Y} = (I_n-P_n)Y$. Recall that Y is a vector of dimensionality n with components (1.14).

We turn now to the nonrandom summands in the r.h.s. of (1.16). In view of Lemma 1.1 formula (10) for $f = f_0$ and relation (21) we have

(30)
$$\log \det B_{f_0}^{(n)} = \log \det B_{f_0}^{(n+q)}$$
$$+ \log \det(V_{1-q, n}^*[B_{f_0}^{(n+q)}]^{-1}V_{1-q, n})$$
$$- \log|\det V_{1-q, 0}|^2 = \frac{n}{2\pi} \int_{-\pi}^{\pi} \log[2\pi f_0(\lambda)] d\lambda$$
$$+ O(\log n).$$

The assertion of Lemma 2 together with the relations (28) and (30) allow us to state the following theorem.

Theorem 2. *Under the conditions of Lemma 2, relation (1) is valid, where L_n is the logarithm of the likelihood function (cf. Theorem 1.1) and*

(31)
$$\tilde{L}_n = -\frac{n}{2} \left\{ \log 2\pi + \frac{1}{2\pi} \int_{-\pi}^{\pi} \log[2\pi f_0(\lambda)] d\lambda \right.$$

$$+ \frac{1}{2\pi} \int_{-\pi}^{\pi} \frac{I_n(\lambda, \tilde{Y})}{f_0(\lambda)}\, d\lambda \Bigg\}.$$

We now present some examples of the applications of formula (31).

Example 1. If X_t is a random process considered in Example 2 of the preceding section, then it is easy to show that

$$\tilde{L}_n = -\frac{n}{2}\left\{ \log 2\pi\sigma^2 + \frac{1}{n\sigma^2}\left[\sum_{k=1}^{n} \left| Y_k - \frac{1}{n} \sum_{j=1}^{n} Y_j \bar{z}_1^{-j} \right|^2 \right]\right\},$$

where

$$Y_j = \sum_{k=1}^{j} z_1^{j-k} X_k, \qquad j = 1, \ldots, n.$$

In particular, for $z_0 = 1$ we have

(32) $$\tilde{L}_n = -\left\{ \log 2\pi\sigma^2 + \frac{1}{n\sigma^2}\left[\sum_{k=1}^{n} \left[Y_k - \frac{1}{n} \sum_{j=1}^{n} Y_j \right]^2 \right]\right\},$$

where $Y_j = X_1 + \ldots + X_j$, $j = 1, \ldots, n$.

Example 2. Let the spectral density f of the process X_t be of the form

$$f(\lambda) = f_1(\lambda) = \frac{\sigma^2}{2\pi}|1-z|^2 \cdot |1-\Theta z|^{-2}, \qquad z = e^{i\lambda},$$

where $-1 < \Theta < 1$, $\sigma^2 > 0$. Since $q = 1$, $\lambda_1 = 0$ here, and

$$f_0(\lambda) = \frac{\sigma^2}{2\pi}|1-\Theta z|^{-2}, \qquad z = e^{i\lambda},$$

it follows that

$$\frac{1}{2\pi}\Bigg| \int_{-\pi}^{\pi} \log[2\pi f_0(\lambda)]d\lambda = \log \sigma^2,$$

$$\frac{1}{2\pi} \int_{-\pi}^{\pi} \frac{I_n(\lambda, \tilde{Y})}{f_0(\lambda)}d\lambda = \frac{1}{\sigma^2 n}\left[(1+\Theta^2)\sum_{j=1}^{n} \tilde{Y}_j^2 - 2\Theta \sum_{j=1}^{n-1} \tilde{Y}_j \tilde{Y}_{j+1} \right],$$

where $\tilde{Y}_j = Y_j - \frac{1}{n}\sum_{k=1}^{n} Y_k$ and $Y_j = X_1 + \cdots + X_j$.

Consequently, we obtain from (31),

(33) $$\tilde{L}_n = -\frac{n}{2}\left\{\log 2\pi\sigma^2 + \frac{1}{\sigma^2}[(1+\theta^2)r_{0,y} - 2\theta r_{1,y}]\right\},$$

where

(34) $$r_{k,y} = \frac{1}{n}\sum_{j=1}^{n-k} Y_j Y_{j+k} - \left[\frac{1}{n}\sum_{j=1}^{n} Y_j\right]^2.$$

Example 3. Let $q = 1$, $\lambda_1 = 0$, and

(35) $$f_1(\lambda) = f_0(\lambda)|1-z|^2, \qquad z = e^{i\lambda},$$

where

(36) $$f_0(\lambda) = \frac{\sigma^2}{2\pi}|1-\theta z|^2, \qquad z = e^{i\lambda}, \quad \sigma^2 > 0, \quad -1 < \theta < 1.$$

Then

$$\frac{1}{2\pi}\int_{-\pi}^{\pi} \log[2\pi f_0(\lambda)]d\lambda = \log \sigma^2,$$

$$\frac{1}{2\pi}\int_{-\pi}^{\pi} \frac{I_n(\lambda, \tilde{Y})}{f_0(\lambda)}d\lambda = \frac{1}{\sigma^2}\int_{-\pi}^{\pi} \frac{I_n(\lambda, \tilde{Y})}{|1-\theta z|^2} d\lambda$$

$$= \frac{1}{\sigma^2(1-\theta^2)}\left\{\frac{1}{n}\sum_{j=1}^{n} \tilde{Y}_j^2 + 2\sum_{k=1}^{n-1}\frac{\theta^k}{n}\sum_{j=1}^{n-k}\tilde{Y}_j\tilde{Y}_{j+k}\right\},$$

and thus in view of (31) and (34) we have

(37) $$\tilde{L}_n = -\frac{n}{2}\left\{\log 2\pi\sigma^2 + \frac{1}{q^2(1-\theta^2)}\left[r_{0,y} + 2\sum_{k=1}^{n-1}\theta^k r_{k,y}\right]\right\}$$

3. The Asymptotic Differentiability of Gaussian Distributions with Spectral Densites Separated from Zero

3.1.   In this Section we shall return to the case where X_t, $t =$ ..., $-1,0,1$, ... is a Gaussian random process with zero mathematical expectation and spectral density $f(\lambda)$, $-\pi \leqslant \lambda \leqslant \pi$, which satisfies the condition $m \leqslant f(\lambda) \leqslant M$.

Consider a sequence of spectral densities $g_n(\lambda) = f(\lambda)(1 + a_n(\lambda)/\sqrt{n})$, $n = 1,2, ...$, where $a_n(\lambda)$, $n = 1,2, ...$, is a sequence of functions on $-\pi \leqslant \lambda \leqslant \pi$ convergent as $n \to \infty$ in the mean square to a square integrable function a on $-\pi \leqslant \lambda \leqslant \pi$.

Denote as above by $P_n(f)$ the Gaussian distribution which corresponds to the spectral density f. Then the following theorem is valid.

Theorem 1. *Sequences of probability measures* $P_n(f)$, $n = 1,2, \ldots$ *and* $P_n(f + n^{-1/2}af)$, $n = 1,2, \ldots$, *are contiguous.*

Proof. As it is well-known (cf., e.g., [110]) it is sufficient to show that the log of the likelihood ratio

$$\Lambda(f,g) = \log \frac{dP_n(g)}{dP_n(f)} = \frac{1}{2} \{\log \det(B_f)$$

$$- \log \det(B_g) + X'(B_f^{-1} - B_g^{-1})X\}$$

(where B_f is as above a Toeplitz matrix associated with f, $X = (X_1, \ldots, X_n)$, and $g = f(1 + n^{-1/2}a)$) is bounded in both $P_n(f)$ and $P_n(g)$ probability.

Since

$$(1) \qquad B_g = B_f + n^{-1/2}B_{fa}$$

we have, evidently,

$$B_f^{-1} - B_g^{-1} = B_f^{-1}(B_g - B_f)B_g^{-1} = n^{-1/2}B_f^{-1}B_{fa}B_g^{-1},$$

$$(2) \qquad \log \det(B_f) - \log \det(B_g) = -\log \det(B_g B_f^{-1})$$

$$= -\log \det(I_n + n^{-1/2}B_{fa}B_f^{-1})$$

and

$$(3) \qquad \Lambda(f,g) = -\frac{1}{2} \{\log \det(I_n + n^{-1/2}B_{fa}B_f^{-1})$$

$$- n^{-1/2}X'B_f^{-1}B_{fa}B_g^{-1}X\}.$$

First we show the boundedness in $P_n(g)$ probability. In view of (2.3), (2.4), and (3) we have

$$(4) \qquad E_g(\Lambda(f,g)) = -\frac{1}{2} \{\log \det(I_n + n^{-1/2}B_{fa}B_f^{-1})$$

$$- n^{1/2}\text{tr}(B_f^{-1}B_{fa})\} = -\frac{1}{2} \{U_n(n^{-1/2}B_{fa}B_f^{-1})$$

$$- \frac{1}{2n}\mathrm{tr}[(B_{\mathrm{fa}}B_{\mathrm{f}}^{-1})^2]\},$$

(5) $$D_{\mathrm{g}}(\Lambda(f,g)) = \frac{1}{2n} \ \mathrm{tr}[(B_{\mathrm{f}}^{-1}B_{\mathrm{fa}})^2],$$

where the index g under E and D -- the symbols of expectation and variance, respectively -- designates that the averaging is carried out with respect to the measure $P_{\mathrm{n}}(g)$; besides

(6) $$U_{\mathrm{n}}(A) = \log \ \det(I_{\mathrm{n}}+A) - \mathrm{tr}(A) + \frac{1}{2} \ \mathrm{tr}(A^2).$$

From Lemma A1.1, assertions 6)-8) of Lemma A1.2 in Appendix 1, and formulas (4) and (5) we obtain

(7) $$E_{\mathrm{g}}(\Lambda(f,g)) \rightarrow \gamma^2/4$$

and

(8) $$D_{\mathrm{g}}(\Lambda(f,g)) \rightarrow \gamma^2/2$$

where

$$\gamma^2 = \frac{1}{2\pi} \ \int_{-\pi}^{\pi} a^2(\lambda)d\lambda.$$

Boundedness with respect to $P_{\mathrm{n}}(g)$ is thus proved. It follows from (2.3), (2), (3), and (6) that

$$E_{\mathrm{f}}(\Lambda(f,g)) = - \frac{1}{2}\{\log \ \det(I_{\mathrm{n}} + n^{-1/2}B_{\mathrm{fa}}B_{\mathrm{f}}^{-1})$$

$$- n^{-1/2}\mathrm{tr}(B_{\mathrm{fa}}B_{\mathrm{g}}^{-1}) \} = - \frac{1}{2} \left\{ U_{\mathrm{n}} (n^{-1/2}B_{\mathrm{fa}} B_{\mathrm{f}}^{-1}) \right.$$

$$+ \frac{1}{n}\mathrm{tr}(B_{\mathrm{fa}}B_{\mathrm{f}}^{-1}B_{\mathrm{fa}}B_{\mathrm{g}}^{-1}) - \frac{1}{2n} \ \mathrm{tr}[(B_{\mathrm{fa}}B_{\mathrm{f}}^{-1})^2] \Big\}$$

$$= - \frac{1}{2} \left\{ U_{\mathrm{n}}(n^{-1/2}B_{\mathrm{fa}}B_{\mathrm{f}}^{-1} + \frac{1}{2n}\mathrm{tr}[(B_{\mathrm{fa}}B_{\mathrm{f}}^{-1})^2] \right.$$

$$- n^{-3/2} \ \mathrm{tr}[(B_{\mathrm{fa}}B_{\mathrm{f}}^{-1})^2 B_{\mathrm{fa}}B_{\mathrm{g}}^{-1}] \Big\}.$$

From here, Lemma A1.1, and assertions 8) and 11) of Lemma A1.2 in Appendix 1 we have

(9) $$E_{\mathrm{f}}(\Lambda(f,g)) \rightarrow -\gamma^2/4.$$

In view of (2.4), (2), (3), and assertions 8), 11), and 12) of Lemma A1.2 in Appendix 1 we obtain

(10)
$$D_f(\Lambda(f,g)) = \frac{1}{2n} \text{tr}[(B_{fa}B_g^{-1})^2] = \frac{1}{2n} \{\text{tr}[(B_{fa}B_f^{-1})^2$$
$$- n^{-1/2}\text{tr}[(B_{fa}B_f^{-1})^2 B_{fa}B_g^{-1}]$$
$$- n^{-1/2}\text{tr}[B_{fa}B_f^{-1}(B_{fa}B_g^{-1})^2)] \to \gamma^2/2.$$

We have thus verified the boundednesss of $\Lambda(f,g)$ with respect to $P_n(f)$ as well. □

Theorem 2. *Under the conditions stated above*

$$\Lambda(f,g_n) - \Lambda(f,g) \to 0$$

in $P_n(f)$ *probability as* $n \to \infty$.

Proof. Since

$$\Lambda(f,g_n) - \Lambda(f,g) = \Lambda(g,g_n)$$
$$= \frac{1}{2}\{\log \det(I_n + n^{-1/2}B_{f(a_n-a)} B_g^{-1})$$
$$- n^{-1/2}X'B_{g_n}^{-1}B_{f(a_n-a)}B_g^{-1}X\}$$

it follows that

$$E_f(\Lambda(f,g_n) - \Lambda(f,g))$$
$$= \frac{1}{2}\{\log \det(I_n + n^{-1/2}B_{f(a_n-a)}B_g^{-1})$$
$$- n^{-1/2}\text{tr}(B_f B_{g_n}^{-1}B_{f(a_n-a)}B_g^{-1})\}$$
$$= \frac{1}{2}\{U_n(n^{-1/2}B_{f(a_n-a)}B_g^{-1})$$
$$+ \frac{1}{n}\text{tr}(B_{fa_n}B_{g_n}^{-1}B_{f(a_n-a)}B_g^{-1})$$
$$- \frac{1}{2n}\text{tr}\{(B_{f(a_n-a)}B_g^{-1})^2\}\}$$

and

$$|E_f(\Lambda(f,g_n)-\Lambda(f,g))| \leqslant \frac{1}{2}\left\{|U_n(n^{-1/2}B_{f(a_n-a)}B_g^{-1})|\right.$$

(11)
$$+ n^{-1/2}|B_{f(a_n-a)}B_g^{-1}|(n^{-1/2}|B_{fa_n}B_g^{-1}|$$

$$\left.+\frac{1}{2}|B_{f(a_n-a)}B_g^{-1}|)\right\} \to 0$$

in view of Lemma A1.1, assertions 2) and 5) of Lemma A1.2
in Appendix 1. Analogously, one shows that

$$D_f(\Lambda(f,g_n)-\Lambda(f,g)) = \frac{1}{2n}\mathrm{tr}[(B_fB_{g_n}^{-1}B_{f(a_n-a)}B_g^{-1})^2]$$

(12)
$$= \frac{1}{2n}\mathrm{tr}\{B_{f(a_n-a)}B_g^{-1}\cdot n^{-1/2}(B_{fa_n}B_{g_n}^{-1}B_{f(a_n-a)}B_g^{-1})]^2\}$$

$$\leqslant \frac{1}{2n}|B_{f(a_n-a)}B_g^{-1}|^2(1+n^{-1/2}|B_{fa_n}B_{g_n}^{-1}|)^2 \to 0.$$

The proof of Theorem 2 now follows from (11), (12), and
Chebyshev's inequality.   □

Theorem 3. *Under the above stipulated conditions*

(13) $$\Lambda(f,g) - \frac{1}{2}\left\{n^{-1/2}[X'B_f^{-1}B_{a/2\pi}X - \mathrm{tr}(B_{a/2\pi})] - \frac{\gamma^2}{2}\right\} \to 0$$

in $P_n(f)$ probability as $n \to \infty$.

Proof. In view of (2.3) and (9) the mathematical expectation
on the l.h.s. of (13) converges to 0, while the variance, which
is equal to

$$\frac{1}{2n}\mathrm{tr}[(B_{fa}B_g^{-1} - B_{a/2\pi})^2]$$

in view of (2.4), converges to 0 by virtue of (2), assertions 5)
and 9) of Lemma A1.2 in Appendix 1, and the inequality

$$\mathrm{tr}[(B_{fa}B_f^{-1} - B_{a/2\pi})^2]$$

$$\leqslant |B_{fa}B_f^{-1} - B_{a/2\pi} - n^{-1/2}B_{fa}B_f^{-1}B_fB_g^{-1}|^2$$

$$\leqslant [|B_{fa}B_f^{-1} - B_{a/2\pi}| + n^{-1/2}|B_{fa}B_f^{-1}B_{fa}B_g^{-1}|]^2.$$ □

Lemma 1. *Let the Fourier coefficients*

$$\beta(j) = \int_{-\pi}^{\pi} f(\lambda)e^{i\lambda j}d\lambda, \qquad \rho(j) = \frac{1}{2\pi}\int_{-\pi}^{\pi}\frac{a(\lambda)}{f(\lambda)}e^{i\lambda j}d\lambda$$

of functions f and a/2πf respectively satisfy the conditions

$$\sum_{j=1}^{\infty} j|\beta(j)|^2 < \infty, \qquad \sum_{j=1}^{\infty} j|\rho(j)|^2 < \infty.$$

Then under the above stated conditions the expectation and the variance of the quadratic form $\mathbf{X}'(B_f^{-1}B_a - B_{a/2\pi f})\mathbf{X}$ *are bounded.*

The proof follows easily from the formulas (2.3) and (2.4) and the assertion of Lemma A1.4 in Appendix 1.
  Since

$$\mathbf{X}'B_{a/2\pi f}\mathbf{X} = \sum_{k,\,j=1}^{n} X_k X_j \frac{1}{2\pi} \int_{-\pi}^{\pi} \frac{a(\lambda)}{f(\lambda)} e^{i\lambda(k-j)} d\lambda$$

$$= n \int_{-\pi}^{\pi} \frac{I_n(\lambda)}{f(\lambda)} a(\lambda) d\lambda,$$

where

$$I_n(\lambda) = I_n(\lambda, X) = \frac{1}{2\pi n} \left| \sum_{j=1}^{n} X_j e^{i\lambda j} \right|^2$$

is a periodogram of X_t and

$$\mathrm{tr}(B_a) = n \int_{-\pi}^{\pi} a(\lambda) d\lambda$$

the results of Theorems 1-3 and Lemma 1 imply that the following theorem is valid.

Theorem 4. *Let the spectral density f of a Gaussian random process* X_t, $t = ..., -1,0,1, ...$ *satisfy the condition* $m \leqslant f(\lambda) \leqslant M$, *where m and M are positive numbers.*
  Let a_n, $n = 1,2, ...$ *be a sequence of square integrable functions convergent in the mean square as* $n \to \infty$ *to a square integrable function a, i.e.,*

$$(16) \qquad \int_{-\pi}^{\pi} |a_n(\lambda) - a(\lambda)|^2 d\lambda \to 0$$

as $n \to \infty$, *where* a_n *and a are such that* $g_n = f(1 + n^{-1/2}a_n)$ *and* $g = f(1 + n^{-1/2}a)$ *are nonnegative integrable functions.*
  Then the following assertions are valid.

1) *The sequences of Gaussian measures* $P_n(f)$, $n = 1,2, ...$ *and* $P_n(g)$, $n = 1,2, ...$ *are contiguous.*

2) $\Lambda(f,g) - \Lambda(f,g_n) \to 0$ in $P_n(f)$ probability as $n \to \infty$.

3) $\Lambda(f,g) - \dfrac{1}{2n^{1/2}} \left\{ \mathbf{X}' B_f^{-1} B_{a/2\pi} \mathbf{X} - \dfrac{n}{2\pi} \int_{-\pi}^{\pi} a(\lambda) d\lambda \right\}$

$$+ \dfrac{1}{8\pi} \int_{-\pi}^{\pi} a^2(\lambda) d\lambda \to 0$$

in $P_n(f)$ probability as $n \to \infty$, where $\mathbf{X} = (X_1, ..., X_n)'$.

4)   Let the conditions of Lemma 1 be fulfilled also. Then

$$\Lambda(f,g) - \dfrac{n^{1/2}}{4\pi} \int_{-\pi}^{\pi} \dfrac{I_n(\lambda) - f(\lambda)}{f(\lambda)} a(\lambda) d\lambda + \dfrac{1}{8\pi} \int_{-\pi}^{\pi} a^2(\lambda) d\lambda \to 0$$

in $P_n(f)$ probability as $n \to \infty$.

3.2.   We now proceed to consider an example of the applicability of the results obtained in the important case when the spectral density $f(\lambda)$ depends on a vector-valued parameter Θ, belonging to an open set Θ of the space R_p, i.e., $f = f_\Theta$, $\Theta \in \Theta$.

Assume that there exist the functions $\dot{\phi}_{k,\Theta} = \dot{\phi}_{k,\Theta}(\lambda)$, square integrable on $-\pi \leqslant \lambda \leqslant \pi$, such that

(17)
$$\int_{-\pi}^{\pi} \left\{ \dfrac{f_{\Theta + \epsilon i}(\lambda) - f_\Theta(\lambda)}{\epsilon f_\Theta(\lambda)} - \dot{\phi}_{k,\Theta}(\lambda) \right\}^2 d\lambda \to 0 ,$$

$$\Theta \in \Theta, \quad k = 1, ..., p,$$

as $\epsilon \to 0$ in which i is a vector whose k-th component equals 1, while all the other components are zero. The function $\dot{\phi}_{k,\Theta}$ defined in this manner is clearly a derivative of the function $\log f_\Theta$ in the L_2 sense with respect to the k-th component of the vector Θ.

Denote

(18) $a_n = a_{n,\Theta} = n^{1/2} \dfrac{f_{\Theta + n^{-1/2} \mathbf{h}_n} - f_\Theta}{f_\Theta}, \qquad n = 1, 2, ... ,$

where $\mathbf{h}_n$, $n = 1, 2, ...$, is a sequence of vectors such that $\Theta + n^{-1/2} \mathbf{h}_n \in \Theta$ and $\mathbf{h}_n \to \mathbf{h}$ as $n \to \infty$. It follows from (17)

(19) $\int_{-\pi}^{\pi} [a_n(\lambda) - \mathbf{h}' \dot{\phi}_\Theta(\lambda)]^2 \to 0$

as $n \to \infty$, where $\dot{\phi}_\Theta = \dot{\phi}_\Theta(\lambda)$ is a p-dimensional column vector, the k-th component of which equals $\dot{\phi}_{k,\Theta}$.

The following corollary from Theorem 4 is valid.

Corollary 1. *Let X_t, $t = ..., -1,0,1, ...$ be a Gaussian random process with spectral density f_Θ such that $m \leqslant f_\Theta(\lambda) \leqslant M$, $-\pi \leqslant \lambda \leqslant \pi$, $\Theta \in \Theta$, where m and M are positive numbers. Now let the square integrable functions $\dot\phi_{k,\Theta}$ exist satisfying condition (17). Then*

1) *the sequence of Gaussian measures $P_n(f_\Theta)$, $n = 1,2, ...$, and $P_n(f_{\Theta+n^{-1/2}h})$ $n = 1,2, ...$, are contiguous;*

2) *$\Lambda(f_\Theta, f_{\Theta+n^{-1/2}h}) - \Lambda(f_\Theta, f_{\Theta+n^{-1/2}h_n}) \to 0$ in $P_n(f_\Theta)$ probability.*

   Next let the conditions of Lemma 1 be satisfied. Then

3) $\Lambda(f_\Theta, f_{\Theta+n^{-1/2}h}) - \dfrac{n^{1/2}}{4\pi} \displaystyle\int_{-\pi}^{\pi} \dfrac{I_n(\lambda)-f(\lambda)}{f(\lambda)} \mathbf{h}'\dot\phi_\Theta(\lambda)d\lambda$

$$+ \frac{1}{8\pi} \int_{-\pi}^{\pi} [\mathbf{h}'\dot\phi_\Theta(\lambda)]^2 d\lambda \to 0,$$

   in $P_n(f)$ probability as $n \to \infty$.

Proof. Using the notation of this Subsection the functions f and g appearing in the statement of Theorem 4 are clearly of the form

(20) $g_n = f_\Theta(1 + n^{-1/2}a_n) = f_{\Theta+n^{-1/2}h_n}$

and in view of (18)

(21) $g = f_\Theta(1 + n^{-1/2}a) = f_\Theta(1 + n^{-1/2}\mathbf{h}'\dot\phi_\Theta)$

also since here $a = \mathbf{h}'\dot\phi_\Theta$.

Assertion (1) of Theorem 4 implies that the sequence of Gaussian measures $P_n(f_\Theta)$, $n = 1,2, ...$, and $P_n(g)$, $n = 1,2, ...$ where g is determined by equation (21), are contiguous. Since contiguity is a transitive property (cf., e.g., Remark 2.1 on page 8 of the book [110]) to prove assertion 1) it is sufficient to show that the sequence of measures $P_n(g)$, $n = 1,2, ...$, and $P_n(f_{\Theta+n^{-1/2}h})$, $n = 1,2, ...$ are contiguous, or clearly

(22) $\Lambda(f_{\Theta+n^{-1/2}h}, g) = \dfrac{1}{2} \{\log \det(B_{f_{\Theta+n^{-1/2}h}}) - \log \det(B_g)$

$$+ \mathbf{X}'(B_{f_{\Theta+n^{-1/2}h}}^{-1} - B_g^{-1})\mathbf{X}]$$

is bounded in $P_n(g)$ as well as in $P_n(f_{\Theta+n^{-1/2}h})$ probabilities.

In view of (20) and assertion 2) of Theorem 4, in order to prove (2) it is sufficient to show that $\Lambda(f_{\Theta+n}-1/2_h, g) \to 0$ in $P_n(f)$ probability. The validity of 3) is then a corollary of assertion 2), formula (20), and assertion 3) of Theorem 4.

Thus, in view of the arguments presented above, the proof of Corollary 1 follows from the validity of Lemma A2.1 presented in Appendix 2. □

The results of this Subsection imply that the family of Gaussian distributions $P_n(f_\Theta)$, $\Theta \in \Theta$ of the Gaussian random process X_t with spectral density f_Θ considered here satisfy conditions (D1)-(D3) presented in the Introduction on page 21 for $\tau_n = \sqrt{n}$,

$$(23) \qquad \Delta_{n,\Theta} = \frac{n^{1/2}}{4\pi} \int_{-\pi}^{\pi} \frac{I_n(\lambda) - f_\Theta(\lambda)}{f_\Theta(\lambda)} \dot{\Phi}_\Theta(\lambda) d\lambda$$

and

$$(24) \qquad \Gamma_\Theta = \frac{1}{4\pi} \int_{-\pi}^{\pi} \dot{\Phi}_\Theta(\lambda) \dot{\Phi}_\Theta'(\lambda) d\lambda,$$

provided only that the conditions of Corollary 1 are fulfilled. If, moreover, one requires that the covariance function $B_\Theta(\tau)$ of the process X_t be a continuous function of Θ for $\Theta \in \Theta$, then the family of distributions $P_n(f_\Theta)$, $\Theta \in \Theta$ will clearly satisfy condition (4) presented on page 21 also.

Thus we have the following

Theorem 5. *Under the conditions stated above the family of distributions $P_n(f_\Theta)$, $\Theta \in \Theta$ is asymptotically differentiable in the sense of definitions presented in the Introduction on page 21, where $\tau_n = \sqrt{n}$ and the random vector $\dot{\Delta}_{n,\Theta}$ and matrix Γ_Θ are given by the formulas (23) and (24) respectively.*

4. The Asymptotic Differentiability of Gaussian Distributions with Spectral Densities Possessing Fixed Zeros

4.1. We now return to the case when the spectral density $f = f_q$ of a Gaussian random process X_t, $t = ..., -1, 0, 1, ...$ can be represented in the form (2.11) where f_0 satisfies the condition

$$m \leqslant f_0 \leqslant M.$$

Consider the sequences of spectral densities

(1)
$$g_{n,j}(\lambda) = f_j(\lambda)(1 + n^{-1/2}a_n(\lambda)),$$

$$n = 1,2, \dots, \quad j = 0,1, \dots, q,$$

and

(2)
$$g_j(\lambda) = f_j(\lambda)(1 + n^{-1/2}a(\lambda)),$$

$$n = 1,2, \dots, \quad j = 0,1, \dots, q,$$

where the $f_j(\lambda)$, $j = 0,1,2, \dots$ satisfy relations (1.8) and a_n, $n = 1,2, \dots$, is a sequence of functions on $[-\pi,\pi]$ convergent in $L_2[-\pi,\pi]$ to a function a (cf. Subsection 3.1). For simplicity, we set $g = g_q$ and $g_n = g_{n,q}$.

The basic task of this subsection is to prove the theorem which actually generalizes the results of Theorem 4 in Section 3 to the case considered herein.

Theorem 1. *Under the above-stated conditions the following assertions are valid:*

1) *The sequences of Gaussian distributions $P_n(f)$, $n = 1,2,\dots$ and $P_n(g)$, $n = 1,2, \dots$ are contiguous.*

2) $\Lambda(f,g) - \Lambda(f,g_n) \to 0$ *in $P_n(f)$ probability as $n \to \infty$.*

3)
(3)
$$\Lambda(f,g) - \frac{1}{2}\left\{n^{-1/2}[X^*B_f^{-1}B_{a/2\pi}X - \mathrm{tr}(B_{a/2\pi})]\right.$$
$$\left. - \frac{1}{4\pi}\int_{-\pi}^{\pi}a^2(\lambda)d\lambda\right\} \to 0$$

in $P_n(f)$ probability as $n \to \infty$.

Assume also that the Fourier coefficients $\beta_0(\tau)$ and $\rho_0(\tau)$ of the function f_0 and respectively $r_0 = a/f_0$ satisfy the conditions

(4')
$$\sum_{\tau=1}^{\infty}\tau|\beta_0(\tau)^2| < \infty, \quad \sum_{\tau=1}^{\infty}\tau|\rho_0(\tau)|^2 < \infty.$$

If, moreover, $\lambda_1, \dots, \lambda_q$ are points of continuity of the functions f_0 and r_0, then, by Fejér's theorem ([65] p. 89) we have

(4")
$$|\sigma_n(\lambda_j,f_0)-f_0(\lambda_j)| \to 0 \quad \text{and} \quad |\sigma_n(\lambda_j,r_0)-r_0(\lambda_j)| \to 0$$

for all λ_j, $j = 1, \dots, q$, where as above, $\sigma_n(\lambda,r_0)$, $n = 1,2, \dots$ are

Fejér's sums of the function r_0.
4) *Then*

(5) $\Lambda(f,g)$ $-\dfrac{1}{2}\left\{\dfrac{n^{1/2}}{2\pi}\int_{-\pi}^{\pi}[I_n(\lambda,\tilde{Y}) - f_0(\lambda)]r_0(\lambda)d\lambda\right.$

$\left.-\dfrac{1}{4\pi}\int_{-\pi}^{\pi}a^2(\lambda)d\lambda\right\} \to 0$

in $P_n(f)$ probability as $n \to \infty$, where $I_n(\lambda,\tilde{Y})$ is given by the formula (2.29).

Proof.   1) It is sufficient to verify that

$$\Lambda(f,g) \;=\; \frac{1}{2}\{\log\det(B_f)\text{-}\log\det(B_g)$$

$$+\; \mathbf{X}^*(B_f^{-1} - B_g^{-1})\mathbf{X}\}$$

is bounded in $P_n(f)$ as well as $P_n(g)$ probabilities.
   In view of the formulas

(6) $E_f[\Lambda(f,g)] = -\dfrac{1}{2}\{\log\det(H_{g_0}H_{f_0}^{-1}) - \mathrm{tr}(I_n-H_{f_0}H_{g_0}^{-1})\}$,

(7) $E_g[\Lambda(f,g)] = -\dfrac{1}{2}\{\log\det(H_{g_0}H_{f_0}^{-1}) - \mathrm{tr}(I_n-H_{g_0}H_{f_0}^{-1})\}$,

(8) $D_f[\Lambda(f,g)] = \dfrac{1}{2}\mathrm{tr}[(I_n - H_{f_0}H_{g_0}^{-1})^2]$,

(9) $D_g[\Lambda(f,g)] = \dfrac{1}{2}\mathrm{tr}[(I_n - H_{g_0}H_{f_0}^{-1})^2]$,

where H_{f_0} is the matrix on the l.h.s. of (1.13) and H_{f_0} is obtained

from H_f with f_0 being replaced by g_0. Recall that in the case under consideration the matrix $V_{1-q,n}$ in (1.13) consists of columns $\mathrm{col}(z_j^{1-q}, ..., z_j^n)$.
   From (6)-(9), (3.7)-(3.10), and the assertions of Lemma A2.2 of Appendix 2 we have

(10) $E_f[\Lambda(f,g)] \to -\gamma^2/4$,

(11) $E_g[\Lambda(f,g)] \to \gamma^2/4$,

(12) $D_f[\Lambda(f,g)] \to \gamma^2/2$,

(13) $D_g[\Lambda(f,g)] \to \gamma^2/2$,

where, as above,

$$y^2 = \frac{1}{2\pi} \int_{-\pi}^{\pi} a^2(\lambda)d\lambda.$$

Assertion 1) is thus proved.

2) In order to prove that

$$\Lambda(f,g_n) - \Lambda(f,g) = \Lambda(g,g_n)$$

$$= \frac{1}{2} \{\log \det(B_g) - \log \det(B_{g_n})$$

$$+ X'(B_g^{-1}) - B_{g_n}^{-1})X\}$$

converges to zero in $P_n(f)$ probability it is sufficient to show that

(14)
$$E_f[\Lambda(g,g_n)] = -\frac{1}{2} \{\log \det(H_{g_{n,0}} H_{g_0}^{-1})$$
$$- tr[H_{f_0}(H_{g_0}^{-1} - H_{g_{n,0}}^{-1})]\} \to 0$$

and

(15)
$$D_f[\Lambda(g,g_n)] = \frac{1}{2} tr\{[H_{f_0}(H_{g_0}^{-1} - H_{g_{n,0}}^{-1})]^2\} \to 0.$$

Relations (14) and (15) follow from (3.11) and (3.12), and the assertions 2), 6), and 8) of Lemma A2.2 in Appendix 2.

3) In view of (1.13), (1.19), (2.3), and (2.4) the mathematical expectation and the variance on the l.h.s. of (3) are equal to

(16)
$$\frac{1}{2} \left[\log \det(H_{f_0} H_{g_0}^{-1}) - tr(I_n - H_{f_0} H_{g_0}^{-1}) + \frac{y^2}{2}\right]$$

and to

(17)
$$\frac{1}{2} tr\left[(I_n - H_{f_0} H_{g_0}^{-1} - n^{-1/2}B_{a/2\pi})^2\right]$$

respectively. Relations (6) and (10) imply that (16) converges to zero. As far as (17) is concerned, see assertion 3) of Lemma A2.2 which states that this expression converges to zero. To complete the proof of assertion 3) it remains only to apply Chebyshev's inequality.

4) From (3), (2.25), (2.28), and (2.29) and formulas (2.3) and (2.4), for mathematical expectation and variance of a quadratic form in normal variables, it follows that it is sufficient to show that

$$\mathrm{tr}[B_a\text{-}B_{r_0/2\pi}(I_n\text{-}P_n)B_{f_0}(I_n\text{-}P_n)]^i < \infty, \qquad i = 1,2.$$

This is, however, a corollary of Lemma A1.4 in Appendix 1 and the following generalizations of the relations (2.26) and (2.27):

(18)
$$\mathrm{tr}[B_a\text{-}B_{r_0/2\pi}(I_n\text{-}P_n)B_{f_0}(I_n\text{-}P_n)] - \mathrm{tr}[B_a\text{-}B_{r_0/2\pi}B_{f_0}]$$

$$= \mathrm{tr}[B_{r_0/2\pi}(P_n B_{f_0}+B_{f_0}P_n-P_n B_{f_0}P_n)]$$

$$\to 2\pi \,\mathrm{diag}\{a(\lambda_1), ..., a(\lambda_q)\}$$

and

(19)
$$\mathrm{tr}[B_a\text{-}B_{r_0/2\pi}(I_n\text{-}P_n)B_{f_0}(I_n\text{-}P_n)]^2 - \mathrm{tr}[(B_a\text{-}B_{r_0/2\pi}B_{f_0})^2]$$

$$= 2\mathrm{tr}(B_a\text{-}B_{r_0/2\pi}B_{f_0})[B_{r_0/2\pi}(P_n B_{f_0}+B_{f_0}P_n-P_n B_{f_0}P_n)]$$

$$+ \mathrm{tr}[B_{r_0/2\pi}(P_n B_{f_0}+B_{f_0}P_n-P_n B_{f_0}P_n)]^2$$

$$\to (2\pi)^2\mathrm{diag}\{a^2(\lambda_1), ..., a^2(\lambda_q)\}.$$

It is easy to see that in the particular case $a = 1/2\pi$, (18) and (19) coincide with (2.26) and (2.27) respectively. The proof of (18) and (19) is also based on utilizing formula (2.19). Theorem 1 is thus proved.   □

4.2. Assume now that the spectral density $f = f_q$ of the process X_t depends on a vector-valued parameter θ belonging to an open set Θ of the space R_p, i.e., $f_j = f_{j,\theta}$, $\theta \in \Theta$ where $j = 0,1, ..., q$. Assume also that the functions $\dot\phi_{k,\theta} = \dot\phi_{k,\theta}(\lambda)$, square integrable on $-\pi \leqslant \lambda \leqslant \pi$, exist such that

(20)
$$\int_{-\pi}^{\pi} \left\{\frac{f_{j,\theta+\varepsilon i} - f_{j,\theta}(\lambda)}{\varepsilon f_{j,\theta}(\lambda)} - \dot\phi_{k,\theta}(\lambda)\right\}^2 d\lambda \to 0, \qquad \theta \in \Theta,$$

as $\varepsilon \to 0$, in which i (analogously to the situation in Subsection 3.2) is a vector with the k-th component equal to 1 and with all its other components being 0. The function $\dot{\phi}_{k,\theta}$ is a derivative of the function $\log f_{j,\theta}$ in the L_2 sense with respect to the k-th element of the vector θ. (In view of (18) it does not depend on j.)

It follows from (26) that -- in particular -- the relation (3.19) is valid where

$$a_n = a_{n,\theta} = n^{1/2} \frac{f_{j,\theta+n^{-1/2}h_n} - f_{j,\theta}}{f_{j,\theta}}, \qquad n = 1,2, ...,$$

and $h_n \to h$ as $n \to \infty$.

It also follows from the arguments analogous to those presented in the course of the proof of Corollary 1 in Section 3 that in view of Theorem 1 and Lemma A2.4 of Appendix 2 the following corollary is valid.

Corollary 1. *Let the spectral density $f = f_q$ of a Gaussian random process X_t, $t = ..., -1,0,1,...$ be of the form (2.11), where f_0 satisfies the condition $m \leqslant f_0 \leqslant M$. Next let a function $\dot{\phi}_{k,\theta}$ exist satisfying the relations (20). Then*

1) *the sequences of Gaussian distributions $P_n(f_{q,\theta})$, $n = 1,2, ...,$ and $P_n(f_{q,\theta+n^{1/2}h})$, $n = 1,2, ...$ are contiguous.*

2)
$$\Lambda(f_{q,\theta}, f_{q,\theta+n^{-1/2}h}) - \Lambda(f_{q,\theta}, f_{q,\theta+n^{-1/2}h}) \to 0$$

in $P_n(f_\theta)$ probability as $n \to \infty$.

3) *Let the function $r_{k,\theta} = \dot{\phi}_{k,\theta}/f_{0,\theta}$, for all $\theta \in \Theta$ and $k = 1, ..., p$, satisfy the same conditions as the function r_0 in assertion 4) of Theorem 1.*
   Then

(21)
$$\Lambda(f_{q,\theta}, f_{q,\theta+n^{-1/2}h}) - h'\Delta_{n,\theta} + \frac{1}{2}h'\Gamma_\theta h \to 0$$

in $P_n(f_{q,\theta})$ probability as $n \to \infty$, where $\Delta_{n,\theta}$ is a p-dimensional random vector, whose k-th component is of the form

(22)
$$\frac{n^{1/2}}{4\pi} \int_{-\pi}^{\pi} [I_n(\lambda, \tilde{Y}) - f_{0,\theta}(\lambda)] r_{k,\theta}(\lambda) d\lambda,$$

and Γ_θ is a (p×p)-matrix whose (k×ℓ)-th entry equals

(23) $\dfrac{1}{4\pi} \displaystyle\int_{-\pi}^{\pi} \dot{\Phi}_{k,\,\Theta}(\lambda)\dot{\Phi}_{\boldsymbol{\ell},\,\Theta}(\lambda)d\lambda.$

4)   *Furthermore, let the covariance function $\beta_{q,\,\Theta}(\tau)$ of the process X_t be continuous in $\Theta \in \boldsymbol{\Theta}$. Then the family of distributions $P_n(f_{q,\,\Theta})$, $\Theta \in \boldsymbol{\Theta}$ is asymptotically differentiable for $\tau_n = \sqrt{n}$ in the sense of the definition presented in the Introduction on page 21.*

Appendix 1

Let A be an $(n\times n)$-matrix. Define the Euclidean norm of the matrix A by the equation

$$|A| = [\operatorname{tr}(AA^*)]^{1/2} = \left[\sum_{i=1}^{n}\sum_{k=1}^{n}|a_{ik}|^2\right]^{1/2},$$

where a_{ik} are the entries of the matrix A (A^* denotes a conjugate transpose of A).

It is convenient here to introduce the sup norm for the matrix A defined by the equation

$$\|A\| = \sup\{|Ax|: |x| = 1\}$$

where x is an n-dimensional column vector.

As it is known for an arbitrary $(n\times n)$-matrix A, the inequality $\|A\| \leqslant |A| \leqslant n^{1/2}\|A\|$ is valid.

In this Chapter we shall also utilize the following inequalities which are valid for arbitrary $(n\times n)$-matrices A and B:

(A1.1) $|\operatorname{tr}(AB)| \leqslant |A| \cdot |B|$

and

(A1.2) $|AB| \leqslant \|A\| \cdot |B|.$

If $\|A\| < 1$ then the inequality

$$|\log \det(I_n+A) - \operatorname{tr}(A) + \frac{1}{2}\operatorname{tr}(A^2)|$$

(1) $\leqslant \dfrac{1}{3}\|A\| \cdot |A|^2(1 - \|A\|)^{-3}$

is also valid (cf. assertion (V) of Appendix II in [61]). It follows from (1) that in particular the following lemma is valid.

Lemma A1.1. *Let A be an (n×n)-matrix such that* $|A| < \infty$ *and* $\|A\| \to 0$; *then* $|U_n(A)| \to 0$, *where*

$$U_n(A) = \log \det(I_n + A) - \operatorname{tr}(A) + \frac{1}{2}\operatorname{tr}(A^2).$$

Lemma A1.2. *Let f and* $g = f(1 + n^{-1/2}a)$ *be spectral densities where a is a square integrable function*

$$\frac{1}{2\pi}\int_{-\pi}^{\pi} a^2(\lambda)d\lambda = \gamma^2 < \infty.$$

Then the following assertions are valid:

1) *if* $f \leqslant M$, *then* $\|B_f\| \leqslant 2\pi M$,

2) *if* $m \leqslant f$, *then* $\|B_f^{-1}\| \leqslant 1/2\pi m$,

3) $n^{1/2}\|B_{a/2\pi}\| \to 0$,

4) $n^{-1/2}|B_{a/2\pi}| \to \gamma$,

5) *if* $f \leqslant M$, *then* $n^{-1/2}|B_{fa} - B_{a/2\pi}B| \to 0$; *if, moreover,* $m \leqslant f$, *then* $n^{-1/2}|B_{fa}B_f^{-1} - B_{a/2\pi}| \to 0$.

Under the condition $m \leqslant f \leqslant M$ *the following assertions also hold:*

6) $n^{1/2}\|B_{fa}B_f^{-1}\| \to 0$,

7) $n^{1/2}|B_{fa}B_f^{-1}| \to \gamma$,

8) $n^{1}\operatorname{tr}[(B_{fa}B_f^{-1})^2] \to \gamma^2$,

9) $n^{1}|B_{fa}B_f^{-1}B_{fa}B_g^{-1}| \leqslant n^{-1/2}|B_{fa}B_f^{-1}|n^{-1/2}\|B_{fa}B_f^{-1}\|$

   $\times\ (1 - n^{-1/2}\|B_{fa}B_f^{-1}\|)^{-1} \to 0$,

10) $n^{1/2}|B_{fa}B_f^{-1}| \to \gamma$,

11) $n^{-3/2}\operatorname{tr}[(B_{fa}B_f^{-1})^2B_{fa}B_g^{-1}] \to 0$,

12) $n^{3/2} \mathrm{tr}[B_{fa}B_f^{-1}(B_{fa}B_g^{-1})^2] \to 0$.

The proofs of assertions 1)-4) and the first assertion of 5) can be found, for example, in [61] (cf. Lemma 2.1 (i)-(iii), (v), (vi)). The second assertion of 5) follows from the first, the inequality $|B_{fa}B_f^{-1}-B_{a/2\pi}| \leqslant |B_{fa}-B_{a/2\pi}B_f| \cdot \|B_f^{-1}\|$, and assertion 2); assertion 6) follows from 2), 5), and the inequality

$$n^{-1/2}\|B_{fa}B_f^{-1}\| \leqslant n^{-1/2}|B_{fa}B_f^{-1}-B_{a/2\pi}| + n^{-1/2}\|B_{a/2\pi}\|;$$

assertion 7) follows from 4), 5), and the inequality

$$|n^{-1/2}|B_{fa}B_f^{-1}| - \gamma| \leqslant n^{-1/2}|B_{fa}B_f^{-1} - B_{a/2\pi}|$$
$$+ |n^{-1/2}|B_{a/2\pi}| - \gamma|.$$

In view of 4), in order to prove assertion 8) it is sufficient to show that

$$|n^{-1}\mathrm{tr}[(B_{fa}B_f^{-1})^2 - (B_{a/2\pi})^2]|$$
$$\leqslant n^{-1}|B_{fa}B_f^{-1}-B_{a/2\pi}|(|B_{fa}B_f^{-1}| + |B_{a/2\pi}|) \to 0.$$

This follows from 4), 5), and 7).
The inequality 9) is valid since

$$|B_{fa}B_f^{-1}B_{fa}(B_f+n^{-1/2}B_{fa})^{-1}| \leqslant |(B_{fa}B_f^{-1})^2|$$
$$+ n^{-1/2}|(B_{fa}B_f^{-1})^2B_{fa}(B_f+n^{-1/2}B_{af})^{-1}|$$
$$\leqslant \|B_{fa}B_f^{-1}\|(|B_{fa}B_f^{-1}|$$
$$+ n^{-1/2}|B_{fa}B_f^{-1}B_{fa}(B_f+n^{-1/2}B_{fa})^{-1}|.$$

The convergence to zero on the r.h.s. of the assertion 9) follows from 6) and 7). Finally assertion 10) is a corollary of 7), 9), and the inequality

$$|n^{-1/2}|B_{fa}(B_f+n^{-1/2}B_{fa})^{-1}| - \gamma| \leqslant |n^{-1/2}|B_{fa}B_f^{-1}| - \gamma|$$
$$+ n^{-1}|B_{fa}B_f^{-1}B_{fa}(B_f + n^{-1/2}B_{fa})^{-1}|,$$

11) and 12) follow from the inequalities

$$\mathrm{tr}[(B_{\mathrm{fa}}B_{\mathrm{f}}^{-1})^2 B_{\mathrm{fa}}(B_{\mathrm{f}}+n^{-1/2}B_{\mathrm{fa}})^{-1}]$$

$$\leqslant \|B_{\mathrm{fa}}B_{\mathrm{f}}^{-1}\| \|B_{\mathrm{fa}}B_{\mathrm{f}}^{-1}\| \cdot |B_{\mathrm{fa}}(B_{\mathrm{f}}+n^{-1/2}B_{\mathrm{f\,a}})^{-1}|,$$

$$\mathrm{tr}\{B_{\mathrm{fa}}B_{\mathrm{f}}^{-1}[B_{\mathrm{fa}}(B_{\mathrm{f}}+n^{-1/2}B_{\mathrm{fa}})^{-1}]^2\}$$

$$\leqslant \|B_{\mathrm{fa}}B_{\mathrm{f}}^{-1}\| \cdot |B_{\mathrm{fa}}(B_{\mathrm{f}}+n^{-1/2}B_{\mathrm{fa}})^{-1}|^2$$

and assertions 6), 7), and 10).

Lemma A1.3. *Let the conditions of Lemma A1.2 be fulfilled. Assume that $a_n(\lambda)$, $n = 1,2, ...$, is a sequence of functions such that*

$$(2) \qquad \int_{-\pi}^{\pi} [a_n(\lambda) - a(\lambda)]^2 d\lambda \to 0;$$

the functions $g_n = f(1+n^{-1/2}a_n)$ and $g = f(1+n^{-1/2}a)$ are spectral densities. Then the following assertions are valid:

1) $n^{1/2}|B_{f(a_n-a)}| \to 0,$

2) $n^{1/2}|B_{f(a_n-a)}B_g^{-1}| \to 0,$

3) $n^{1/2}\|B_{\mathrm{fa}_n}B_{\mathrm{f}}^{-1}\| \to 0,$

4) $n^{1/2}|B_{\mathrm{fa}_n}B_{\mathrm{f}}^{-1}| \to \gamma,$

5) $n^{1/2}|B_{\mathrm{fa}_n}B_{g_n}^{-1}| \to \gamma.$

Proof. Since

$$n^{-1/2}|B_{f(a_n-a)}| = \left[\int_{-\pi}^{\pi} f^2(\lambda)[a_n(\lambda)-a(\lambda)]^2 d\lambda\right]^{1/2} + o(1),$$

assertion 1) follows from (2) and the boundedness of f.
From 1), the inequality

$$|B_{f(a_n-a)}(B_{\mathrm{f}}+n^{-1/2}B_{\mathrm{a}})^{-1}| \leqslant |B_{f(a_n-a)}B_{\mathrm{f}}^{-1}|$$

$$+ n^{-1/2}|B_{f(a_n-a)}B_{\mathrm{f}}^{-1}B_{\mathrm{fa}}(B_{\mathrm{f}}+n^{-1/2}B_{\mathrm{fa}})^{-1}|$$

$$\leqslant |B_{f(a_n-a)}| \cdot \|B_{\mathrm{f}}^{-1}\|(1+n^{-1/2}|B_{\mathrm{fa}}(B_{\mathrm{f}}+n^{-1/2}B_{\mathrm{fa}})^{-1}|),$$

and the assertions 2) and 10) of Lemma A1.2 assertion 2) follows. Applying assertions 2), 6), 7) of Lemma A1.2 assertion 1) of Lemma A1.3 and the inequality

$$\|B_{fa_n} B_f^{-1}\| \leqslant B_{fa} B_f^{-1}\| + |B_{f(a_n-a)}| \cdot \|B_f^{-1}\|,$$

$$|B_{fa_n} B_f^{-1}| \leqslant |B_{fa} B_f^{-1}| + |B_{f(a_n-a)}| \cdot \|B_f^{-1}\|,$$

it is easy to verify the validity of assertions 3) and 4).

Assertion 5) is proved analogously to assertion 10) of Lemma A1.2.   □

Lemma A1.4. *Let the following inequalities*

$$\sum_{j=1}^{\infty} j|\beta(j)|^2 < \infty, \qquad \sum_{j=1}^{\infty} j|\rho(j)|^2 < \infty$$

be valid, where $\beta(j)$ *and* $\rho(j)$ *are the Fourier coefficients of the functions* f *and* $r = a/2\pi f$ *respectively. Then*

$$|\mathrm{tr}(B_a - B_r B_f)| < \infty, \qquad \mathrm{tr}((B_a - B_r B_f)^2) < \infty.$$

Proof. Since the (k,ℓ)-th entry of the matrix B_a can be represented in the form $\sum_{j=-\infty}^{\infty} \beta(j)\rho(j-k+\ell)$, then the (k,ℓ)-th entry of the matrix $B_a - B_r B_f$ is of the form

$$\sum_{j=-\infty}^{k-n} \beta(j)\rho(j-k+\ell) + \sum_{j=k+1}^{\infty} \beta(j)\rho(j-k+\ell).$$

Consequently,

$$(3) \qquad \mathrm{tr}(B_a - B_r B_f) = 2 \sum_{j=0}^{\infty} \min(n, j)\beta(j)\rho(j),$$

and the first assertion of Lemma A1.4 follows from the obvious inequality

$$\left| \sum_{j=1}^{\infty} \min(n, j)\beta(j)\rho(j) \right|^2 \leqslant \sum_{j=1}^{\infty} j|\beta(j)|^2 \sum_{j=1}^{\infty} j|\rho(j)|^2 < \infty.$$

The proof of the second assertion follows from the inequality

$$\mathrm{tr}[(B_a - B_r B_f)^2] \leqslant |B_a - B_r B_f|^2$$

$$= \sum_{k,\, \mathbf{l}=0}^{n-1} \left\{ \sum_{j=-\infty}^{k-n} \beta(j)\rho(j-k+\mathbf{l}) + \sum_{j=k+1}^{\infty} \beta(j)\rho(j-k+\mathbf{l}) \right\}^2$$

$$\leqslant \sum_{k,\, \mathbf{l}=0}^{n-1} \left\{ \sum_{j=-\infty}^{k-n} \beta^2(j) + \sum_{j=k+1}^{\infty} \beta^2(j) \right\}$$

$$\times \left\{ \sum_{j=n}^{\infty} \rho^2(j-\mathbf{l}) + \sum_{j=1}^{\infty} \rho^2(j+\mathbf{l}) \right\}$$

$$= 2 \sum_{j=1}^{\infty} \min(n,\, j)\beta^2(j) \; 2 \sum_{j=1}^{\infty} \min(n,\, j)\rho^2(j) < \infty. \qquad \square$$

Remark. It is easy to verify that for $a(\lambda) = \text{const}$ and $f(\lambda) > m$ (m is a positive number) condition $\sum_{j=1}^{\infty} j|\beta(j)|^2 < \infty$ implies $\sum_{j=1}^{\infty} j|\rho(j)|^2 < \infty$.

To prove this assertion we first show that the condition $\sum_{j=1}^{\infty} j|\beta(j)|^2 < \infty$ is equivalent to the condition

(4) $$\sum_{n=1}^{\infty} \omega_f^2(1/n) < \infty,$$

where

$$\omega_f(t) = \left\{ \sup_{|h| \leqslant t} \int_{-\pi}^{\pi} |f(\lambda+h) - f(\lambda)|^2 d\lambda \right\}^{1/2}$$

(cf., e.g., Lemma 7 on page 131 of the book [68]). Next we check that condition (4) implies the inequality $\sum_{n=1}^{\infty} \omega_{1/f}^2(1/n) < \infty$.

Appendix 2

1. Retaining the notation utilized in Section 3 we shall prove the following lemma here.

Lemma A2.1. *Under the conditions of Corollary 1 of Section 3, $\Lambda(f_{\Theta+n^{-1/2}h}, g) \to 0$ as $n \to \infty$ in $P_n(\hat{g})$ probability where $\hat{g}$ equals either $f_{\Theta+n^{-1/2}t}$ (where t equals either 0 or h) or g.*

Proof. In view of (3.21)

(1) $$\log \det(B_{f_{\Theta+n^{-1/2}h}}) - \log \det(B_g)$$

$$= -\log \det(I_n - n^{-1/2} B_{\psi_n} B^{-1}_{f_{\Theta+n^{-1/2}h}}),$$

where

(2) $\qquad \psi_n = n^{1/2}(f_{\Theta+n^{-1/2}h} - f_\Theta) - h'\dot\phi_\Theta f_\Theta.$

It follows from (3.18) and (3.19) that

$$\int_{-\pi}^{\pi} \psi_n^2(\lambda)d\lambda \leqslant M^2 \int_{-\pi}^{\pi} \psi_n^2(\lambda)f_\Theta^{-2}(\lambda)d\lambda \to 0.$$

Consequently,

(3) $\qquad n^{-1/2}|B_{\psi_n}| \to 0$

and

(4) $\qquad n^{-1/2}|B_{\psi_n} B^{-1}_{f_{\Theta+n^{-1/2}h}}| \leqslant n^{-1/2}|B_{\psi_n}| \cdot \|B^{-1}_{f_{\Theta+n^{-1/2}h}}\| \to 0$

since $\|B^{-1}_{f_{\Theta+n^{-1/2}h}}\| < \infty$ (cf. [61], Corollary 3.3). In view of

(3.21), (3.22), and (1)

(5)
$$\Lambda(f_{\Theta+n^{-1/2}h}, g) = -\frac{1}{2}|\log \det(I_n - n^{-1/2}B_{\psi_n} B^{-1}_{f_{\Theta+n^{-1/2}h}})$$
$$+ n^{-1/2}X'B^{-1}_g B_{\psi_n} B^{-1}_{f_{\Theta+n^{-1/2}h}} X|.$$

Applying the formulas (2.3) and (2.4) we obtain

(6)
$$E^\wedge_g(\Lambda(f_{\Theta+n^{-1/2}h}, g)) = -\frac{1}{2} \left\{ U_n(n^{-1/2}B_{\psi_n} B^{-1}_{f_{\Theta+n^{-1/2}h}}) \right.$$
$$- n^{-1/2}\mathrm{tr}[(I_n - B^\wedge_g B^{-1}_g)B_{\psi_n} B^{-1}_{f_{\Theta+n^{-1/2}h}}]$$
$$\left. - \frac{1}{2n}\mathrm{tr}[(B_{\psi_n} B_{f_{\Theta+n^{-1/2}h}})^2] \right\},$$

and

(7) $\qquad D^\wedge_g(\Lambda(f_{\Theta+n^{-1/2}h}, g)) = \frac{1}{2n}\mathrm{tr}[(B^\wedge_g B^{-1}_g B_{\psi_n} B^{-1}_{f_{\Theta+n^{-1/2}h}})^2].$

It is easy to verify that if $\hat g = g$, the proof of the Lemma
follows from the inequality

(8)
$$\mathrm{tr}[(B_{\psi_n} B_{f_{\Theta+n^{-1/2}h}}^{-1})^2] \leqslant |B_{\psi_n} B_{f_{\Theta+n^{-1/2}h}}^{-1}|^2,$$

formula (4) and the assertion of Lemma A1.1 in Appendix 1.

For $\hat{g} = f_{\Theta+n^{-1/2}t}$, $t = 0$ or h, we have

(9)
$$n^{-1/2}|\mathrm{tr}[(I_n - B_{\hat{g}}B_{\hat{g}}^{-1})B_{\psi_n} B_{f_{\Theta+n^{-1/2}h}}^{-1}]\|$$
$$\leqslant n^{-1/2}|B_{\psi_{n,t}} B_{\hat{g}}^{-1}|n^{-1/2}|B_{\psi_n} B_{f_{\Theta+n^{-1/2}h}}^{-1}| \to 0,$$

where

$$\psi_{n,t} = n^{1/2}(f_{\Theta+n^{-1/2}t} - f_\Theta) - f_\Theta \mathbf{h'}\dot{\phi}_\Theta,$$

since for $t = 0$

(10')
$$n^{-1/2}|B_{\psi_{n,t}} B_{\hat{g}}^{-1}| = n^{-1/2}|B_{f_\Theta \mathbf{h'}\dot{\phi}_\Theta} B_{\hat{g}}^{-1}|$$
$$\to \left[\frac{1}{2\pi}\int_{-\pi}^{\pi}[\mathbf{h'}\dot{\phi}_\Theta(\lambda)]^2 d\lambda\right]^{1/2}$$

(cf. assertion 10) of Lemma A1.2 in Appendix 1), and for $t = h$

(10")
$$n^{-1/2}|B_{\psi_{n,t}} B_{\hat{g}}^{-1}| = n^{-1/2}|B_{\psi_n} B_{\hat{g}}^{-1}|$$
$$= n^{-1/2}|B_{\psi_n} B_{f_\Theta}^{-1}(I_n - n^{-1/2}B_{f_\Theta \mathbf{h'}\dot{\phi}_\Theta} B_{\hat{g}}^{-1})|$$
$$\leqslant n^{-1/2}|B_{\psi_n}| \cdot \|B_{f_\Theta}^{-1}\|(1+n^{-1/2}|B_{f_\Theta \mathbf{h'}\dot{\phi}_\Theta} B_{\hat{g}}^{-1}|) \to 0$$

in view of formula (3) and assertions 2) and 10) of Lemma A1.2 in Appendix 1.

From (8), (9), and Lemma A1.1 follows the convergence to zero as $n \to \infty$ of the mathematical expectation of the expression (5) (represented by formula (6) for $\hat{g} = f_{\Theta+n^{-1/2}h}$)

and the convergence to zero of the variance (given by formula (7)) from (4), (10), and the inequality

$$\mathrm{tr}[(B_{\hat{g}}B_{\hat{g}}^{-1}B_{\psi_n} B_{f_{\Theta+n^{-1/2}h}}^{-1})^2] \leqslant |B_{\hat{g}}B_{\hat{g}}^{-1}B_{\psi_n} B_{f_{\Theta+n^{-1/2}h}}^{-1}|^2$$
$$\leqslant |B_{\psi_n} B_{f_{\Theta+n^{-1/2}h}}^{-1}|^2 (1+n^{-1/2}|B_{\psi_{n,t}} B_{\hat{g}}^{-1}|)^2.$$

Applying Chebyshev's inequality now we obtain the proof
of the lemma.   □

2.   Here we shall prove the general assertion (2.19)
presented in Section 2, Remark 3 and then deduce from it the
required corollary.

Proposition A2.1. *Let λ_1, ..., λ_q be q distinct points of continuity
of some even positive functions f_j, $j = 1$, ..., m, possessing
Fourier coefficients $\beta_j(\tau)$ such that*

(11) $\sum\limits_{\tau=1}^{\infty} \tau |\beta_j(\tau)|^2 < \infty, \quad j = 1, ..., m.$

   Then

(12)
$$\frac{1}{n} V^* B_{f_1} \cdots B_{f_m} V \to (2\pi)^m \mathrm{diag}\{f_1(\lambda_j) \cdots f_m(\lambda_j),$$
$$j = 1, ..., q\},$$

*where V is an $(n \times q)$-matrix consisting of columns $v_j = \mathrm{col}(z_j, ...,
z_j^n)$, $z_j = e^{i\lambda_j}$, $j = 1, ..., q$.*

Proof.   Analogously to [37], Section 11.7 (cf. also [163] or
[164]), we shall consider only the special case $m = 2$ which is
typical[9] for the general situation.   Following the arguments
presented therein we shall show that

(13)
$$v_k^* B_{f_1} B_{f_2} v_\ell = \sum\limits_{n-M(k,\ell)+\mu(k,\ell)>0} R_{k,\ell}^{(n)}(\overline{z}_k z_\ell) z_k^{j_1+j_2} \cdot$$
$$\cdot \beta_1(j_1)\beta_2(j_2),$$

where $M(k,\ell) = \max(0,k,k+\ell)$ and $\mu(k,\ell) = \min(0,k,k+\ell)$ and

(14) $R_{k,\ell}^{(n)}(z) = \begin{cases} n-M(k,\ell)+\mu(k,\ell), & \text{for } k = \ell \\[2mm] \dfrac{z}{1-z}[z^{M(k,\ell)}-z^{n+\mu(k,\ell)}], & \text{for } k \neq \ell. \end{cases}$

   Indeed, utilizing the characteristic function $\psi(i)$ of the set
of values $i = 1, ..., n$ (which is -- as usual -- equal to 1 for the

---

[9]When dealing with the general case it may be useful to trace the related
argument presented in [157], Sections 3.4 and 4.1.

indicated values and zero otherwise) we write

$$v_k^* B_{f_1} B_{f_2} v_\ell = \sum_{i_1, i_2, i_3 = 1}^{n} \bar{z}_k^{-i_1} z_\ell^{i_3} B_1(i_1 - i_2) B_2(i_2 - i_3)$$

(15)
$$= \sum \psi(i_1)\psi(i_2)\psi(i_3)\bar{z}_k^{-i_1} z_\ell^{i_3} B_1(i_1 - i_2) B_2(i_2 - i_3)$$

$$= \sum \psi(j_1)\psi(j_1 + j_2)\psi(j_1 + j_2 + j_3)\bar{z}_k^{-j_1} z_\ell^{j_1 + j_2 + j_3} .$$

$$\cdot B_1(j_2) B_2(j_3)$$

(here we apply the transformation of variables $j_1 = i_1$, $j_2 = i_2 - i_1$ and $j_3 = i_3 - i_2$).

Thus (13) and (14) follow from (15) and the easily verified relation that

$$\sum_{j_1} \psi(j_1)\psi(j_1 + j_2)\psi(j_1 + j_2 + j_3) z^{j_1 + j_2 + j_3}$$

$$= \sum_{j = M(j_2, j_3) + 1}^{n + \mu(j_2, j_3)} z^j = R_{j_2, j_3}^{(n)} (z)$$

provided only $n + 1 - M(j_2, j_3) + \mu(j_2, j_3)$ is positive; otherwise the sum is zero.

Now from (13) and (14) for $k \neq \ell$ we have

$$|v_k^* B_{f_1} B_{f_2} v_\ell| \leq \frac{2}{|1 - z_\ell \bar{z}_k|} \sum_{|j_1| \leq n} |B_1(j_1)| \sum_{|j_2| \leq n} |B_2(j_2)|$$

$$\leq \frac{2}{|1 - z_\ell \bar{z}_k|} \left\{ B_1(0) + 2 \left[\sum_{j=1}^{n} \frac{1}{j} \sum_{j_1=1}^{n} j_1 |B_1(j_1)|^2 \right]^{1/2} \right\}.$$

$$\cdot \left\{ B_2(0) + 2 \left[\sum_{j=1}^{n} \frac{1}{j} \sum_{j_2=1}^{n} j_2 |B_2(j_2)|^2 \right]^{1/2} \right\}$$

$$= O(\log n),$$

in view of (11). This clearly indicates that the non-diagonal entries of the matrix in the l.h.s. of (12) converge to zero (for $m = 2$).

As far as the diagonal entries are concerned, in view of the convergence of Fejer's sums $\sigma_n(\lambda, f_j)$ and the partial sums $s_n(\lambda, f_j)$ of the Fourier series of functions f_j, $j = 1, ..., m$ at the continuity points $\lambda = \lambda_1, ..., \lambda_q$ (cf. (2.12)-(2.16)) it is sufficient to verify that

$$\frac{1}{n} \sum_{j_1, j_2, j_3 = 1}^{n} e^{i\lambda(j_3 - j_1)} \beta_1(j_1 - j_2)\beta_2(j_2 - j_3)$$

$$= 4\pi^2 [\sigma_n(\lambda, f_1) s_{n-1}(\lambda, f_2) + \sigma_n(\lambda, f_2) s_{n-1}(\lambda, f_1)$$

$$- s_{n-1}(\lambda, f_1) s_{n-1}(\lambda, f_2)]$$

(16)
$$+ 2\mathrm{Re}\ \frac{1}{n} \left\{ \sum_{\ell=1}^{n-1} e^{-i\lambda\ell} \beta_1(\ell) \sum_{k=1}^{\ell} k e^{i\lambda k} \beta_2(k) \right.$$

$$+ \sum_{\ell=2}^{n-1} e^{i\lambda\ell} \beta(\ell_2) \sum_{k=1}^{\ell-1} k e^{-i\lambda k} \beta\ (k)$$

$$+ \left. \sum_{\ell=1}^{n-1} \beta_1(\ell) e^{-i\lambda\ell} \sum_{k=1}^{\ell-1} k e^{-i\lambda(n-\ell+k)} \beta_2(n-\ell+k) \right\}$$

and that the quantity whose real part appears in the second term absolutely converges to zero. But this is indeed evident from the following two inequalities

$$\frac{1}{n} \sum_{\ell=1}^{n} |\beta_1(\ell)| \sum_{k=1}^{\ell} |\beta_2(k)|$$

$$\leqslant \frac{1}{n} \sum_{\ell=1}^{n} |\beta_1(\ell)| \left\{ \sum_{k=1}^{\ell} k \sum_{k=1}^{\ell} k |\beta_2(k)|^2 \right\}^{1/2}$$

$$\leqslant \frac{1}{n} \sum_{\ell=1}^{n} \ell |\beta_1(\ell)| \cdot \left\{ \sum_{k=1}^{\ell} k |\beta_2(k)|^2 \right\}^{1/2}$$

and

$$\frac{1}{n} \sum_{\ell=1}^{n} |\beta_1(\ell)| \sum_{k=0}^{\ell-1} k |\beta_2(n-\ell+k)|$$

$$\leqslant \frac{1}{n} \sum_{\ell=1}^{n-1} \ell |\beta_1(\ell)| \left\{ \sum_{k=0}^{\ell-1} k |\beta_2(n-\ell+k)|^2 \right\}^{1/2}$$

$$\leqslant \frac{1}{n} \sum_{\ell=1}^{n} \ell |\beta_1(\ell)| \left\{ \sum_{k=1}^{\infty} k |\beta_2(k)|^2 \right\}^{1/2},$$

and noting that the condition $\sum_{k=1}^{\infty} k |\beta_1(k)|^2 < \infty$ assures that $(1/n)\sum_{k=1}^{n} k |\beta(k)| \to 0$ (cf. [65], page 79).

To complete the proof we shall verify the validity of the relation (16). Here we can bypass the relation (13) and start from the equality

$$\frac{1}{n} \sum_{j_1, j_2, j_3 = 1}^{n} e^{i\lambda(j_3 - j_1)} \beta_1(j_1 - j_2)\beta_2(j_2 - j_3)$$

$$- 4\pi^2 \sigma_n(\lambda, f_1) s_{n-1}(\lambda, f_2)$$

$$= -\frac{1}{n} \sum_{j_1, j_2 = 1}^{n} e^{i\lambda(j_2 - j_1)} \beta_1(j_1 - j_2) S_{j_2}^{(n)}(f_2)$$

(17)

$$= -\frac{1}{n} \sum_{|k| < n} e^{-i\lambda k} \beta_1(k) \Sigma_k^{(n)}(f_2)$$

$$= -\frac{2\pi}{n} s_{n-1}(\lambda, f_1) \Sigma_0^{(n)}(f_2)$$

$$- \frac{2}{n} \mathrm{Re} \sum_{k=1}^{n-1} e^{-i\lambda k} \beta_1(k) [\Sigma_0^{(n)}(f_2) - \Sigma_k^{(n)}(f_2)],$$

where

$$S_j^{(n)}(f) = \sum_{k=j}^{n-1} \beta(k)e^{-i\lambda k}(1 - \delta_{jn}) + \sum_{k=n-j+1}^{n-1} \beta(k)e^{i\lambda k}(1 - \delta_{j1}),$$

and

$$\Sigma_k^{(n)}(f) = \sum_{j=1-\min(0,k)}^{n-\max(0,k)} S_j^{(n)}(f), \qquad -n < k < n,$$

so that

$$\Sigma_k^{(n)}(f) = \overline{\Sigma}_{-k}^{(n)}(f),$$

(18)

$$\Sigma_0^{(n)}(f) = \sum_{k=1}^{n-1} k\beta(k)[e^{i\lambda k} + e^{-i\lambda k}]$$

$$= 2\pi n[s_{n-1}(\lambda, f) - \sigma_n(\lambda, f)]$$

and

$$\Sigma_0^{(n)}(f) - \Sigma_k^{(n)}(f) = S_n^{(n)}(f) + \cdots + S_{n-k+1}^{(n)}(f)$$

$$= \sum_{\ell=1}^{n-1} \min(\ell, k)\beta(\ell)e^{i\lambda \ell} + \sum_{\ell=0}^{k-1} \ell\beta(n-k+\ell)e^{-i\lambda(n-k+\ell)}.$$

It follows from the last relation that

$$\sum_{k=1}^{n-1} e^{-i\lambda k}\beta_1(k)[\Sigma_0(f_2) - \Sigma_k(f_2)]$$

$$= \sum_{\ell=1}^{n-1} e^{-i\lambda\ell}\beta_1(\ell)\sum_{k=1}^{\ell} k\beta_2(k)e^{i\lambda k}$$

(19)

$$+ \sum_{\ell=2}^{n-1} \beta_2(\ell)e^{i\lambda\ell}\sum_{k=1}^{\ell-1} k\beta_1(k)e^{-i\lambda k}$$

$$+ \sum_{\ell=1}^{n-1} e^{-i\lambda\ell}\beta_1(\ell)\sum_{k=0}^{\ell-1} k\beta_2(n-\ell+k)e^{-i\lambda(n-\ell+k)}.$$

Equations (17)-(19) imply (16). The assertion is proved. □

Corollary A2.1. *Let* $\lambda_1, ..., \lambda_q$ *be points of continuity of the spectral density* f *satisfying conditions* 1) *and* 2) *of Lemma* 2.1. *Then:*

1) $(1/n)V^*B_f^{-1}V \to \mathrm{diag}\{1/2\pi f(\lambda_1), ..., 1/2\pi f(\lambda_q)\}.$

Furthermore, let the spectral densities $f, g = f(1+n^{-1/2}a)$ *and* $g_n = f(1+n^{-1/2}a_n)$ *satisfy the conditions preceding the assertions* 1)-3) *of Theorem* 3.4. *Then*

2) $(1/n)V^*B_g^{-1}V \to \mathrm{diag}\{1/2\pi f(\lambda_1), ..., 1/2\pi f(\lambda_q)\},$

3) $(1/n)V^*B_{g_n}^{-1}V \to \mathrm{diag}\{1/2\pi f(\lambda_1), ..., 1/2\pi f(\lambda_q)\},$

4) $(1/n)V^*B_f^{-1}B_gB_f^{-1}V \to \mathrm{diag}\{1/2\pi f(\lambda_1), ..., 1/2\pi f(\lambda_q)\},$

5) $(1/n)V^*B_g^{-1}B_fB_g^{-1}V \to \mathrm{diag}\{1/2\pi f(\lambda_1), ..., 1/2\pi f(\lambda_q)\},$

6) $(1/n)V^*B_{g_n}^{-1}B_fB_{g_n}^{-1}V \to \mathrm{diag}\{1/2\pi f(\lambda_1), ..., 1/2\pi f(\lambda_q)\},$

7) $(1/n)V^*B_f^{-1}B_gB_f^{-1}B_gB_f^{-1}V \to \mathrm{diag}(1/2\pi f(\lambda_1), ..., 1/2\pi f(\lambda_q)\},$

8) $(1/n)V^*B_g^{-1}B_fB_g^{-1}B_fB_g^{-1}V \to \mathrm{diag}\{1/2\pi f(\lambda_1), ..., 1/2\pi f(\lambda_q)\},$

9) $(1/n)V^*B_{g_n}^{-1}B_fB_{g_n}^{-1}B_fB_{g_n}^{-1}V \to \mathrm{diag}\{1/2\pi f(\lambda_1), ..., 1/2\pi f(\lambda_q)\}.$

Proof. Since

$$B_f^{-1} = B_h + (I_n - B_hB_f)B_h + (I_n - B_hB_f)B_f^{-1}(I_n - B_fB_h),$$

where $h = 1/(2\pi)^2f$, assertion 1) follows from Proposition

A2.1 and the inequality

$$\frac{1}{n}V^*(I_n\text{-}B_hB_f)B_f^{-1}(I_n\text{-}B_fB_h)V$$

$$\leqslant \|B_f^{-1}\|\frac{1}{n}V^*(I_n\text{-}B_hB_f)(I_n\text{-}B_fB_h)V$$

(cf. assertion 2) of Lemma A1.2).

Since

$$B_g^{-1} = B_f^{-1}\text{-}n^{-1/2}B_f^{-1}B_{af}B_f^{-1} + n^{-1}B_f^{-1}B_{af}B_f^{-1}B_{af}B_g^{-1},$$

assertion 2) follows from assertion 1), the inequality

$$\frac{1}{n}|V^*(B_g^{-1}\text{-}B_f^{-1})V| \leqslant qn^{-1/2}\|B_f^{-1}\|\cdot\|B_{af}B_f^{-1}\|$$

$$+ qn^{-1}\|B_f^{-1}\|\cdot\|B_{af}B_f^{-1}B_{af}B_g^{-1}\|$$

and assertions 2), 6) and 9) of Lemma A1.2.

Assertion 3) is proved analogously by utilizing assertion 3) of Lemma A1.3 (instead of 6) of Lemma A1.2) and also the inequality

$$n^{-1}\|B_{fa_n}B_f^{-1}B_{fa_n}B_{g_n}^{-1}\|$$

$$\leqslant n^{-1}\|B_{fa_n}B_f^{-1}\|^2(1\text{-}n^{-1/2}\|B_{fa_n}B_f^{-1}\|)^{-1},$$

which is analogous to 9) of Lemma A1.2.

Now it is clear that the proofs of the remaining assertions follow along the lines indicated above. For example, assertions 4) and 5) follow from the obvious equalities

$$B_f^{-1}B_gB_f^{-1} = B_f^{-1} + n^{-1/2}B_f^{-1}B_{fa}B_f^{-1}$$

and

$$B_g^{-1}B_fB_g^{-1} = B_f^{-1}\text{-}2n^{-1/2}B_f^{-1}B_{af}B_f^{-1}(I_n\text{-}n^{-1/2}B_{af}B_g^{-1})$$

$$+ B_f^{-1}[n^{-1/2}B_{af}B_f^{-1}(I_n\text{-}n^{-1/2}B_{af}B_g^{-1})]^2,$$

which imply the inequalities

$$\frac{1}{n}|V^*(B_f^{-1}B_gB_f^{-1}\text{-}B_f^{-1})V| \leqslant qn^{-1/2}\|B_f^{-1}\|\cdot\|B_{fa}B_f^{-1}\|$$

and

$$\frac{1}{n}|V^*(B_g^{-1}B_f B_g^{-1}-B_f^{-1})V| \leqslant q\|B_f^{-1}\|(2n^{-1/2}\|B_{af}B_f^{-1}\|$$

$$+ 2n^{-1}\|B_{af}B_f^{-1}B_{af}B_g^{-1}\|+ \|n^{-1/2}B_{af}B_f^{-1}$$

$$+ n^{-1}B_{af}B_f^{-1}B_{af}B_g^{-1}\|^2).$$

The assertions of Corollary A2.1 are used for the proof of the following lemma.

Lemma A2.2. *Let the spectral densities f, g and g_n satisfy the conditions of the Corollary A2.1. As in Section 4 denote*

(20) $\qquad H_f = [-V_{1,n}V_{1-q,0}^{-1},I_n]B_f^{(n+q)}[-V_{1,n}V_{1-q,0}^{-1},I_n]^*$

where $V_{s,t}$ consists of columns $\mathrm{col}(z_j^s, ..., z_j^t)$, $j = 1, ..., q$, $s \leqslant t$. Then, as $n \to \infty$ the following assertions are valid.

1) $|\mathrm{tr}\{(I_n-H_g H_f^{-1})^i\} - \mathrm{tr}\{(I_{n+q}-B_g^{(n+q)}[B_f^{(n+q)}]^{-1})^i\}| \to 0$, $\quad i = 1,2$,

2) $|\mathrm{tr}\{(I_n-H_{\hat{f}}H_g^{-1})^i\} - \mathrm{tr}\{(I_{n+q}-B_{\hat{f}}^{(n+q)}[B_g^{(n+q)}]^{-1})^i\}| \to 0$, $\quad i = 1,2$ *where $\hat{g} = g$ or g_n,*

3) $|\mathrm{tr}\{(I_n-H_f H_g^{-1} - n^{-1/2}B_{a/2\pi})^2\}| \to 0$,

4) $|\log \det(H_g H_f^{-1}) - \log \det(B_g^{(n+q)}[B_f^{(n+q)}]^{-1})| \to 0$, *where $\hat{f} = f$ or g_n.*

Proof.   Since H_f^{-1} is the lower right-angular submatrix in the corresponding subdivision of the matrix $[B_f^{(n+2q)}]^{-1}$ into blocks (cf. (1.12) and (1.13)) it satisfies the following relation

$$[-V_{1,n}V_{1-q,0}^{-1},I_n]^*H_f^{-1}[-V_{1,n}V_{1-q,0}^{-1},I_n] = [B_f^{(n+q)}]^{-1}$$

(21) $\qquad - [B_f^{(n+q)}]^{-1}V_{1-q,n}(V_{1-q,n}^*[B_f^{(n+q)}]^{-1}V_{1-q,n})^{-1}$

$$\times V_{1-q,n}^*[B_f^{(n+q)}]^{-1}$$

(the matrix in the r.h.s. of (21) was introduced in (1.15) and was denoted by $D_f^{(n+q)}$; when deriving (21) it is necessary to utilize the fact that $D_f^{(n+q)}V_{1-q,n} = 0$).

From (20) and (21) we have, in particular, that for a pair

of spectral densities f_1 and f_2

$$\text{tr}(H_{f_2} H_{f_1}^{-1}) = \text{tr}(B_{f_2}^{(n+q)} D_{f_1}^{(n+q)}) = \text{tr}(B_{f_2}^{(n+q)}[B_{f_1}^{(n+q)}]^{-1})$$

(22)
$$- \text{tr}\{V_{1-q, n}^*[B_{f_1}^{(n+q)}]^{-1} B_{f_2}^{(n+q)}[B_{f_1}^{(n+q)}]^{-1}$$

$$\cdot V_{1-q, n}(V_{1-q, n}^*[B_{f_1}^{(n+q)}]^{-1} V_{1-q, n})^{-1}\}$$

and

$$\text{tr}[(H_{f_2} H_{f_1}^{-1})^2] = \text{tr}[(B_{f_2}^{(n+q)}[B_{f_1}^{(n+q)}]^{-1})^2]$$

$$-2\text{tr}\{V_{1-q, n}^*[B_{f_1}^{(n+q)}]^{-1} B_{f_2}^{(n+q)}[B_{f_1}^{(n+q)}]^{-1} B_{f_2}^{(n+q)}$$

(23)
$$\cdot [B_{f_1}^{(n+q)}]^{-1} V_{1-q, n}(V_{1-q, n}^*[B_{f_1}^{(n+q)}]^{-1} V_{1-q, n})^{-1}\}$$

$$+ \text{tr}\{(V_{1-q, n}^*[B_{f_1}^{(n+q)}]^{-1} B_{f_2}^{(n+q)}[B_{f_1}^{(n+q)}]^{-1} V_{1-q, n}$$

$$\cdot (V_{1-q, n}^*[B_{f_1}^{(n+q)}]^{-1} V_{1-q, n})^{-1})^2\}.$$

The results obtained in Section 1 also yield

$$\det H_{f_2} H_{f_1}^{-1} = \det B_{f_2}^{(n+q)}[B_{f_1}^{(n+q)}]^{-1}$$

(24)
$$\cdot \det(V_{1-q, n}^*[B_{f_2}^{(n+q)}]^{-1} V_{1-q, n})/\det(V_{1-q, n}^*$$

$$\cdot [B_{f_1}^{(n+q)}]^{-1} V_{1-q, n}).$$

Applying (22)-(24) to the spectral densities f, g, and g_n and taking into account the assertions of Corollary A2.1 it is easy to arrive at the relations 1), 2), and 4).

As far as the remaining relation 3) is concerned, in view of

$$I_n - H_f H_g^{-1} = n^{-1/2} H_{af} H_g^{-1},$$

this relation is equivalent to

(25) $\qquad n^{-1}\text{tr}[(H_{af} H_g^{-1} - B_{a/2\pi})^2] \to 0,$

and we shall now verify the latter.

First observe that the matrix $D_f^{(n+q)}$ appearing in the r.h.s. of (21) satisfies the relation

(26) $\qquad D_f^{(n+q)}[0, I_n]'[-V_{1, n} V_{1-q, 0}^{-1}, I_n] = D_f^{(n+q)}.$

This can be easily verified taking the fact that $D_f^{(n+q)}V_{1-q,n} = 0$ into account. From (21) and (26) we obtain that

$$H_f^{-1} = [0,I_n]D_f^{(n+q)}[0,I_n]',$$

and this, in turn, implies that

$$H_{af}H_g^{-1} = [-V_{1,n}V_{1-q,0}^{-1}, I_n]B_{af}^{(n+q)}D_g^{(n+q)}[0,I_n]'.$$

Therefore the l.h.s. of (25) can be written in the form

(27)
$$n^{-1}\mathrm{tr}[(B_{a/2n}^{(n)})^2] + n^{-1}\mathrm{tr}[(B_{af}^{(n+q)}D_g^{(n+q)})^2]$$
$$- n^{-1}2\mathrm{tr}(B_{af}^{(n+q)}D_g^{(n+q)}B_{a/2n}^{(n+q)})$$
$$+ n^{-1}2\mathrm{tr}\{B_{af}^{(n+q)}D_g^{(n+q)}(B_{a/2n}^{(n+q)}$$
$$- [0,I_n]'B_{a/2n}^{(n)}[-V_{1,n}V_{1-q,0}^{-1}, I_n])\}.$$

The first term in expression (27) converges to γ^2 (Lemma A1.2, assertion 4)). In view of assertion 10) of Lemma A1.2 and the corresponding assertions in Corollary A2.1, the second terms also possesses the limit:

$$\frac{1}{n}\,\mathrm{tr}\{[B_{af}(B_g^{-1}-B_g^{-1}V(V^*B_g^{-1}V)^{-1}V^*B_g^{-1})]^2\} \to \gamma^2,\quad V=V_{1,n}$$

The limit of the third term is $-2\gamma^2$, since

$$\frac{1}{n}\,|\mathrm{tr}\{[B_{af}(B_g^{-1}-B_g^{-1}V(V^*B_g^{-1}V)^{-1}V^*B_g^{-1})-B_{a/2n}]B_{a/2n}\}|$$
$$\leqslant \frac{1}{n}\,|B_{af}B_f^{-1}-B_{a/2n}|\cdot|B_{a/2n}|$$
$$+ \frac{1}{n}\,|B_{af}B_f^{-1}B_{fa}B_g^{-1}|\cdot n^{-1/2}|B_{a/2n}|$$
$$+ \frac{1}{n}|\mathrm{tr}\{(V^*B_g^{-1}V)^{-1}V^*B_g^{-1}B_{a/2n}B_{af}B_g^{-1}V\}| \to 0,$$

in view of the Corollary A2.1 and the assertions 4), 5), and 9) of Lemma A1.1. Thus for the proof of the required relation (25) it is sufficient to show that the last summand in expression (27) converges to 0. However, using the obvious notation

$$B_{a/2\pi}^{(n+q)} = \begin{bmatrix} B_{a/2\pi}^{(q)} & B_{a/2\pi}^{(q\times n)} \\ B_{a/2\pi}^{(n\times q)} & B_{a/2\pi}^{(n)} \end{bmatrix}$$

we have

$$B_{a/2\pi}^{(n+q)} - [0,I_n]'B_{a/2\pi}^{(n)}[-V_{1,n}V_{1-q}^{-1}, 0, I_n]$$

$$= \begin{bmatrix} B_{a/2\pi}^{(q)} & B_{a/2\pi}^{(q\times n)} \\ B_{a/2\pi}^{(n\times q)} & 0 \end{bmatrix} + \begin{bmatrix} 0 \\ B_{a/2\pi}^{(n)}V_{1,n}V_{1-q,0}^{-1} \end{bmatrix}[I_q, 0],$$

thus the convergence of (27) to 0 follows from the fact that in view of (A1.1)

$$n^{-1}\left|\text{tr}\left\{B_{af}^{(n+q)}D_g^{(n+q)}\begin{bmatrix} B_{a/2\pi}^{(q)} & B_{a/2\pi}^{(q\times n)} \\ B_{a/2\pi}^{(n\times q)} & 0 \end{bmatrix}\right\}\right|$$

(28)
$$\leqslant n^{-1}\left|B_{af}^{(n+q)}D_g^{(n+q)}\right|\left(\left|B_{a/2\pi}^{(q)}\right|^2 + 2\left|B_{a/2\pi}^{(q\times n)}\right|^2\right)^{1/2}$$

$$\leqslant 4\pi q\gamma n^{-1}\left|B_{af}^{(n+q)}D_g^{(n+q)}\right| \to 0$$

and

$$n^{-1}\left|\text{tr}\left\{[I_q,0]B_{af}^{(n+q)}D_g^{(n+q)}\begin{bmatrix} 0 \\ B_{a/2\pi}^{(n)}V_{1,n}V_{1-q,0}^{-1} \end{bmatrix}\right\}\right|$$

(29)
$$\leqslant \left|[I_q,0]B_{af}^{(n+q)}\right| \cdot \left\|D_g^{(n+q)}\right\| \cdot (q/n)^{1/2}\left\|B_{a/2\pi}^{(n)}\right\| \cdot$$

$$\cdot \left|V_{1-q,0}^{-1}\right|$$

$$\leqslant q2\pi M\gamma\left\|D_g^{(n+q)}\right\| \cdot \left|V_{1-q,0}^{-1}\right|n^{-1/2}\left\|B_{a/2\pi}^{(n)}\right\| \to 0$$

(recall that M and γ are constants appearing in the formulation of Lemma A1.2). When deriving (28) the fact that

$$| B_{af}(B_g^{-1}-B_g^{-1}V(V^*B_g^{-1}V)^{-1}V^*B_g^{-1}) | \leqslant \| B_{af}B_g^{-1} \|$$

$$\times \ (1 \ + \{\mathrm{tr}[V^*V(V^*B_g^{-1}V)^{-1}V^*B_g^{-2}V(V^*B_g^{-1}V)^{-1}]\}^{1/2}),$$

$$V = V_{1,n},$$

together with assertion 10) of Lemma A1.2 and Corollary A2.1 are utilized, while in order to derive (29) we use the inequality

$$\| D_g \| = \| B_g^{-1}-B_g^{-1}V(V^*B_g^{-1}V)^{-1}V^*B_g^{-1} \| \ \leqslant \ \| B_g^{-1} \|$$

$$(1 + \{\mathrm{tr}[V^*V(V^*B_g^{-1}V)^{-1}V^*B_g^{-2}V(V^*B_g^{-1}V)^{-1}]\}^{1/2}),$$

together with

$$\| B_g^{-1} \| \ \leqslant \ \| B_f^{-1} \|(1 + n^{-1/2}|B_{af}B_g^{-1}|)$$

and the assertions 2), 3) and 10) of Lemma A1.2.

Thus relation (25) and hence the assertion 3) are verified which completes the proof of Lemma A2.2.

3. As in Subsection 4.2 we shall consider the case when f_0 depends on a vector-valued parameter θ, i.e., $f_0 = f_{0,\theta}$.

It is easy to verify that the formula (22) as applied to the pair of spectral densities $g_0 = f_{0,\theta}(1 + n^{-1/2}a)$ (where $a = h'\dot\phi_\theta$) and $f_{0,\theta+n^{-1/2}h}$ yields the following relation:

$$\mathrm{tr}(H_{\psi_{0,n}} H_{f_{0,\theta+n^{-1/2}h}}^{-1}) = \mathrm{tr}(B_{\psi_{0,n}}^{(n+q)} [B_{f_{0,\theta+n^{-1/2}h}}^{(n+q)}]^{-1})$$

$$- \ \mathrm{tr}\{V_{1-q,n}^* [B_{f_{0,\theta+n^{-1/2}h}}^{(n+q)}]^{-1} B_{\psi_{0,n}}^{(n+q)} [B_{f_{0,\theta+n^{-1/2}h}}^{(n+q)}]^{-1} \cdot$$

$$\cdot V_{1-q,n}(V_{1-q,n}^* [B_{f_{0,\theta+n^{-1/2}h}}^{(n+q)}]^{-1} V_{1-q,n})^{-1}\},$$

where (cf. (2))

$$\psi_{0,n} = n^{1/2}(f_{0,\theta+n^{-1/2}h} - f_{0,\theta}) - f_{0,\theta}a.$$

Applying analogously the formulas (23) and (24) (to the above indicated pair of spectral densities as well as to the other pair appearing below) we obtain that the following lemma is valid.

Lemma A2.3. *Let* $\hat{g}_0$ *be equal to either* $f_{0,\Theta+n^{-1/2}t}$ *or to* $g_0 = f_0(1+n^{-1/2}a)$, *where* $t = 0$ *or* h. *Then*

1)
$$\lim_{n\to\infty} \text{tr}\{[\hat{H}(H_{g_0}^{-1}-H_{f_{0,\Theta+n^{-1/2}h}}^{-1})]^k$$
$$- [B_{\hat{g}_0}(B_{g_0}^{-1}-B_{f_{0,\Theta+n^{-1/2}h}}^{-1})]^k\} = 0, \quad k = 1,2,$$

where $\hat{H}$ equals either $H_{f_{0,\Theta+n^{-1/2}t}}$ or H_{g_0}.

2)
$$\lim_{n\to\infty} \log \det(H_{g_0} H_{f_{0,\Theta+n^{-1/2}h}}^{-1} B_{g_0}^{-1}B_{f_{0,\Theta+n^{-1/2}h}}) = 0.$$

A direct application of Lemma A2.3 yields the following result.

Lemma A2.4. *Under the conditions of Corollary 1 in Section 4*

$$\Lambda(f_{q,\Theta+n^{-1/2}h}, g_q) \to 0$$

in $P_n(g_q)$ *probability as* $n \to \infty$. *(Recall that* $g_q(\lambda) = g(\lambda) = f_q(\lambda)(1 + n^{-1/2}a(\lambda))$, *according to (4.2).)*

Proof. In view of (2.3), (2.4), and Lemma A2.3

$$E\{\Lambda(f_{q,\Theta+n^{-1/2}h},g_q)\} = \frac{1}{2}\{\log \det(H_{g_0} H_{f_{0,\Theta+n^{-1/2}h}}^{-1})$$
$$+ \text{tr}[\hat{H}(H_{g_0}^{-1} -H_{f_{0,\Theta+n^{-1/2}h}}^{-1})]\}$$
$$= \frac{1}{2} \{\log \det(B_{g_0}^{-1}B_{f_{0,\Theta+n^{-1/2}h}})$$
$$+ \text{tr}[B_{\hat{g}_0}(B_{g_0}^{-1} - B_{f_{0,\Theta+n^{-1/2}h}}^{-1})]\} + o(1)$$

and
$$D\{\Lambda(f_{q,\Theta+n^{-1/2}h},g_q)\}$$
$$= \frac{1}{2} \text{tr}\{[B_{\hat{g}_0}(B_{g_0}^{-1} - B_{f_{0,\Theta+n^{-1/2}h}}^{-1})]^2\} + o(1),$$

where the mathematical expectation and variance are taken with respect to the measure corresponding to the function $\hat{g}_0$. To complete the proof it remains only to apply the results of Lemma A2.1 to $f_\Theta = f_{0,\Theta}$.

Appendix 3. Remarks and Bibliography

Section 1

1. The problem of obtaining an explicit expression for a matrix which is the inverse of the Toeplitz matrix B_f (associated with the spectral density f) in the case of an autoregressive process is discussed in the papers [111,118] and also in [30,101]. The case of a moving average process is discussed in the papers [3,133,148], while the case of a mixed autoregressive moving average process is discussed in [84,87,92,93,113,149]. More recent references can be found in [152].

2. In Subsection 1.2 the general case is discussed when the spectral density f can be represented in the form $f(\lambda) = f_0(\lambda)|(z-z_1) \cdots (z-z_q)|^2$, $z = e^{i\lambda}$, $|z_i| \leqslant 1$, where f_0 is a nonnegative summable function. We obtain an explicit expression for B_f^{-1} and $\det(B_f)$ in terms of $B_{f_0}^{-1}$ and $\det(B_{f_0})$.

This allows us to write the final formulas in a relatively compact form suitable for further applications. We note in passing the role played by the above-stated representation of f in investigating the regularity condition of Gaussian processes in the sense of [68].

3. In the case when X_t, $-\infty < t < \infty$, is a random process with continuous time t observed on the time interval $[0,T]$, "the probability density in a functional space of functions $x(t)$, $0 \leqslant t \leqslant T$", $p_T\{x(t), 0 \leqslant t \leqslant T\} = (dP_T/dP_T^{(0)})\{x(t), 0 \leqslant t \leqslant T\}$ plays the role of a finite-dimensional probability density $p_n(x_1, ..., x_n)$. This probability density is the Radon-Nikodym derivative of the measure $P_T\{x(t)\}$ corresponding to the process X_t, $0 \leqslant t \leqslant T$, with respect to some similar "standard measure" $P_T^{(0)}\{x(t)\}$ (the derivative $dP_T/dP_T^{(0)}$ exists only under certain restrictions imposed on the "standard measure" $P_T^{(0)}$). These restrictions in the Gaussian case are discussed, for example, in [108]. In the particular case when X_t, $-\infty < t < \infty$, is a Gaussian autoregressive process of the first order, an explicit expression for p_T was obtained in the paper [112]. In the papers [8,30,102] this expression is generalized for the case of an autoregressive process of a finite order. In [30]

(cf. also [116]) an algorithm is also presented for computing the expression p_T in the case of a general Gaussian process X_t, $-\infty < t < \infty$, with a rational spectral density. Based on this algorithm the author, in his Ph.D. dissertation ("Estimation of parameters of a spectrum of a Gaussian stationary process with a rational spectral density," Moscow, 1971) produced a very cumbersome explicit expression for p_T in the last case. Later in [50] the problem of obtaining p_T in the case when X_t, $-\infty < t < \infty$, is a generalized Gaussian process with asymptotically (as $|\lambda| \to 0$) constant rational spectral density $f = f(\lambda)$, $-\infty < \lambda < \infty$ is considered. We also note the report [114] and the paper [117] where this problem is considered for a vector-valued Gaussian process X_t, $-\infty < t < \infty$, with a rational spectral density. Several further references may be found in [152] where a few specific examples are presented. Especially interesting is the Example 4.7, where X_t is a generalized process with a degenerating (for $\lambda = 0$) spectral density of the form $f(\lambda) = f_0(\lambda)\lambda^2$, $-\infty < \lambda < \infty$ with f_0 being the spectral density corresponding to an Ornstein-Uhlenbeck process (in this connection see also a short remark below).

Section 2

1. Conditions for the validity of formula (7) presented in [103] actually coincide with the conditions utilized in [37] for the proof of the refined Szegö theorem (see the theorem on page 101). In the paper [67] this theorem is proved again under more general conditions which for $f > 0$ coincide with condition (2) of Lemma 1 (see also [153,157]). In the note [104] a simple and lucid derivation for formula (7) is presented in particular in the case of a mixed autoregressive-moving average process. The expression for the "principal part" of the logarithm of the likelihood function related to expression (7) for the case of the multidimensional vector-valued process X_t is given in [123-125] and for the case of a random field X_t (t is a vector-valued parameter) in [32,33,159,180].

2. For a Gaussian process X_t, $-\infty < t < \infty$, with continuous time and a general rational spectral density $f = f(\lambda) > 0$, $-\infty < \lambda < \infty$, the formula for the "principal part" of the logarithm of the likelihood function $\log p_T\{x(t), 0 \leqslant t \leqslant T\}$ related to (7) is

given in [103]. It is shown in [50] that this formula is valid also in the case of a generalized process X_t, $-\infty < t < \infty$, with asymptotically (as $|\lambda| \to \infty$) constant rational spectral density f. There is, however, no doubt that the results of the paper [103] can be carried over to a substantially wider class of Gaussian processes (for a mathematically nonrigorous discussion of this problem (cf. [78], where the case when $-\infty < -\lambda_1 \leqslant \lambda \leqslant \lambda_1 < \infty$ is considered; cf. also note [158]).

In the concluding paragraph of Subsection 5.5 of the paper [152] the possibility of extending the results of Subsection 2.2 to the case of continuous time processes is discussed. It is indicated therein that this can be accomplished at least for the simple case mentioned above with $f(\lambda) = f_0(\lambda)\lambda^2$, f_0 being the spectral density of the Ornstein-Uhlenbeck process.

3. Return to the above-mentioned refinement of Szegö's theorem which states that the formula (10) can be refined:

$$\log \det(B_f) = \frac{n}{2\pi}\int_{-\pi}^{\pi} \log[2\pi f(\lambda)]d\lambda$$
$$+ \frac{1}{2} \sum_{k=1}^{\infty} k|(\widetilde{\log f})_k|^2 + o(1),$$

where the tilde over the function of λ indicates the corresponding Fourier coefficient of this function. Observe that

$$\frac{1}{2} \sum_{k=1}^{\infty} k|(\widetilde{\log f})_k|^2 = \frac{1}{2\pi}\iint_{|z|\leqslant 1} \left|\frac{A'(z)}{A(z)}\right|^2 d\sigma,$$

where $d\sigma$ denotes the area element of the unit circle and $A(z)$ = $\sum_{s=0}^{\infty} g_s z^s$ is the exterior function of representation of f in the form (II.6.2)-(II.6.3) under the condition (II.1.3).

Moreover, it is also easy to refine the first assertion of Lemma A1.4 (cf. (A1.3)):

$$\frac{1}{n}\,\mathrm{tr}(B_a\text{-}B_r B_f) = \frac{2}{n} \sum_{j=1}^{\infty} \min(n,j)\beta(j)\rho(j)$$
$$= \sum_{j=-\infty}^{\infty} \beta(j)\rho(j) - \sum_{j=-n}^{n} \left[1 - \frac{|j|}{n}\right]\beta(j)\rho(j)$$
$$= \int_{-\pi}^{\pi} a(\lambda)d\lambda - \iint_{-\pi}^{\pi} K_n(\lambda\text{-}\mu)a(\lambda)\frac{f(\mu)}{f(\lambda)}d\lambda d\mu$$

where $K_n(\lambda) = [\sin(1/2)n\lambda/\sin(1/2)\lambda]^2/2\pi n$ is Fejér's kernel.

These two refinements taken together result in sharpening the assertion of Theorem 1 in the sense that $\tilde{L}_n$ defined by formula (7) not only satisfies the relation (1) but also possesses the following properties: up to a term o(1)

$$E[L_n-\tilde{L}_n] = -\frac{1}{4}\sum_{k=1}^{\infty} k|(\overline{\log f})_k|^2 - \sum_{k=1}^{\infty} k\beta(k)\rho(k)$$

$$= -\frac{1}{2}\left\{ n - \frac{n}{2\pi}\int\int_{-\pi}^{\pi} K_n(\lambda-\mu)\frac{f(\mu)}{f(\lambda)}d\lambda d\mu \right.$$

$$\left. + \frac{1}{2\pi}\int\int_{|z|\leqslant 1}\left|\frac{A'(z)}{A(z)}\right|^2 d\sigma \right\}$$

and as it was already established (Theorem 1)

$$E[L_n - \tilde{L}_n]^2 < \infty$$

(cf. [38,154,160]).

4. A similar refinement is also valid in the case of spectral densities with fixed zeros studied in Subsection 2.2.[10]   In particular, in this case

$$E[L_n-\tilde{L}_n] = -\frac{1}{4}\sum_{k=1}^{\infty} k|(\overline{\log f_0})|^2 - \sum_{k=1}^{\infty} k\beta_0(k)\rho_0(k) - \frac{q}{2}$$

$$- \frac{1}{2}\sum_{j=1}^{\infty} \log(n\sigma^2/2\pi f_0(\lambda_j))$$

$$+ \frac{1}{2}(1-\delta_{1q})\sum_{k<l} \log|1-e^{i(\lambda_k-\lambda_l)}|^2,$$

in view of Remark 2 in Subsection 1.2 and formulas (2.23) and (2.26) ($\rho_0(k)$ is the k-th Fourier coefficient of the function $h_0 = 1/(2\pi)^2 f_0$).

5. Under the Assumption 1 stipulated in Subsection 2.2 the spectral density f_q can evidently be written in the form (1.10) with polynomial $Q_q(z)$ possessing q distinct roots $z_j = e^{i\lambda_j}$, $j = 1, ..., q$. The restriction to simple roots is motivated only by the desire to avoid further complications in an already cumbersome exposition of the material in Subsection 2.2

---

[10]Cf. [152] -- there are unfortunate misprints on pages 48 and 49.

and Section 4. Actually, the results of Subsection 1.2 allow us to conjecture that the formula (31) will be valid also in the general case of spectral densities of the form (1.10) with a polynomial $Q_q(z)$ admitting multiple roots with a unit modulus provided only the matrices S and V are modified accordingly -- when constructing $\mathbf{Y}_n$ (in accordance with Remark 1 and the expression for the V preceding it as presented in 1.2).

When carrying out such a generalization it is necessary to take into account in particular that in view of the well-known results of linear regression analysis[11] in place of the formulas (17) and (23) we have

$$\lim_{n \to \infty} D_n^{-1} V^* B_{f_0} V D_n^{-1} = \text{diag}\{2\pi f_0(\lambda_j) H_j, \quad j = 1, ..., r\}$$

and

$$\lim_{n \to \infty} D_n^{-1} V^* B_{f_0}^{-1} V D_n^{-1} = \text{diag}\{H_j / 2\pi f_0(\lambda_j), \quad j = 1, ..., r\},$$

where

$$V = V_{1,n} = [V_{1,n}^{(1)}, ..., V_{1,n}^{(r)}],$$

$$V_{1,n}^{(j)} = [k^{\ell} \zeta_j^k, \quad k = 1, ..., n; \quad \ell = 0, ..., q_j-1]$$

(as in Subsection 1.2, $\zeta_1, ..., \zeta_r$ are $r \leqslant q$ distinct roots of multiplicity $q_1, ..., q_r$ with $q_1 + ... + q_r = q$ of the polynomial $Q_q(z)$) and

$$D_n = \text{diag}\{D_n^{(1)}, ..., D_n^{(r)}\},$$

$$D_n^{(j)} = \sqrt{n} \, \text{diag}\{n^{\ell}; \quad \ell = 0, ..., q_j-1\},$$

while $H_j = [1/k+\ell+1; \quad k,\ell = 0,1, ..., q_j-1]$ are the Hilbert matrices (positively definite since

$$x^* H_j x = \int_0^1 \left| \sum_{k=0}^{q_j-1} x_k \xi^k \right|^2 d\xi > 0$$

for any nonzero $x = \text{col}(x_0, ..., x_{q_j-1})$).

---

[11]Cf., e.g., [35, 161] where these results are derived under the assumption that $f_0 > 0$ is a piecewise-continuous function, but it is also noted therein that only the absence of jumps at the points $\lambda_1, ..., \lambda_r$ of the function f_0 is essential.

We also note that $\det(V^* B_{f_0}^{-1} V) = O(n^{q_1^2 + \cdots + q_r^2})$ since

$$\det D_n^{(j)} = O(n^{q_j^2/2}).$$

6. It is of interest to trace the application of the formulas presented herein to another problem of time series analysis namely the problem of estimating regression coefficients, say $\eta = \text{col}(\eta_1, ..., \eta_s)$ in a linear model of polynomial-trigonometric regression in which the observations $X = \text{col}(X_1, ..., X_n)$ are characterized by the mathematical expectation of the form

$$EX = \Phi\eta, \qquad \Phi = [\Phi^{(1)}, ..., \Phi^{(m)}]$$

with

$$\Phi^{(j)} = [k^{\ell} e^{i\omega_j k}; \quad k = 1, ..., n; \quad \ell = 0, ..., s_j{-}1]$$

$(s_1 + \cdots + s_m = s)$ and covariance matrix B_f -- a Toeplitz matrix associated with the spectral density f (continuous at least at the regression spectrum consisting of a finite number of distinct cyclic frequencies $\{\omega_1, ..., \omega_m\}$).

As it is well-known (cf., e.g., [35], Chapter 7) in the case of everywhere positive spectral density $f > 0$ the finiteness of the regression spectrum $\{\omega_1, ..., \omega_m\}$ assures the asymptotic efficiency of the least square estimator $\eta_{LS} = [\Phi^*\Phi]^{-1}\Phi^*X$ of parameter η in the sense that it has exactly the same limiting variance (as $n \to \infty$) as the best linear unbiased estimator (BLUE)

$$\eta_0 = (\Phi^* B_f^{-1} \Phi)^{-1} \Phi^* B_f^{-1} X.$$

Namely, from the formulas presented in the preceding remark,

$$E(\eta_{LS}{-}\eta)^{\otimes 2} = [\Phi^*\Phi]^{-1}\Phi^* B_f \Phi[\Phi^*\Phi]^{-1}$$

and $E(\eta_0{-}\eta)^{\otimes 2} = (\Phi^* B_f^{-1}\Phi)^{-1}$ possess the same principal parts represented in the form $\text{diag}\{2\pi f(\omega_j)[D_n^{(j)} H_j D_n^{(j)}]^{-1}, j = 1, ..., m\}$ where $D_n^{(j)} = \sqrt{n} \, \text{diag}\{n^{\ell}, \ell = 0, ..., s_j{-}1\}$ now,

However, in the case when the spectral density f can degenerate being of the form (11) or more generally of the form (1.10) with a polynomial $Q_q(z)$ possessing r $(r \leqslant q)$

distinct roots $e^{i\lambda j}$, $j = 1, ..., r$, the situation is not as simple: in general an LSE preserves the property of asymptotic efficiency in the above-mentioned sense only if the sets $\{\omega_1, ..., \omega_m\}$ and $\{\lambda_1, ..., \lambda_r\}$ do not intersect. On the other hand another estimator possesses this property irrespectively of the last condition. This is the so-called pseudo best estimator η_{PB} corresponding to the pseudo spectral density $|Q_q(z)|^2/2\pi$ (in the terminology of [68], Chapter VII). This estimator is constructed in the same manner as η_0 but in place of B_f a Toeplitz matrix associated with the pseudo spectral density is chosen. We observe that as a result of the simple transformations utilized in Subsection 1.2, one can arrive at the following representation $\eta_{PB} = [\psi^*\psi]^{-1}\psi^*\tilde{Y}_{n+q}$, where

$$\tilde{Y}_{n+q} = (I_{n+q} - P_{n+q})Y_{n+q} \quad \text{and} \quad \psi = (I_{n+q} - P_{n+q})\begin{bmatrix} 0 \\ S\Phi \end{bmatrix};$$

here the previous notation is retained:

$$P_{n+q} = V_{1-q, n}(V^*_{1-q, n} V_{1-q, n})^{-1} V^*_{1-q, n}.$$

It has a clear interpretation: in the presence of a polynomial factor in the expression for spectral density, in place of LSE being constructed directly from the observed X, we construct an estimator which preserves the LSE form, but now with the transformed statistic $\tilde{Y}_{n+q}$ and the transformed regressor ψ (cf. [107]). The following example illustrates the above discussion.

Example. Let a stationary time series X_t possess unknown mathematical expectation $EX_t = \eta$ and spectral density $f = f_1$ of the same form as in Example 4 presented in Section 1 but with a specific $z_1 = e^{i\lambda_1}$.

The LSE which is the sample mean $\eta_{LS} = (1/n)\Sigma_{t=1} X_t$ has the variance

$$D(\eta_{LS}) = \frac{2}{n} \sum_{-n}^{n} \left[1 - \frac{|t|}{n}\right] b(t)$$

$$= \frac{2\pi f(0)}{n}\left[1 + \frac{1-\theta^n}{n} \frac{1+\theta^2}{1-\theta^2}\right] + \frac{2\sigma^2}{n^2} \cdot \frac{1-\theta^n}{1-\theta^2},$$

since here $b(t) = 2b_0(t) - \bar{z}_1 b_0(t+1) - z_1 b_0(t-1)$, $b_0(t) = \sigma^2\theta^{|t|}/(1-\theta^2)$.

Utilizing the formulas presented in Example 1.4 it is also not difficult to verify that $D(\eta_0) = (v'B_f^{-1}v)^{-1}$ where

$$\eta_0 = (v'B_f^{-1}v)^{-1}v'B_f^{-1}X, \quad v = \mathrm{col}(1, ..., 1)$$

possesses the principal part $2\pi f(0)/n$ for $\lambda_1 \neq 0$ and $2\pi f_0(0)\cdot 12/n^3$ for $\lambda_1 = 0$.

We have thus seen that LSE is asymptotically efficient in the sense that $D(\eta_0)/D(\eta_{LS}) \to 1$ only in the case when $\lambda_1 \neq 0$; however, in the case of degeneracy of the spectral density on the regression spectrum $\{0\}$ (i.e., in the case when $\lambda_1 = 0$) it possesses a zero asymptotic efficiency:

$$\frac{D(\eta_0)}{D(\eta_{LS})} \sim \frac{12\sigma^2}{n^3(1-\theta)^2} \Big/ \frac{2\sigma^2}{n^2(1-\theta^2)} \to 0.$$

In both cases the estimator

$$\eta_{PB} = \left\{ \sum_{j=1}^{n} \left| \sum_{k=0}^{j-1} z_1^k \right|^2 - (n+1) \left| \sum_{j=1}^{n} z_1^j \left[1 - \frac{j}{n+1} \right] \right|^2 \right\}^{-1} \cdot$$

$$\cdot \left\{ \sum_{j=1}^{n} Y_j \sum_{k=0}^{j-1} \overline{z}_1^k - \sum_{k=1}^{n} \overline{z}_1^k Y_k \cdot \sum_{j=1}^{n} z_1^j \left[1 - \frac{j}{n+1} \right] \right\}$$

is an asymptotically efficient estimator. (Recall that by definition η_{PB} is constructed in the same manner as η_0 above, but in place of B_f the Toeplitz matrix associated with $|e^{i\lambda}_1 - e^{i\lambda}|^2/2\pi$ is substituted.)

7. One can grasp the connection between the approximation (7) and an explicit expression for the logarithm of the conditional probablity density of variables $X_1, ..., X_n$ under the condition that $X_s = 0$, where $s \leqslant 0$ or $s > n$ (this expression can be found from the results of the paper [185] devoted to Gibbs' description of Gaussian homogeneous fields; cf. [152], formula (5.13)): simply, they are equal up to an unessential deterministic term.

8. If in (7) the integral from $-\pi$ to π with respect to $d\lambda$ is replaced by its Riemann sum, for example, by $2\pi/n$ times the sum of the integrand at $\lambda_j = 2\pi j/n$, $j = 0,\pm 1, ..., \pm[(n-1)/2]$, we get the following discrete approximation to the log likelihood L_n:

$$\tilde{L}_n^{(d)} = -\sum_{j=0}^{[(n-1)/2]} \left\{ \log f(\lambda_j) + \frac{I_n(\lambda_j)}{f(\lambda_j)} \right\} + \text{const.},$$

$$\lambda_j + \frac{2\pi j}{n}.$$

Clearly, if n is large, this expression will give essentially the same asymptotic results as did Whittle's approximation (7). It should be noted also that the replacement of the integrals by their Riemann sums is necessary if the maximization of the approximate likelihood is to be performed numerically in a computer. Moreover, for large values of n it is often even feasible to replace the Riemann sum by a sum corresponding to the division of the integration interval into considerably less than n subintervals (cf. [140, p. 381, Remark 4]).

These considerations allow one to get yet another simple interpretation of Whittle's approximation to the function L_n ([186]; cf. also [187]). For this let us recall that under sufficiently mild conditions the random variables $L_n(2\pi j/n)$, $j = 1, ..., [(n-1)/2]$, $[(n-1)/2] = (n-1)/2$ if n is odd and $= (n/2) - 1$ if n is even, are asymptotically mutually independent and identically distributed; the limiting distribution is exponential with expectation $f(\lambda_j) = f(2\pi j/n)$ and variance $f^2(2\pi j/n)$ ([188]; cf. also [4,26,140]). Instead of the probability density p_n of the variables $X_1, ..., X_n$, let us consider now the proability density $p_I(I(\lambda_j))$, $j = 1, ..., [(n-1)/2]$, of the complex variables $I(\lambda_j)$, $j = 1, ..., [(n-1)/2]$. Then, asymptotically (for large n's),

$$p_I = \prod_{j=0}^{[(n-1)/2]} [f(\lambda_j)]^{-1} \exp \frac{-I_n(\lambda_j)}{f(\lambda_j)},$$

$$L_I = \log p_1 = \sum_{j=0}^{[(n-1)/2]} \left[\log f(\lambda_j) + \frac{I_n(\lambda_j)}{f(\lambda_j)} \right],$$

where $\lambda_j = 2\pi j/n$ (cf. $\tilde{L}_n^{(d)}$ above). Note that these considerations do not employ the Gaussian assumption since the asymptotic properties of the periodogram values $I_n(\lambda_j)$ mentioned above are valid under much broader conditions on X_t, in particular, under the conditions of Sections 6 and 7 of the next chapter.

Chapter II
ESTIMATION OF PARAMETERS BY MEANS OF P. WHITTLE'S METHOD

1. Asymptotic Maximum Likelihood Estimators

Let X_t, $t = ..., -1,0,1, ...$ be a Gaussian process with zero expectation and spectral density f depending on an unknown vector-valued parameter θ so that $f = f_\theta$, $\theta \in \Theta$, where Θ is a subset in R_p. Assume furthermore, that it is required to estimate the value of the unknown parameter θ based on a sequence of observations from the random process X_t, for $t = 1, ..., n$.

The formula (I.1.3) allows us to consider the maximum likelihood estimator $\bar{\theta}$ of the parameter θ determined by the condition

$$(1) \qquad L_n(\bar{\theta}) = \max_\theta L_n(\theta),$$

where $L_n(\theta) = L_n$. This estimator in the regular case is determined as the root of a system of p equations:

$$(2) \qquad \frac{\partial}{\partial \theta_k} L_n(\theta) = -\frac{1}{2} \frac{\partial}{\partial \theta_k} [\log \det(B_{f_\theta}) + X' B_{f_\theta}^{-1} X] = 0, \qquad k = 1, ..., p$$

(here θ_k is the k-th element of the vector θ and $X = (X_1, ..., X_n)'$). To be able to solve the system of equations (2) it is required to have an explicit expression for the determinant

$\det(B_{f_\Theta})$ and the matrix $B_{f_\Theta}^{-1}$. The results presented in Section

1 of Chapter I allow us in principle to obtain these explicit expressions for a wide class of spectral densities (which includes all rational densities). However, as it is clearly illustrated by the examples presented in Section 1 of Chapter I, even in those cases when one is able to write an expression for $L_n(\Theta)$ explicitly, equation (2) may be so cumbersome that it becomes, in a majority of cases, a hopeless task to determine its roots. However, since the maximum likelihood estimator $\bar{\Theta}$ may be considered to be optimal only in the limit as $n \to \infty$, Mann and Wald [86], in the case of an autoregressive process and later Whittle [121] in the general case, suggested overcoming this difficulty by replacing $\bar{\Theta}$ with the estimator $\tilde{\Theta}$ of the parameter Θ which equals the value of Θ maximizing the "principal part" of $L_n(\Theta) = L_n$ (rather than the whole quantity $L_n(\Theta)$). This principal part satisfies the condition $[L_n(\Theta) - \tilde{L}_n(\Theta)]/\sqrt{n} \to 0$ as $n \to \infty$ (in the sense of convergence in probability). It was established in Section 2 of Chapter I that under general conditions on f_Θ the quantity $\tilde{L}_n(\Theta)$ can be chosen in a manner such that its expression is much simpler than that of $L_n(\Theta)$; at the same time, the estimator $\tilde{\Theta}$ turns out to be asymptotically equivalent to the maximum likelihood estimator $\bar{\Theta}$ (i.e., it possesses the same asymptotic properties as the estimator $\bar{\Theta}$).

The asymptotic properties of the estimators $\tilde{\Theta}$ (which we shall, for convenience, refer to as asymptotic m.l. estimators) were studied by Whittle [121] (see also [131]) under a number of assumptions including, in particular, the one that the spectral density f_Θ satisfies the inequality $f_\Theta > 0$ and one of the elements of the vector Θ is the parameter

$$(3) \qquad \sigma^2 = 2\pi \exp\left\{ \frac{1}{2\pi} \int_{-\pi}^{\pi} \log f(\lambda) d\lambda \right\} > 0.$$

Moreover, in [121] it was shown that under some very general regularity conditions on the function f_Θ, the asymptotic m.l. estimator $\tilde{\Theta}$ is also a consistent, asymptotically normal, and asymptotically efficient estimator.

It should, however, be noted that in many important particular situations, the conditions presented in [121] mentioned above may not be fulfilled. For example, both the normalized spectral density f/σ^2 as well as the parameter σ^2 may depend on the unknown parameter Θ. We shall see below that this is indeed the case often encountered in practice when the parameter Θ, appearing in the expression for spectral density f_Θ of the process S_t, is estimated by means of the observed values of the process $X_t = S_t + N_t$, where N_t is a sequence of independent Gaussian random variables with expectation zero and unknown variance (which evidently must also be estimated; see Section 5 of this Chapter for more details).[1] In the next section we shall only slightly modify the arguments presented in [131] and generalize the results of [121,131] dealing with properties of asymptotic m.l. estimators $\tilde\Theta$ to a more general case when the parameter σ^2 is not necessarily an element of Θ (but still under the assumption that $f > 0$); the case when f satisfies Proposition 1 presented in Section 2 of Chapter 1 (i.e., when it can vanish at certain points) will be considered in Section 3.

2. Properties of Asymptotic Maximum Likelihood Estimators in the Case of Strictly Positive Spectral Density

2.1. Assume that the true value Θ_0 of the parameter Θ belongs to a closed set Θ contained in an open set S of the p-dimensional Euclidean space R_p.

For $\Theta \in S$ let the function $f = f_\Theta$ satisfy the following condition:

(A) If Θ_1 and Θ_2 are two different values of the parameter belonging to Θ, then $f_{\Theta_1} \neq f_{\Theta_2}$ for almost all λ.

Denote

$$(1) \qquad U_n(\Theta) = U_n(\Theta;X) = \frac{1}{4\pi} \int_{-\pi}^{\pi} \left[\log f_\Theta(\lambda) + \frac{I_n(\lambda)}{f_\Theta(\lambda)} \right] d\lambda,$$

where $I_n(\lambda) = I_n(\lambda;X)$ is the periodogram of the process X_t [see (I.2.8)] and determine the estimator $\tilde\Theta$ from the condition

$$(2) \qquad U_n(\tilde\Theta) = \min_\Theta U_n(\Theta).$$

(In view of formula (I.2.7) this estimator is called the

---

[1]A discussion of this problem is also given in the recent papers [39,165].

asymptotically m.l. estimator here.) The following theorem is valid.

Theorem 1. *Let the continuous derivatives $(\partial/\partial\theta_k)f_\theta^{-1}$ of the function f_θ^{-1} exist for all elements θ_k of the vector θ, where $\theta \in S$ and $\lambda \in [-\pi,\pi]$. Then under the conditions stipulated above the asymptotically m.l. estimator $\tilde{\theta}$ is consistent (i.e., $\tilde{\theta} \to \theta$ in probability $P_{n,\theta} = P_n(f_\theta)$ as $n \to \infty$).*

Proof. Let θ_1 and θ_2 be values of θ such that for $\theta_1 \in \theta$ and $\theta_2 \in S$ and $|\theta_1-\theta_2| < \delta$ where δ may depend on θ_1. Taking (1) and the obvious inequality

$$(3) \qquad \log\frac{f_{\theta_1}}{f_{\theta_2}} \leqslant \frac{f_{\theta_1}}{f_{\theta_2}} - 1$$

into account we obtain

$$(4) \qquad |U_n(\theta_2)-U_n(\theta_1)| \leqslant \frac{1}{4\pi}\int_{-\pi}^{\pi}[I_n(\lambda) + f_{\theta_1}(\lambda)]|f_{\theta_2}^{-1}(\lambda)$$
$$- f_{\theta_1}^{-1}(\lambda)\; d\lambda.$$

Introducing the notation

$$H(\theta_1,\delta(\theta_1)) = \sup\{|(\partial/\partial\theta_k)f_\theta^{-1}(\lambda)|\colon -\pi \leqslant \lambda \leqslant \pi,$$

$$|\theta_1-\theta| \leqslant \delta(\theta_1),\;\; k = 1, ..., p\}$$

where $\delta(\theta_1) > \delta$ is chosen in such a manner that the set $\{\theta\colon |\theta-\theta_1| \leqslant \delta(\theta_1)\}$ is contained in S, in view of the mean value theorem we obtain from (4):

$$(5) \qquad |U_n(\theta_2) - U_n(\theta_1)| \leqslant H_{\delta,n}(\theta_1,X),$$

where

$$(6) \qquad H_{\delta,n} = \frac{\delta}{4\pi}H(\theta_1,\delta(\theta_1))\int_{-\pi}^{\pi}[I_n(\lambda) + f_{\theta_1}(\lambda)]d\lambda.$$

It follows from the formulas (I.2.3) and (I.2.4) that one has

$$(7) \qquad \lim_{\delta\to 0} E(H_{\delta,n}) = 0$$

uniformly in n, and

(8) $\qquad \lim_{n\to\infty} D(H_{\delta, n}) = \left[\frac{\delta}{4\pi} H(\theta_1, \delta(\theta_1))\right]^2 \lim_{n\to\infty} \frac{2}{n^2} |B_f|^2 = 0$

for all δ.

Consider now the difference

$$U_n(\theta_0) - U_n(\theta) = \frac{1}{4\pi} \int_{-\pi}^{\pi} \left\{ \log \frac{f_{\theta_0}(\lambda)}{f_\theta(\lambda)} \right.$$

$$\left. + I_n(\lambda) \left[\frac{1}{f_{\theta_0}(\lambda)} - \frac{1}{f_\theta(\lambda)} \right] \right\} d\lambda,$$

where $\theta \neq \theta_0$, $\theta \in \Theta$. In view of the formula (I.2.3)

(9) $\qquad \begin{aligned} &\lim_{n\to\infty} E(U_n(\theta_0) - U_n(\theta)) \\ &= \frac{1}{4\pi} \int_{-\pi}^{\pi} \left[\log \frac{f_{\theta_0}(\lambda)}{f_\theta(\lambda)} + 1 - \frac{f_{\theta_0}(\lambda)}{f_\theta(\lambda)} \right] d\lambda < 0 \end{aligned}$

since it follows from condition (A) that for $\theta_1 = \theta_0$, $\theta_2 = \theta$ and $(\theta \neq \theta_0)$ the inequality (3) is strict for all λ. Since, moreover, θ_0 and θ belong to the closed set Θ there exists a positive number $K(\theta_0, \theta)$ for which the inequality

(10) $\qquad \lim_{n\to\infty} E(U_n(\theta_0) - U_n(\theta)) < -K(\theta_0, \theta)$

is valid. Applying the formula (I.2.4) we obtain

(11) $\qquad \begin{aligned} &\lim_{n\to\infty} D[n^{1/2}(U_n(\theta_0) - U_n(\theta))] \\ &= \frac{1}{4\pi} \int_{-\pi}^{\pi} \left[1 - \frac{f_{\theta_0}(\lambda)}{f_\theta(\lambda)} \right]^2 d\lambda < \infty. \end{aligned}$

Thus, in view of Chebyshev's inequality

(12) $\qquad \lim_{n\to\infty} P\{U_n(\theta_0) - U_n(\theta) < -K(\theta_0, \theta)\} = 1, \quad \theta_0 \neq \theta.$

The proof of Theorem 1 now follows from the validity of the following lemma (cf. [131], Lemma 2, p. 368).

Lemma 1. *Let $U_n(\theta)$ be a random functional satisfying the relations (5) and (12). Then $\tilde{\theta} \to \theta$ in probability $P_{n, \theta}$ (as $n \to \infty$) where $\tilde{\theta}$ is determined from condition (2).*

2.2.   To investigate further properties of the estimator $\tilde{\theta}$ we shall return to the case considered in Subsection 3.2 of the preceding chapter (retaining however the conditions for the consistency of $\tilde{\theta}$; cf. Theorem 1).   Under the last conditions the vector $\dot{\phi}_\theta$ in (I.3.19) is a gradient vector of continuous derivatives of the logarithm of f_θ; its k-th element is equal to $\dot{\phi}_{k,\theta} = (\partial/\partial\theta_k)\log f_\theta$.

Observe now that in this case the square integrable function $a = a_\theta = \mathbf{h}'\dot{\phi}_\theta$, $\mathbf{h} = \mathrm{col}(h_1, ..., h_p)$ is equal to

(13) $a(\lambda) = a_\theta(\lambda) = \overset{p}{\underset{k=1}{\Sigma}} h_k \dfrac{\partial}{\partial\theta_k} \log f_\theta(\lambda).$

By assuming that it is different from zero for almost all λ for any nonvanishing vector $\mathbf{h}$, we shall assure the positive definiteness of the matrix

(14)
$$\Gamma_\theta = \left[\frac{1}{4\pi} \int_{-\pi}^{\pi} \frac{\partial}{\partial\theta_k} \log f_\theta(\lambda)\frac{\partial}{\partial\theta_\ell} \log f_\theta(\lambda)d\lambda, \right.$$
$$\left. k,\ell = 1, ..., p \right]$$

(cf. (I.3.24)) which is actually the limit as $n \to \infty$ of the Fisher's information matrix

(15) $\dfrac{1}{n} \left[E\left[\dfrac{\partial}{\partial\theta_k}L_n(\theta)\dfrac{\partial}{\partial\theta_\ell} L_n(\theta) \right], \ k,\ell = 1, ..., p \right]$

since

$$\frac{1}{n} D\left[\overset{p}{\underset{k=1}{\Sigma}} h_k \frac{\partial}{\partial\theta_k} L_n(\theta) \right] = \frac{1}{2n} \mathrm{tr}\left[\left[B_{f_\theta} \overset{p}{\underset{k=1}{\Sigma}} h_k\frac{\partial}{\partial\theta_k} B_{f_\theta}^{-1} \right]^2 \right]$$

$$= \frac{1}{2n} \mathrm{tr}\left[\left[B_{f_\theta}^{-1} B_{\Sigma h_k(\partial/\partial\theta_k) f_\theta} \right]^2 \right] \to \frac{1}{4\pi}\int_{-\pi}^{\pi} a_\theta^2(\lambda)d\lambda$$

in view of (I.1.3), (I.1.4) and the assertion 8) of Lemma A1.2 of Appendix 1 to Chapter I.

In view of the assertion 3) of Corollary I.3.1 the log of the likelihood ratio satisfies the relation

(16) $\Lambda(f_\theta, f_{\theta+n^{-1/2}\mathbf{h}}) - \mathbf{h}'\Delta_{n,\theta} + \dfrac{1}{2} \mathbf{h}'\Gamma_\theta\mathbf{h} \to 0$ as $n \to \infty$

in $P_n(f_\theta)$ probability, if only the condition of Lemma I.3.1 on the Fourier coefficients of functions f_θ and $a_\theta/f_\theta = -\Sigma_{k=1}^{p} h_k(\partial/\partial\theta_k)f_\theta^{-1}$ are satisfied.   Here $\Delta_{n,\theta}$ is the gradient vector of the functional $U_n(\theta)$ defined by formula (2.1) multiplied by $-n^{1/2}$, i.e.,

$$\Delta_{n,\Theta} = -n^{1/2} \left[\frac{\partial}{\partial\Theta}\right] U_n(\Theta)$$

(17)
$$= \text{col}\left\{\frac{n^{1/2}}{4\pi} \int_{-\pi}^{\pi} \frac{I_n(\lambda)-f_\Theta(\lambda)}{f_\Theta(\lambda)} \frac{\partial}{\partial\Theta_k}\log f_\Theta(\lambda)d\lambda,\right.$$
$$\left. k = 1, ..., p\right\}$$

(cf. (I.3.23)).

Moreover, as it was established in Corollary A1.1 of Appendix 1 to this Chapter, under the last conditions, the family of Gaussian distributions $\{P_n(f_\Theta), \Theta \in \Theta\}$ is locally asymptotically normal in the sense that in the representation (16) the vector $\Delta_{n,\Theta}$ as $n \to \infty$ is distributed according to the normal distribution $N(0,\Gamma_\Theta)$ with zero expectation and covariance matrix Γ_Θ:

(18) $\{\Delta_{n,\Theta} \mid P_{n,\Theta}\}$ $N(0,\Gamma_\Theta)$.

In view of this property of local asymptotic normality (LAN) of the family $\{P_n(f_\Theta), \Theta \in \Theta\}$ one would expect[2] that the asymptotic normality and asymptotic efficiency of the estimator $\tilde{\Theta}$ is valid in the sense that

(19) $\{n^{1/2}(\Theta-\tilde{\Theta}) \mid P_n(f_\Theta)\}$ $N(0,\Gamma_\Theta^{-1})$,

where (as we have shown above) Γ_Θ is the limit as $n \to \infty$ of Fisher's information matrix (14). Indeed, the following theorem is valid.

---

[2] The definition of the LAN property for a family of distributions $\{P_{n,\Theta}, \Theta \in \Theta\}$ in a general sequence of experiments and certain notions and results connected with it may be found in a recent book [70]. Following the ideas of this book, one could investigate more subtle properties of $\tilde{\Theta}$ and not be limited by Fisher's efficiency presented herein. Indeed, such an investigation is certainly of theoretical interest, however, its practical value is limited by the necessary assumption of the Gaussian nature of the observed series. We shall not dwell on this problem here and will only disuss certain corollaries of the LAN condition in Appendix 2 of this chapter (see also Remarks 2 and 3 to this section in Appendix 3). In Sections 6 and 7 we shall concentrate on relaxing the assumption about the Gaussian nature of the observed series.

Theorem 2. *Let the above stated conditions[3] be satisfied. Moreover let the spectral density f_Θ be a twice differentiable function of $\Theta \in \Theta$ and the second derivatives be continuous in $\lambda \in [-\pi,\pi]$. Then*

1) (20) $\Gamma_\Theta n^{1/2}(\tilde\Theta - \Theta) - \Delta_{n,\Theta} \to 0$

   in $P_n(f_\Theta)$ probability as $n \to \infty$.
2) (19) *holds.*

Proof. Only assertion 1) requires a proof since this assertion together with the relation (18) yields (19). In view of (13) and (17)

$$
\begin{aligned}
(21) \quad & \mathbf{h}'\Delta_{n,\tilde\Theta} - \mathbf{h}'\Delta_{n,\Theta} \\
&= \frac{n^{1/2}}{4\pi} \int_{-\pi}^{\pi} \left[\frac{a_{\tilde\Theta}(\lambda)}{f_{\tilde\Theta}(\lambda)} - \frac{a_\Theta(\lambda)}{f_\Theta(\lambda)}\right][I_n(\lambda) - f_\Theta(\lambda)]d\lambda \\
&\quad - \frac{n^{1/2}}{4\pi}\int_{-\pi}^{\pi} a_{\tilde\Theta}(\lambda)\left[1 - \frac{f_\Theta(\lambda)}{f_{\tilde\Theta}(\lambda)}\right]d\lambda.
\end{aligned}
$$

It follows from (1), (2), and (17) that clearly $\Delta_{n,\tilde\Theta} = 0$. By the mean value theorem, the second summand on the r.h.s. of (21) equals

$$
(22) \quad -\frac{n^{1/2}}{4\pi}\sum_{k=1}^{P}(\tilde\Theta_k - \Theta_k)\int_{-\pi}^{\pi}\left[\frac{\partial}{\partial\Theta_k}f_\Theta(\lambda)\right]_{\Theta=\Theta'}\frac{a_{\tilde\Theta}(\lambda)}{f_{\tilde\Theta}(\lambda)}d\lambda
$$

where $\Theta' \in [\tilde\Theta, \Theta]$, and consequently the integral in (22) with the multiplier $1/4\pi$ converges in $P_n(f_\Theta)$ probability to the k-th element of the vector $\Gamma_\Theta \mathbf{h}$. Thus (21) implies (20) provided the convergence to zero in $P_n(f_\Theta)$ probability of the first

---

[3]Above, the validity of Lemma I.3.1 was required in particular which is equivalent to Lemma A1.4 of Chapter I. However, in the literature on this subject often conditions alternative to those of Lemma I.3.1 (or Lemma A1.4, Chapter I) are met. Cf. Lemma A1.1 of Chapter II (also the subsequent Corollary A1.3) for a hint on the alternative possibilities.

summand in the r.h.s. of (21) is proved. More precisely, since $\tilde{\Theta}$ is consistent, in view of the conditions on the second derivatives of f_Θ and the mean value theorem, it is sufficient to verify that

(23)
$$\frac{1}{2\pi} \int_{-\pi}^{\pi} [I_n(\lambda) - f_\Theta(\lambda)]a_\Theta(\lambda)d\lambda$$
$$= \frac{1}{n} [X'B_a X - \text{tr } B_{2\pi fa}] \to 0,$$
$$a = a_\Theta = \sum_{k,\,j=1}^{p} h_k h_j \frac{\partial^2}{\partial \Theta_k \partial \Theta_j} f_\Theta^{-1}$$

in $P_n(f_\Theta)$ probability. However, from the formulas (I.2.3) and (I.2.4) relation (23) follows from

$$n^{-1}|\text{tr}(B_f B_a - B_{2\pi fa})| \leqslant n^{-1/2} |B_f B_a - B_{2\pi fa}| \to 0$$

and

$$n^{-2}\text{tr}[(B_f B_a)^2] \leqslant n^{-2} |B_f B_a|^2 \to 0$$

by virtue of the inequality (A1.1) and assertion 5) of Lemma A1.2 in Appendix 1 to the preceding chapter.   ◻

3. Consistency, Asymptotic Normality, and Asymptotic Efficiency of the Estimator $\tilde{\Theta}$ in the Case of Spectral Density Possessing Fixed Zeros

3.1. In this subsection we shall consider the case when the spectral density $f = f_q$ of a Gaussian random process X_t, $t =$ ..., $-1,0,1$, ... satisfies the assumption 1 presented in Section 2 of Chapter I (i.e., possesses fixed simple zeros).

Assume as above that the value of the parameter Θ appearing in the expression for the spectral density $f_q = f_{q,\Theta}$ (or more precisely, in the expression of the function $f_0 = f_{0,\Theta}$ appearing in the formula (I.2.11)) belongs to a closed set $\bar{\Theta}$ contained in an open set S of p-dimensional Euclidean space R_p. Next, let the spectral density $f = f_q = f_{q,\Theta}$ satisfy condition (A) of the preceding section.

Retaining the notation of Chapter I, introduce a random variable

$$U_n(\Theta, X) = \frac{1}{4\pi} \int_{-\pi}^{\pi} \left[\log f_{0,\Theta}(\lambda) + \frac{I_n(\lambda, \widetilde{Y})}{f_{0,\Theta}(\lambda)} \right] d\lambda$$

(1)

$$= \frac{1}{4\pi} \int_{-\pi}^{\pi} \log f_{0,\Theta}(\lambda) d\lambda + \frac{1}{2n} \widetilde{Y}^* B_{h_0,\Theta} \widetilde{Y},$$

where $I_n(\lambda, \widetilde{Y})$ is defined by the formula (I.2.29), while $h_0 = 1/(2\pi)^2 f_0$. Recall that the n-dimensional vector $\widetilde{Y}$ is related to observations $X_1, \ldots, X_n$ as it is indicated in the assertion of Lemma I.2.2.

Then in view of the formula (I.2.31) the asymptotic maximum likelihood estimator $\tilde{\Theta}$ is again determined from the condition (2.2), where, however, $U_n(\Theta)$ is determined now by formula (1).

When proving the consistency of the estimator $\tilde{\Theta}$ we shall basically follow the arguments presented in Section 1 of the previous chapter. Here some assertions are required whose content is presented in the following two lemmas.

Lemma 1. *Let* $\Theta_1 \in \Theta$ *be such that* $\Theta_1 \neq \Theta_0$, *where* Θ_0 *is the true value of the parameter* Θ. *Then there exists a positive number* $K(\Theta_0, \Theta_1)$ *such that*

(2) $$\lim_{n \to \infty} P\{U_n(\Theta_0) - U_n(\Theta_1) < -K(\Theta_0, \Theta_1)\} = 1.$$

Proof. Using the method utilized in the proof of Lemma I.2.2 it follows that

$$E[U_n(\Theta_0) - U_n(\Theta_1)]$$

(3)

$$= \frac{1}{4\pi} \int_{-\pi}^{\pi} \left[\log \frac{f_{0,\Theta_0}(\lambda)}{f_{0,\Theta_1}(\lambda)} d\lambda + 1 - \frac{f_{0,\Theta_0}(\lambda)}{f_{0,\Theta_1}(\lambda)} \right] d\lambda < 0.$$

(We basically follow the considerations leading to the formula (2.9) of the previous section, taking also into account that terms like $n^{-2} \mathrm{tr}(V^* B_{f_0} B_{h_0} V)$ vanish as $n \to \infty$; of course, V is

the same as in (I.2.19).)

Analogous considerations lead to also the limiting expression for the variance which is of form (2.11) with f_0 instead of f.

Since Θ_1 and Θ_2 belong to a closed set Θ, in view of Chebyshev's inequality this fact and (4) imply (3).   □

Lemma 2. *Let there exist the continuous derivatives* $\partial f_{0,\Theta}^{-1}/\partial \Theta_k$,

$k = 1, ..., p$ for $\theta \in \Theta$ while $\theta_1 \in \Theta$ and $\theta_2 \in S$ are chosen in such a manner that $|\theta_2 - \theta_1| < \delta$ (δ may depend on θ_1). Then there exists a positive number n_0 such that for $n \geqslant n_0$

(4) $|U_n(\theta_2) - U_n(\theta_1)| \leqslant H_{\delta, n}(\theta_1, X),$

where $H_{\delta, n} = H_{\delta, n}(\theta, X)$ is a random variable such that one has

(5) $\lim_{\delta \to 0} E(H_{\delta, n}) = 0$

uniformly in n and

(6) $\lim_{n \to \infty} D(H_{\delta, n}) = 0$

for all δ.

Proof. In view of (1), (2.3), and the mean value theorem

(7)
$$|U_n(\theta_2) - U_n(\theta_1)| = \frac{1}{4\pi} \int_{-\pi}^{\pi} \left\{ \log \frac{f_{0, \theta_2}(\lambda)}{f_{0, \theta_1}(\lambda)} \right.$$
$$\left. + I_n(\lambda, \tilde{Y}) \left[\frac{1}{f_{0, \theta_2}(\lambda)} - \frac{1}{f_{0, \theta_1}(\lambda)} \right] \right\} d\lambda$$
$$\leqslant \frac{\delta}{4\pi} H(\theta_1, \delta(\theta_1)) \left\{ \int_{-\pi}^{\pi} [f_{0, \theta_1}(\lambda) + I_n(\lambda, \tilde{Y})] d\lambda,$$

where $H(\theta_1, \delta(\theta_1))$ is chosen in the same manner as in the preceding section (with f_0 in place of f).

It follows from (4)-(7) that to complete the proof of the lemma it remains only to take into consideration that, by (I.2.25),

$$E(n^{-1} Y^* Y) = \int_{-\pi}^{\pi} f_{0, \theta_0}(\lambda) d\lambda < \infty,$$

and that $n^{-1} D(Y^* Y)$ is bounded uniformly in n.   □

Now applying Lemma 1 from the preceding section (which is valid in the case considered in this subsection in view of the assertion of Lemma 1) we obtain from the assertion of Lemma 2 that the following theorem holds:

Theorem 1. *Let the conditions indicated above be fulfilled. Then the asymptotic m.l. estimator $\tilde{\Theta}$ is consistent: $\tilde{\Theta} \to \Theta$ in $P_n(f_{q,\Theta})$ probability.*

3.2.   For the proof of additional properties of the estimators $\tilde{\Theta}$ we shall substantially utilize the assertions stated in the following lemmas.

Lemma 3. *Assume that in view of the condition stipulated in the preceding Subsection as well as in Subsection 4.2 of the preceding chapter, $\dot{\phi}_{k,\Theta} = (\partial/\partial\Theta_k)\log f_{0,\Theta}$ and $r_{k,\Theta} = -(\partial/\partial\Theta_k)f_{0,\Theta}^{-1}$, $k = 1, ..., p$. Then the p-dimensional random vector*

$$\Delta_{n,\Theta} = col\left\{-\frac{n^{1/2}}{4\pi}\int_{-\pi}^{\pi}\frac{\partial}{\partial\Theta_k}\log f_{0,\Theta}(\lambda)\right.$$

(8)
$$\left. \cdot \frac{I_n(\lambda,\tilde{Y})-f_{0,\Theta}(\lambda)}{f_{0,\Theta}(\lambda)}d\lambda, \quad k = 1, ..., p\right\}$$

(cf. (I.4.22)) *as $n \to \infty$ possesses the normal distribution $N(0,\Gamma_\Theta)$, where the covariance matrix Γ_Θ is given by the formula*

(9)
$$\Gamma_\Theta = \left[\frac{1}{4\pi}\int_{-\pi}^{\pi}\frac{\partial}{\partial\Theta_k}\log f_{0,\Theta}(\lambda)\frac{\partial}{\partial\Theta_l}\log f_{0,\Theta}(\lambda)d\lambda,\right.$$
$$\left. k,l = 1, ..., p\right]$$

(cf. (I.4.23)).

The proof is presented in Appendix 1 of this Chapter (assertion 4) of Theorem A1.3).

Lemma 4. *Let all the derivatives $(\partial^2/\partial\Theta_k\partial\Theta_j)f_{0,\Theta}^{-1}(\lambda)$ exist and be continuous in $\lambda \in [-\pi,\pi]$ and $\Theta \in \Theta$. Then under the conditions stipulated above the following relation is valid:*

(10) $\Gamma_\Theta n^{1/2}(\tilde{\Theta}-\Theta) - \Delta_{n,\Theta} \to 0$ *as $n \to \infty$*

in $P_n(f_{q,\Theta})$ probability.

Proof. Since it follows from the condition (2.2) that $\Delta_{n,\tilde{\Theta}} = 0$ where $\Delta_{n,\Theta} = -n^{1/2}(\partial/\partial\Theta)U_n(\Theta)$, the k-th element of the latter vector can be written in the form

$$-n^{1/2}\frac{\partial}{\partial\Theta_k}U_n(\Theta) = \frac{n^{1/2}}{4\pi}\int_{-\pi}^{\pi}[I_n(\lambda,\tilde{Y}) - f_{0,\Theta}(\lambda)]$$

(11)
$$\cdot[r_{k,\tilde{\Theta}}(\lambda) - r_{k,\Theta}(\lambda)]d\lambda$$

$$-\frac{n^{1/2}}{4\pi}\int_{-\pi}^{\pi}\dot{\Phi}_{k,\Theta}(\lambda)[1-f_{0,\Theta}(\lambda)f_{0,\tilde{\Theta}}^{-1}(\lambda)]d\lambda.$$

Proceeding in the same manner as in the course of deriving (2.20) one can easily verify from (2.21) that (11) implies (10) provided only that

$$\left|\frac{1}{2\pi}\int_{-\pi}^{\pi}[I_n(\lambda,\tilde{Y}) - f_{0,\Theta}(\lambda)]a_\Theta(\lambda)d\lambda\right| \to 0$$

in $P_n(f_{q,\Theta})$ probability where

$$a_\Theta = \sum_{k,j=1}^{p}h_k h_j\frac{\partial^2}{\partial\Theta_k\partial\Theta_j}f_{0,\Theta}^{-1}$$

(cf. (2.23)). This is proved in the same manner as Lemma 1. Lemma 4 is proved.   □

The validity of the following theorem follows directly from the assertions of Lemmas 3 and 4.

Theorem 2. *In addition to the conditions of Lemma 4 let the condition $a(\lambda) \neq 0$ be valid for almost all λ, where a is defined by the relation (2.13) for $f_\Theta = f_{0,\Theta}$ so that the matrix Γ_Θ is nonsingular. Then the asymptotic m.l. estimator $\tilde{\Theta}$ is asymptotically normal and asymptotically efficient, i.e., the random vector $n^{1/2}(\tilde{\Theta}-\Theta)$ as $n \to \infty$ possesses the normal distribution $N(0,\Gamma_\Theta^{-1})$ with zero expectation and covariance matrix Γ_Θ^{-1}, where Γ_Θ is the limit as $n \to \infty$ of the Fisher's information matrix (2.15).*

Remark. The fact that Γ_Θ is the limit as $n \to \infty$ of the matrix (2.15) can be verified directly: in the same manner as in the preceding section we have the relation

$$\frac{1}{n} D\left[\sum_{k=1}^{P} h_k \frac{\partial}{\partial \theta_k} L_{n, \theta}\right]$$

$$= \frac{1}{2n} \text{tr}\left[\left[B_{f_0, \theta}^{-1} B_{\Sigma h_k (\partial/\partial \theta_k) f_{0,\theta}}\right]^2\right]$$

$$+ o(1) \rightarrow \frac{1}{4\pi} \int_{-\pi}^{\pi} a^2(\lambda) d\lambda$$

which is valid in view of the assertions 4) of Lemma A2.2 and 8) of Lemma A1.2 of Appendix 1 to Chapter I for

$$a = \sum_{k=1}^{P} h_k \frac{\partial}{\partial \theta_k} \log f_{0, \theta}.$$

The assertion concerning the distribution of the vector $n^{1/2}(\bar{\theta}-\theta)$ follows in an obvious manner from (10), the nonsingularity of Γ_θ, and the assertion of Lemma 3.

4. Examples of Determination of Asymptotic Maximum Likelihood Estimators

4.1. Autoregressive Process

We shall begin by considering the case very often encountered in practical applications when the random process X_t, $t = ...,$ -1,0,1, ..., satisfies the difference equation

(1) $X_t - \iota_1 X_{t-1} - \cdots - \iota_q X_{t-q} = \varepsilon_t,$

where ε_t, $t = ...,-1,0,1, ...$ is a sequence of independent Gaussian random variables with zero expectation $E(\varepsilon_t) = 0$ and a positive variance $E(\varepsilon_t^2) = \sigma^2 > 0$, while the coefficients $\iota_1, ...,$ ι_q are such that all the roots of the characteristic equation

(2) $h_q(z) = 1 - \iota_1 z - \cdots - \iota_q z^2 = 0, \quad z = e^{i\lambda},$

exceed one in their absolute value. As it is known (cf. for example, [58] Chapter 10, Section 10), under the last condition the parameters $\iota_1, ..., \iota_q$ and σ^2 are uniquely determined by f which here is of the form

(3) $f(\lambda) = \frac{\sigma^2}{2\pi}|h_q(z)|^{-2}, \quad z = e^{i\lambda}.$

The covariance matrix $\beta(\tau)$ of an autoregressive process of the q-th order X_t with spectral density (3) satisfies the so-called Yule-Walker equation

$$(4) \qquad \beta(\tau) - \sum_{j=1}^{q} \iota_j \beta(\tau-j) = \delta_{0\tau}\sigma^2, \qquad \tau = 0,1, \dots$$

(cf. [58], Chapter 10, Section 10; or [4], Section 5.2; or [140], Chapter VI, Section 2).

Assume that the parameters $\iota_1, \dots, \iota_q$ and σ^2 are unknown, i.e.,

$$\Theta = (\iota_1, \dots, \iota_q, \sigma^2), \qquad p = q+1.$$

To obtain the maximum likelihood estimators $\bar{\Theta}$ of the parameter Θ it would be necessary to have an explicit expression of the corresponding logarithm of likelihood L_n. However, as it was noted above (cf. Section 1 of Chapter I) the formulas for L_n even in the relatively simple case under consideration turn out to be very cumbersome if only the order of autoregression q exeeds the minimal value $q = 1$. Moreover, even in the simplest case considered in Example I.1.1 of Chapter I where $q = 1$ (and $\iota_1 = \Theta$), the values $\bar{\sigma}^2$ and $\bar{\iota}_1$ of the parameters σ^2 and ι_1, maximizing the expression for L_n (cf. (I.1.7)) turn out to be roots of quite cumbersome equations.

At the same time since for f of the form (3) the expression (I.2.7) for the "principal part" of L_n is of the form

$$\tilde{L}_n = -\frac{n}{2}\left\{ \log 2\pi\sigma^2 + \frac{1}{\sigma^2}\int_{-\pi}^{\pi} I_n(\lambda)|h_q(z)|^2 d\lambda \right\}$$

$$= -\frac{n}{2}\left\{ \log 2\pi\sigma^2 + \frac{1}{\sigma^2}\sum_{k,j=0}^{q} \iota_k \iota_j \beta_n^*(k-j) \right\},$$

where $\iota_0 = 1$ and

$$(5) \qquad \beta_n^*(\tau) = \int_{-\pi}^{\pi} I_n(\lambda)e^{i\lambda\tau}d\lambda = \frac{1}{n}\sum_{j=1}^{n-\tau} X_j X_{j+\tau},$$

$$\tau = 0,1, \dots, n-1$$

is the empirical covariance function, then clearly the asymptotic m.l. estimators $\tilde{\iota}_1, \dots, \tilde{\iota}_q$ are roots relative to $\iota_1, \dots, \iota_q$ of a simple system of linear equations

(6) $\qquad \beta_n^*(\tau) - \sum\limits_{j=1}^{q} \iota_j \beta_n^*(\tau-j) = 0, \qquad \tau = 1, \ldots, q,$

and $\tilde{\sigma}^2$ is determined by the formula

(7) $\qquad \tilde{\sigma}^2 = \beta_n^*(0) - \sum\limits_{j=1}^{q} \tilde{\iota}_j \beta_n^*(j).$

Since

$$\frac{1}{4\pi} \int_{-\pi}^{\pi} \frac{\partial}{\partial \iota_k} \log\left[\frac{\sigma^2}{2\pi}|h_q(z)|^{-2}\right] \frac{\partial}{\partial \iota_\ell} \log\left[\frac{\sigma^2}{2\pi}|h_q(z)|^{-2}\right] d\lambda$$

$$= \frac{1}{4\pi} \int_{-\pi}^{\pi} 2\mathrm{Re}\left[\frac{z^k}{h_q(z)}\right] 2\mathrm{Re}\left[\frac{z^\ell}{h_q(z)}\right] d\lambda$$

$$= \frac{1}{2\pi} \int_{-\pi}^{\pi} \frac{z^{k-\ell}}{|h_q(z)|^2} d\lambda = \beta(k-\ell)/\sigma^2, \quad k, \ell+1, \ldots, q,$$

$$\frac{1}{4\pi} \int_{-\pi}^{\pi} \frac{\partial}{\partial \iota_k} \log\left[\frac{\sigma^2}{2\pi}|h_q(z)|^{-2}\right] \frac{\partial}{\partial \sigma^2} \log\left[\frac{\sigma^2}{2\pi}|h_q(z)|^{-2}\right] d\lambda$$

$$= \frac{1}{2\pi\sigma^2} \int_{-\pi}^{\pi} \mathrm{Re}\left[\frac{z^k}{h_q(z)}\right] d\lambda = 0, \quad k = 1, \ldots, q,$$

$$\frac{1}{4\pi} \int_{-\pi}^{\pi} \left\{\frac{\partial}{\partial \sigma^2} \log\left[\frac{\sigma^2}{2\pi}|h_q(z)|^{-2}\right]\right\}^2 d\lambda = 1/2\sigma^4,$$

the limit of the Fisher's information matrix Γ_Θ is of the form

$$\Gamma_\Theta = \Gamma_{\iota,\sigma} = \begin{bmatrix} \Gamma_\iota^{(q)} & 0 \\ 0 & 1/2\sigma^4 \end{bmatrix},$$

(8)

$$\Gamma_\iota^{(q)} = [\beta(\ell-k)/\sigma^2]_{k,\,\ell=1,\ldots,q},$$

where $\iota = (\iota_1, \ldots, \iota_q)'$. Therefore we have up to a summand of order $o(1/n)$

$$[\mathrm{cov}(\tilde{\iota}_k, \tilde{\iota}_\ell)]_{k,\,\ell=1,\ldots,q} = (n\Gamma_\iota^{(q)})^{-1},$$

$$D(\tilde{\sigma}^2) = 2\sigma^4/n, \quad \mathrm{cov}(\tilde{\sigma}^2, \tilde{\iota}_k) = 0, \quad k = 1, \ldots, q.$$

Example 1. Let $q = 1$, i.e., $X_t - \iota_1 X_{t-1} = \varepsilon_t$ (cf. (I.1.4)). Then, in view of (6) and (7)

(9) $\tilde{\iota}_1 = \beta_n^*(1)/\beta_n^*(0), \qquad \tilde{\sigma}^2 = ([\beta_n^*(0)]^2 - [\beta_n^*(1)]^2)/\beta_n^*(0)$

and

(10) $D(\tilde{\iota}_1) = \dfrac{1-\iota_1^2}{n} + o(1/n).$

4.2. Moving Average Process

Let the process X_t be represented in the form

(11) $X_t = \varepsilon_t - \alpha_1 \varepsilon_{t-1} - \cdots - \alpha_r \varepsilon_{t-r},$

where the coefficients $\alpha_1, ..., \alpha_r$ are such that the roots of the polynomial

(12) $g_r(z) = 1 - \alpha_1 z - \cdots - \alpha_r z^r, \quad z = e^{i\lambda},$

are not less than one in their absolute value. This condition assures (cf., e.g., [58], Chapter X, Section 10) the uniqueness of representation of the density f in the form

(13) $f(\lambda) = \dfrac{\sigma^2}{2\pi} |g_r(z)|^2, \quad z = e^{i\lambda}.$

We shall confine ourself to positive spectral densities f and we shall assume below that all the roots of $g_r(z)$ exceed 1 in their absolute value.

The covariance function $\beta(\tau)$ of a moving average process of r-th order X_t with the spectral density (13) is represented in the form

(14) $\beta(\tau) = \begin{cases} \sigma^2 \displaystyle\sum_{j=0}^{r-|\tau|} \alpha_j \alpha_{j+|\tau|}, & \text{for } |\tau| \leqslant r, \\[2mm] 0 & \text{for } |\tau| > r \end{cases}$

where $\alpha_0 = 1$. Assuming that $\alpha_1, ..., \alpha_r$ and σ^2 are unknown parameters and taking into account that in view of (I.2.7) and (13)

$$\tilde{L}_n = -\frac{n}{2}\left\{\log 2\pi\sigma^2 + \frac{1}{\sigma^2}\int_{-\pi}^{\pi} I_n(\lambda)|g_r(z)|^{-2}d\lambda\right\},$$

we obtain that the asymptotic m.l. estimators $\alpha_1, ..., \tilde{\alpha}_r$ of the parameters $\alpha_1, ..., \alpha_r$ are the roots relative to $\alpha_1, ..., \alpha_r$ of the system of equations

(15)
$$\int_{-\pi}^{\pi} I_n(\lambda)|g_r(z)|^{-4}\left[\cos k\lambda - \sum_{j=1}^{r}\alpha_j\cos(k-j)\lambda\right]d\lambda = 0,$$

$$k = 1, ..., r,$$

and

(16)
$$\tilde{\sigma}^2 = \int_{-\pi}^{\pi} I_n(\lambda)|1-\tilde{\alpha}_1 z - \cdots - \tilde{\alpha}_r z^r|^{-2}d\lambda, \qquad z = e^{i\lambda}.$$

It is easy to verify here that

$$\Gamma_{\Theta} = \Gamma_{\alpha,\sigma^2} = \begin{bmatrix} \Gamma_{\alpha}^{(r)} & 0 \\ 0 & 1/2\sigma^4 \end{bmatrix},$$

(17)
$$\Gamma_{\alpha}^{(r)} = \left[\frac{1}{2\pi}\int_{-\pi}^{\pi} e^{i(\ell-k)\lambda}|g_r(z)|^{-2}d\lambda\right]_{k,\ell=1,...,r},$$

where

$$\Theta = (\alpha_1, ..., \alpha_r, \sigma^2)', \qquad p = r+1, \qquad \alpha = (\alpha_1, ..., \alpha_r)',$$

i.e.,

(18)
$$[\text{cov}(\tilde{\alpha}_k, \tilde{\alpha}_\ell)]_{k,\ell=1,...,r} = (n\Gamma_{\alpha}^{(r)})^{-1} + o(1/n),$$

$$D(\tilde{\sigma}^2) = \frac{2\sigma^4}{n} + o(1/n), \qquad \text{cov}(\tilde{\sigma}^2, \tilde{\alpha}_k) = o(1/n),$$

$$k = 1, ..., r.$$

Instead of the coefficients $\alpha_1, ..., \alpha_r$ and σ^2 one may choose any $r+1$ functions of $\alpha_1, ..., \alpha_r$ and σ^2 as the unknown parameters provided that their values uniquely determine the quantities $\alpha_1, ..., \alpha_r$ and σ^2.

In particular, the covariances $\beta(0), \beta(1), ..., \beta(r)$ defined by the formula (14) can be chosen for these functions. The dependence of f on these new parameters $\beta(k), k = 0,1, ..., r$, is described by the formula

(19) $\qquad f(\lambda) = \dfrac{1}{2\pi}\,\beta(0) + \dfrac{1}{\pi}\sum_{k=1}^{r}\beta(k)\cos k\lambda = \sum_{k=1}^{r}w_k(\lambda)\beta(k),$

where $w_0(\lambda) = 1/2\pi$ and $w_k(\lambda) = \cos k\lambda/\pi$, $k = 1, ..., r$.

This case is evidently a particular case of a somewhat more general situation when the spectral density f depends linearly on the unknown parameters $\Theta = (\Theta_1, ..., \Theta_p)'$ so that

(20) $\qquad f_\Theta(\lambda) = \sum_{k=1}^{p}\Theta_k w_k(\lambda),$

where $w_1(\lambda), ..., w_p(\lambda)$ are some fixed functions of the frequency λ. It follows~from (2.1), (2.2), and (20) that asymptotic m.l. estimators $\tilde\Theta = (\tilde\Theta_1, ..., \tilde\Theta_p)'$ are the roots of the system of equations

(21) $\qquad \displaystyle\int_{-\pi}^{\pi}\dfrac{I_n(\lambda)-f_\Theta(\lambda)}{f_\Theta^2(\lambda)}w_k(\lambda)d\lambda = 0, \qquad k = 1, ..., p.$

Clearly in this case

(22) $\qquad \Gamma_\Theta = \left[\dfrac{1}{4\pi}\displaystyle\int_{-\pi}^{\pi}\dfrac{w_k(\lambda)w_\ell(\lambda)}{f_\Theta^2(\lambda)}d\lambda\right]_{k,\,\ell=1,\ldots,p}.$

Example 2. In particular, for $k = 1$ we have

$$\beta(0) = (1+\alpha_1^2)\sigma^2, \quad \beta(1) = -\alpha_1\sigma^2, \quad \beta(\tau) = 0, \quad \tau = 2,3,...,$$

and

(23) $\qquad f(\lambda) = \dfrac{\sigma^2}{2\pi}(1+\alpha_1^2-2\alpha_1\cos \lambda) = \dfrac{\sigma^2}{2\pi}|1-\alpha_1 z|^2, \quad z = e^{i\lambda}.$

From (15), for $r = 1$, we obtain that $\tilde\alpha_1$ is the root relative to α_1 of the equation

$$\int_{-\pi}^{\pi}I_n(\lambda)|1-\alpha_1 e^{i\lambda}|^{-4}(\cos \lambda-\alpha_1)d\lambda = 0,$$

or, in view of the equality

$$|1-\alpha_1 e^{i\lambda}|^{-4}(\cos \lambda-\alpha_1) = -\dfrac{1}{2}\dfrac{\partial}{\partial\alpha_1}|1-\alpha_1 e^{-i\lambda}|^{-2}$$

(24)

$$= -\dfrac{1}{2}\dfrac{\partial}{\partial\alpha_1}\dfrac{1}{1-\alpha_1^2}\sum_{j=-\infty}^{\infty}\alpha_1^{|j|}e^{i\lambda j},$$

of the equation

(25) $\dfrac{\partial}{\partial\alpha_1}\left\{\dfrac{1}{1-\alpha_1^2}\left[\beta_n^*(0) + 2\sum_{j=1}^{n-1}\alpha_1^j\beta_n^*(j)\right]\right\} = 0.$

The variance of the estimator $\tilde{\alpha}_1$ is of the form

(26) $D(\tilde{\alpha}_1) = \dfrac{1-\alpha_1^2}{n} + o(1/n).$

If the parameter σ^2 is also unknown then its asymptotic m.l. estimator $\tilde{\sigma}^2$ is given by the formula (16) with $r = 1$.

4.3. The Mixed Autoregressive-Moving Average Process

Generalizing the two cases discussed above we shall consider the process X_t satisfying the difference equation

$$X_t - \iota_1 X_{t-1} - \cdots - \iota_q X_{t-q} = \varepsilon_t - \alpha_1\varepsilon_{t-1} - \cdots - \alpha_r\varepsilon_{t-r},$$

where the coefficients $\iota_1, \ldots, \iota_q$ and $\alpha_1, \ldots, \alpha_r$ are such that all the roots of the polynomials

(27)
$$h_q(z) = 1 - \iota_1 z - \cdots - \iota_q z^q,$$

$$g_r(z) = 1 - \alpha_1 z - \cdots - \alpha_r z^r, \qquad z = e^{i\lambda}$$

are greater than 1 in the absolute value.

As it is well-known, the spectral density f of the process X_t is a rational function of the form

(28) $f(\lambda) = \dfrac{\sigma^2}{2\pi}\, |g_r(z)|^2 |h_q(z)|^{-2} > 0,$

and the covariance function $\beta(\tau)$ satisfies the equations

(29) $\beta(t) - \sum_{j=1}^{q}\iota_j\beta(t-j) = 0, \qquad t = r+1, r+2, \ldots$

(cf. [58], Chapter X, Section 10; or [4], Section 5.2; or [140], Chapter VI, Section 2).

Assume that the unknown parameters $\theta = (\theta_1, \ldots, \theta_p)'$ are the coefficients $\iota_1, \ldots, \iota_q, \alpha_1, \ldots, \alpha_r$, and σ^2 ($p = q+r+1$). Then (2.1), (2.2), and (28) imply that the asymptotic m.l. estimators $\tilde{\iota}_1, \ldots, \tilde{\iota}_q, \tilde{\alpha}_1, \ldots, \tilde{\alpha}_r$ of the first $(q+r)$ parameters are the roots of the system of equations

$$(30) \qquad \int_{-\pi}^{\pi} I_n(\lambda)|h_q(z)|^2 |g_r(z)|^{-4} \left[\cos k\lambda \right.$$
$$\left. - \sum_{j=1}^{r} \alpha_j \cos(k-j)\lambda \right] d\lambda = 0, \quad k = 1,..., r,$$

$$(31) \qquad \int_{-\pi}^{\pi} I_n(\lambda)|g_r(z)|^{-2} \left[\cos k\lambda - \sum_{i=1}^{q} \iota_j \cos(k-j)\lambda \right] d\lambda = 0,$$
$$k = 1, ..., q$$

with respect to $\iota_1, ..., \iota_q, \alpha_1, ..., \alpha_r$, and

$$(32) \qquad \tilde\sigma^2 = \int_{-\pi}^{\pi} I_n(\lambda)|1-\tilde\iota_1 z - \cdots - \tilde\iota_q z^q|^2 |1-\tilde\alpha_1 z - \cdots - \tilde\alpha_r z^r|^{-2} d\lambda.$$

It is easy to verify that in this case

$$(33) \qquad \Gamma_\Theta = \begin{bmatrix} \Gamma_\iota^{(q)} & -\Omega^* & 0 \\ -\Omega & \Gamma_\alpha^{(r)} & 0 \\ 0 & 0 & 1/2\sigma^4 \end{bmatrix},$$

where the $(q \times q)$-matrix $\Gamma_\iota^{(q)}$ and the $(r \times r)$-matrix $\Gamma_\alpha^{(r)}$ are given by the formulas (8) and (17) respectively, while the $(r \times q)$-matrix Ω is such that its (k, ℓ)-th element is of the form

$$(34) \qquad \frac{1}{2\pi} \int_{-\pi}^{\pi} \frac{z^{k-\ell}}{h_q(z)g_r(z)} d\lambda, \quad z = e^{i\lambda}.$$

Observe that if the spectral density f of the process X_t is of the form (28) then the process

$$(35) \qquad Y_t = X_t - \sum_{j=1}^{q} \iota_j X_{t-j}$$

will possess the spectral density (13) and its covariance function $\beta_y(\tau) = E(Y_{t+\tau} Y_t)$ will be expressed in terms of

$$\beta(\tau) = E(X_{t+\tau} X_t)$$

by the formula

$$(36) \qquad \beta_y(\tau) = \sum_{j, k=0}^{q} \iota_j \iota_k \beta(\tau+j-k), \quad \iota_0 = -1.$$

As in the case of the preceding subsection we can suggest several other choices of unknown parameters of the spectrum which are in a one-to-one correspondence with the parameters

$\iota_1, ..., \iota_q, \alpha_1, ..., \alpha_r, \sigma^2$. Thus, for example, formula (28) may be rewritten in the form

$$(37) \qquad f(\lambda) = \frac{\beta_y(0) + 2 \sum\limits_{k=1}^{r} \beta_y(k)\cos k\lambda}{2\pi|1 - \sum\limits_{j=1}^{q} \iota_j e^{i\lambda j}|^2},$$

and the quantities $\iota_1, ..., \iota_q, \beta_y(0), \beta_y(1), ..., \beta_y(r)$ can serve as a new system of unknown parameters. In that case (cf. [100]) the asymptotic m.l. estimators

$$\tilde{\iota}_1, ..., \tilde{\iota}_q, \qquad \beta_y(0), ..., \beta_y(r)$$

of these parameters are the roots of the system of equations

$$\int_{-\pi}^{\pi} \frac{I_n(\lambda)}{f_y(\lambda)} \left[\cos j\lambda - \sum\limits_{k=1}^{r} \iota_k \cos(j-k)\lambda\right] d\lambda = 0,$$

$$j = 1, ..., q,$$

$$(38) \qquad \int_{-\pi}^{\pi} \left[\sum\limits_{k=0}^{r} \beta_y(k) w_k(\lambda) - |h_q(z)|^2 I_n(\lambda)\right] w_j(\lambda) f_y^{-2}(\lambda) d\lambda = 0,$$

$$j = 0, 1, ..., r,$$

$$(39) \qquad f_y(\lambda) = \sum\limits_{k=0}^{r} \beta_y(k) w_k(\lambda)$$

$$= \frac{1}{2\pi} \left[\beta_y(0) + 2 \sum\limits_{k=1}^{r} \beta_y(k)\cos k\lambda\right],$$

with respect to $\iota_1, ..., \iota_q, \beta_y(0), ..., \beta_y(r)$, and the limit of Fisher's information matrix is given by the formula

$$(40) \qquad \Gamma_\Theta = \Gamma_{\iota,\beta_y} = \begin{bmatrix} \Gamma_\iota^{(q)} & -\Psi^* \\ -\Psi & \Gamma_{\beta_y}^{(r+1)} \end{bmatrix},$$

where

$$\Theta = (\iota_1, ..., \iota_q; \beta_y(0), \beta_y(1), ..., \beta_y(r))',$$

$$p = q + r + 1, \qquad \iota = (\iota_1, ..., \iota_q)',$$

$$\beta_y = (\beta_y(0), \beta_y(1), ..., \beta_y(r))',$$

$\Gamma_{\iota}^{(q)}$ is the same matrix as in (8), $\Gamma_{\beta_y}^{(r+1)}$ denotes the $(r+1) \times$ $(r+1)$-matrix which differs from the r.h.s. of (22) in that f is replaced by f_y (cf. [131]), and the $(k,\mathbf{l})$-th element of the $(r+1) \times q$-matrix Ψ is of the form

$$\frac{1}{2\pi} \int_{-\pi}^{\pi} \text{Re } \frac{e^{i\lambda\mathbf{l}}w_k(\lambda)}{h_q(\lambda)f_y(\lambda)}\, d\lambda.$$

Example 3.   In the simplest particular case when $q = r = 1$ the process X_t, satisfying the difference equation

$$X_t - \iota_1 X_{t-1} = \varepsilon_t - \alpha_1 \varepsilon_{t-1},$$

where $|\iota_1| < 1$ and $|\alpha_1| < 1$, has a spectral density f of the form

(41) $\qquad f(\lambda) = \dfrac{\sigma^2}{2\pi} |1 - \alpha_1 z|^2 |1 - \iota_1 z|^{-2}$

and the covariance function $\beta(\tau)$ defined by the equalities

$$\beta(0) = (1 + \alpha_1^2 - 2\iota_1\alpha_1)(1 - \iota_1^2)^{-1}\sigma^2$$

$$\beta(k) = (1 - \alpha_1\iota_1)(\iota_1 - \alpha_1)(1 - \iota_1^2)^{-1}\iota_1^k\sigma^2, \quad k = 1,2, \dots .$$

For $q = r = 1$ it follows from (30) and (31) that the asymptotic m.l. estimators $\tilde{\iota}_1$ and $\tilde{\alpha}_1$ are the roots of the equations

(42) $\qquad \int_{-\pi}^{\pi} I_n(\lambda)|1 - \tilde{\alpha}_1 z|^{-2}[\cos \lambda - \tilde{\iota}_1]d\lambda = 0.$

(43) $\qquad \int_{-\pi}^{\pi} I_n(\lambda)|1 - \tilde{\iota}_1 z|^2 |1 - \tilde{\alpha}_1 z|^{-4}[\cos \lambda - \tilde{\alpha}_1]d\lambda = 0,$

and

(44) $\qquad \tilde{\sigma}^2 = \int_{-\pi}^{\pi} I_n(\lambda)|1 - \tilde{\iota}_1 z|^2 |1 - \tilde{\alpha}_1 z|^{-2}d\lambda.$

Since for $q = r = 1$ the formula (33) becomes

(45) $\qquad \Gamma_{\Theta} = \Gamma_{\iota_1,\alpha_1,\sigma^2} = \begin{bmatrix} (1-\iota_1^2)^{-1} & -(1-\iota_1\alpha_1)^{-1} & 0 \\ -(1-\iota_1\alpha_1)^{-1} & (1-\alpha_1^2)^{-1} & 0 \\ 0 & 0 & 1/2\sigma^4 \end{bmatrix},$

the covariance matrix

$$D_\theta = D_{\iota_1, \alpha_1, \sigma^2} = (n \; \Gamma_{\iota_1, \alpha_1, \sigma^2})^{-1} + o(1/n)$$

of the estimators $\tilde{\iota}_1$, $\tilde{\alpha}_1$, $\tilde{\sigma}_2$ is of the form

(46)
$$D_{\iota_1, \alpha_1, \sigma^2} = \frac{1}{n(\iota_1 - \alpha_1)^2}$$

$$\times \begin{bmatrix} (1-\iota_1^2)(1-\iota_1\alpha_1)^2 & (1-\iota_1^2)(1-\alpha_1^2)(1-\alpha_1\iota_1) & 0 \\ (1-\iota_1^2)(1-\alpha_1^2)(1-\alpha_1\iota_1) & (1-\alpha_1^2)(1-\alpha_1\iota_1)^2 & 0 \\ 0 & 0 & 2\sigma^4(\iota_1-\alpha_1)^2 \end{bmatrix} + o(1).$$

4.4. Process with an Exponential Spectral Density

Recently, Bloomfield [23] observed that very often the logarithm of a nonparametric estimator of a spectral density f turns out to be a sufficiently "smooth" function and can therefore be well approximated by a finite Fourier series. In such cases he suggested to utilize the model of the process X_t with spectral density of the form

(47)
$$f(\lambda) = \frac{\sigma^2}{2\pi} \exp\left\{ 2 \sum_{j=1}^{r} \gamma_j \cos j\lambda \right\}, \qquad \sigma^2 > 0,$$

where $\gamma_1, \dots, \gamma_r$ and σ^2 are unknown parameters $((\gamma_1, \dots, \gamma_r, \sigma^2) = \theta, \; p = r+1)$.

It follows from (2.1) and (2.2) that the asymptotic m.l. estimators $\tilde{\gamma}_1, \dots, \tilde{\gamma}_r$ are the roots of the system of equations

(48)
$$\int_{-\pi}^{\pi} I_n(\lambda) \cos k\lambda \; \exp\left[-2 \sum_{j=1}^{r} \tilde{\gamma}_j \cos j\lambda \right] d\lambda = 0,$$
$$k = 1, \dots, r,$$

and

(49)
$$\tilde{\sigma}^2 = \int_{-\pi}^{\pi} I_n(\lambda) \exp\left[-2 \sum_{j=1}^{r} \tilde{\gamma}_j \cos j\lambda \right] d\lambda.$$

It is easy to verify that in this case

(50)
$$\Gamma_\theta = \Gamma_{\gamma, \sigma^2} = \text{diag}\{1, \dots, 1, 1/2\sigma^4\}.$$

4.5. Processes with Spectral Densities Possessing Fixed Zeros

Consider now some simple examples of spectral densities f satisfying the conditions of the preceding section.

Example 4. Let the process X_t be represented in the form X_t $= \varepsilon_t - \varepsilon_{t-1}$ where ε_t, $t = ..., -1,0,1, ...$ is, as above, a sequence of independent random variables with zero expectations $E(\varepsilon_t) = 0$ and positive variances $E(\varepsilon_t^2) = \sigma^2 > 0$ (cf. Examples I.1.2 and I.2.1).

In view of (I.1.20) the maximum likelihood estimator $\overline{\sigma}^2$ of the unknown parameter σ^2 is of the form

$$\overline{\sigma}^2 = \frac{1}{n}\left[\sum_{j=1}^{n}Y_j^2 - \frac{1}{n+1}\left(\sum_{j=1}^{n}Y_j\right)^2\right] = \frac{\sigma^2}{n}\mathbf{Y'}H^{-1}\mathbf{Y},$$

where $Y_j = X_1 + \cdots + X_j$ and $H = E\mathbf{YY'} = \sigma^2(I_n + \mathbf{vv'})$ with $\mathbf{v} =$ col$(1, ..., 1)$. The random variable $n\overline{\sigma}^2/\sigma^2 = \mathbf{Y'}H^{-1}\mathbf{Y}$ possesses χ^2-distribution with n degrees of freedom so that $E(\overline{\sigma}^2) = \sigma^2$, $D(\overline{\sigma}^2) = 2\sigma^4/n$. Moreover, since it follows from (I.1.20) that

$$\frac{\partial L_n}{\partial \sigma^2} = n(\sigma^2 - \overline{\sigma}^2)/2\sigma^4,$$

Fisher's information quantity $E(\partial L_n/\partial\sigma^2)^2$ equals $n/2\sigma^4 = [D(\overline{\sigma}^2)]^{-1}$ here.

Consequently, $\overline{\sigma}^2$ is an unbiased and efficient estimator of the parameter σ^2.

As the maximum likelihood estimator $\overline{\sigma}^2$ itself is quite simple in this case, it is not necessary to carry out further simplification, i.e., it is not required to introduce the asymptotic maximum likelihood estimator $\widetilde{\sigma}^2$. Observe, however, for comparison, that in view of (I.2.32)

$$\widetilde{\sigma}^2 = \frac{1}{n}\sum_{j=1}^{n}Y_j^2 - \left(\frac{1}{n}\sum_{j=1}^{n}Y_j\right)^2 = \frac{1}{n}\mathbf{Y'}\left[I_n - \frac{1}{n}\mathbf{vv'}\right]\mathbf{Y},$$

since

$$\left[I_n - \frac{1}{n}\mathbf{vv'}\right](I_n + \mathbf{vv'})\left[I_n - \frac{1}{n}\mathbf{vv'}\right] = \left[I_n - \frac{1}{n}\mathbf{vv'}\right]$$

is a matrix of rank $(n-1)$, the random variable $n\widetilde{\sigma}^2/\sigma^2$ possesses χ^2-distribution with $n-1$ degrees of freedom (cf. [106], 3c.4) so that

$$E(\tilde{\sigma}^2) = \sigma^2(n-1)/n, \qquad D(\tilde{\sigma}^2) = 3\sigma^4(n-1)/n^2.$$

Example 5. Let the process X_t satisfy the stochastic difference equation

$$(51) \qquad X_t - \iota_1 X_{t-1} = \varepsilon_t - \varepsilon_{t-1}, \qquad -1 < \iota_1 < 1$$

(cf. Examples I.1.4 and 2.2 for $\iota_1 = 0$ and $z_1 = 1$).

If the value of the parameter ι_1 is known and only the parameter $\sigma^2 = E(\varepsilon_t^2)$ is unknown (and must be estimated), we then have (in view of (I.1.22)) that its maximum likelihood estimator $\overline{\sigma}^2$ is of the form

$$\overline{\sigma}^2 = \frac{1}{n}\left[\sum_{j=1}^{n} Y_{j,\iota_1}^2 - (1-\iota_1)(1+\iota_1+n(1-\iota_1))^{-1}\left(\sum_{j=1}^{n} Y_{j,\iota_1}\right)\right]$$

where

$$Y_{1,\iota_1} = X_1, \qquad Y_{j,\iota_1} = (1-\iota_1)\sum_{k=1}^{j-1} X_k + X_j, \qquad j = 2, ..., n.$$

As in the case of the preceding example, it is easy to prove that in this case as well, the estimator $\overline{\sigma}^2$ will be an unbiased and efficient estimator for all values of n. If, however, the parameter ι_1 is unknown or both parameters σ^2 and ι_1 are unknown, then in this case the maximum likelihood estimators of unknown parameters will be only asymptotically efficient. Moreover, if the value of ι_1 is unknown, in view of the formula (I.1.22) for maximum likelihood estimators of unknown parameters, one can obtain only very cumbersome nonlinear equations, involving lengthy analytic expressions and the only possible numerical solution requires substantial effort. Therefore, in the case of unknown ι_1 the maximum likelihood estimators of the parameter ι_1 or parameters ι_1 and σ^2 are of no substantial practical usefulness. On the other hand, the asymptotic m.l. estimators $\tilde{\iota}_1$ and $\tilde{\sigma}^2$ determined as the values of ι_1 and σ^2 maximizing L_n in view of (I.2.33) are very simple in this case:

$$(52) \qquad \tilde{\iota}_1 = r_{1,y}/r_{0,y}, \qquad \tilde{\sigma}^2 = (r_{0,y}^2 - r_{1,y}^2)/r_{0,y},$$

where $r_{j,y}$ is given by the formula (I.2.34) (cf. (9)). As in the case of Example 1 we have -- up to a summand of order $o(1/n)$ -- the following expressions for variances and the covariance

$$D(\tilde{\tau}_1) = (1-\iota_1^2)/n, \quad D(\tilde{\sigma}^2) = 2\sigma^4/n, \quad \text{cov}(\tilde{\tau}_1,\tilde{\sigma}^2) = 0.$$

Example 6. Let X_t be a moving average process represented in the form $X_t = \varepsilon_t - (1-\alpha_1)\varepsilon_{t-1} + \alpha_1\varepsilon_{t-2}$ and thus possessing the spectral density of the form

(53) $$f_1(\lambda) = \frac{\sigma^2}{2\pi}|(1-z)(1-\alpha_1 z)|^2, \quad \sigma^2 = E(\varepsilon_t^2).$$

(cf. (I.2.35) and (I.2.36) for $\alpha_1 = 0$). In view of (I.2.37) an asymptotic m.l. estimator $\tilde{\alpha}_1$ is a root of an equation obtained from (25) in which $\beta_n^*(j)$ is replaced by $r_{j,Y}$ The variance of the estimator $\tilde{\alpha}_1$ is of the form (26). If the parameter σ^2 is also unknown then

$$\tilde{\sigma}^2 = (1-\tilde{\alpha}_1^2)^{-1}\left[r_{0,Y} + 2\sum_{j=1}^{n-1}\tilde{\alpha}_1^j\, r_{j,Y}\right]$$

is its asymptotic m.l. estimator and analogously to the case discussed in Subsection 4.2, the relations (18) are valid here with $r = 1$.

5. Asymptotic Maximum Likelihood Estimator of the Spectrum of Processes Distorted by "White Noise"

5.1.   The examples presented in the preceding Section do not take into account the important fact that in real-world situations, observations of the values of the random process

$$X_t, \quad t = ..., -1,0,1, ... ,$$

are never absolutely accurate but rather they always contain "measurement errors" which often are quite substantial. These "errors," also called "disturbances" or "noise," are usually of the form of disordered fluctuations which are superimposed on the true values of the measured quantity and can be described only statistically. A sufficiently wide class of noise -- occurring in the real world -- can be described by assuming that the given observations are represented as realizations of the sum

(1) $$X_t = S_t + N_t$$

of two mutually independent stationary random processes:

the signal S_t (which is of interest to us) and the "noise" N_t which distorts the signal. From physical considerations it seems reasonable to assume also that N_t is "white noise," i.e., it represents the realization of a sequence of mutually independent and identically distributed random variables.

Below we shall confine ourselves to the case when both the "signal" S_t and the "noise" N_t are Gaussian stationary processes with zero means. As above, we shall assume that the spectral density $f^{(S)}$ of the "signal" S_t depends on the unknown parameters $\theta_1, ..., \theta_{p-1}$ so that $f^{(S)} = f^{(S)}_{\theta_1, ..., \theta_{p-1}}$

It follows from the independence of the summands on the r.h.s. of equation (1) that the spectral density f of the process X_t can be represented as a sum of a function $f^{(S)} = f^{(S)}(\lambda)$ and a constant term $\sigma_N^2/2\pi$ where $\sigma_N^2 = E(N_t^2)$, i.e.,

$$(2) \qquad f = f^{(S)} + \frac{\sigma_N^2}{2\pi}.$$

In the case frequently occurring in practice when the effect of the "noise" cannot be ignored, it is necessary to consider the problem of estimating the unknown parameters $\theta_1, ..., \theta_{p-1}$ appearing in the expression for spectral density $f^{(S)}_{\theta_1, ..., \theta_{p-1}}$ of the "signal" S_t as well as the single unknown parameter σ_N^2 of the "noise" N_t based on the observation values of the sum $X_t = S_t + N_t$ at the n consecutive times $t = 1, 2, ..., n$.

Since the variance σ_N^2 of the "noise" N_t is also assumed to be unknown it is convenient to introduce the notation $\sigma_N^2 = \theta_p$; moreover, as above, $f = f_\theta$ where $\theta = (\theta_1, ..., \theta_{p-1}, \theta_p)'$.

The inequality $f_\theta > 0$ which is valid in view of (2) and the inequalities $f^{(S)} \geqslant 0$ and $\sigma_N^2 > 0$ allow us to determine from condition (2.2) the asymptotic maximum likelihood estimators θ which turn out to be (under general conditions stipulated in the statements of the Theorems 2.1 and 2.2) consistent, asymptotically normal, and asymptotically efficient estimators of parameters θ. It should be noted, however, that since the spectral density f_θ depends on the unknown parameters $\theta_1, ..., \theta_p$ in a specific manner, the asymptotic likelihood equations are in this case more complicated as compared with the examples discussed in Section 4.

5.2. In this paragraph we shall consider especially the

important case when the signal S_t, observed in a noisy background with intensity $\sigma_N^2 = \theta_p$, possesses spectral density of the form

$$(3) \qquad f^{(S)}(\lambda) = \frac{\sigma_S^2}{2\pi} |g_r(z)|^2 |h_q(z)|^{-2},$$

where

$$g_r(z) = 1 - \alpha_1 z - \cdots - \alpha_1 z^r$$

and

$$h_q(z) = 1 - \iota_1 z - \cdots - \iota_q z^q$$

are polynomials whose roots exceed 1 in the absolute value. The spectral density (3) is assumed to depend on the unknown parameters $\iota_1, \ldots, \iota_q, \alpha_1, \ldots, \alpha_r$ and σ_S^2. Then the components of the vector $\theta = (\theta_1, \ldots, \theta_p)'$ introduced in the preceding Section will be the unknown parameters $\iota_1, \ldots, \iota_q, \alpha_1, \ldots, \alpha_r, \sigma_S^2, \sigma_N^2$; here $p = q + r + 2$.

In this case the following lemma is valid.

Lemma 1. *For $0 \leqslant q \leqslant r$ there exist two different choices of values θ_1 and θ_2 of the parameter θ so that $f_{\theta_1}(\lambda) = f_{\theta_2}(\lambda)$ for almost all λ. If, however, $0 \leqslant r < q$, then $f_{\theta_1}(\lambda) \neq f_{\theta_2}(\lambda)$ for almost all λ provided $\theta_1 \neq \theta_2$.*

Proof. To prove the first assertion of Lemma 1 it is clearly sufficient to show that for $0 \leqslant q \leqslant r$ there exist θ_1 and θ_2 ($\theta_1 \neq \theta_2$) such that

$$(4) \qquad \beta_{\theta_1}(\tau) = \beta_{\theta_2}(\tau),$$

for any $\tau = 0, 1, \ldots$, where

$$(5) \qquad \beta_\theta(\tau) = \int_{-\pi}^{\pi} f_\theta(\lambda) e^{i\lambda\tau} d\lambda$$

is the covariance function of the observed process $X_t = S_t + N_t$.

We shall utilize the fact that the Fourier transform $\beta^{(S)}(\tau)$ of a spectral density $f^{(S)}$ of the form (3) -- the covariance function of signal S_t -- satisfies the following difference equations (cf., for example, [24] page 75)

$$\beta^{(S)}(k) - \iota_1\beta^{(S)}(k-1) - \cdots - \iota_q\beta^{(S)}(k-q)$$

(6)

$$= \sigma_S^2(h_{-k} - \alpha_1 h_{1-k} - \cdots - \alpha_r h_{r-k}), \qquad k = 0,1, \ldots \, ,$$

where

(7)
$$h_k = \begin{cases} 0, & \text{for } k < 0, \\ 1, & \text{for } k = 0, \\ c_k - \alpha_1 c_{k-1} - \cdots - \alpha_{k-1} c_1 - \alpha_k, & \text{for } 0 < k \leqslant r, \end{cases}$$

and $c_1, \ldots, c_r$ are uniquely determined by the equality

$$(1 - \iota_1 z - \cdots - \iota_q z^q)^{-1} = 1 + c_1 z + \cdots + c_r z^r + \cdots$$

(and hence $c_1, \ldots, c_r$ depend only on $\iota_1, \ldots, \iota_q$). In view of (6) and the self-evident equality

$$\beta_\Theta(k) = \beta_\Theta^{(S)}(k) + \sigma_N^2 \delta_{0k},$$

we obtain

$$\beta_\Theta(k) - \iota_1\beta_\Theta(k-1) - \cdots - \iota_q\beta_\Theta(k-q)$$

$$- \sigma_S^2(h_{-k} - \alpha_1 h_{1-k} - \cdots - \alpha_r h_{r-k})$$

(8)

$$= \begin{cases} \sigma_N^2, & \text{for } k = 0, \\ -\iota_k \sigma_N^2, & \text{for } k = 1, \ldots, q, \\ 0, & \text{for } k > q. \end{cases}$$

If $q \leqslant r$, then in view of (7) the equation (8) for $k > r$, does not depend on the parameters $\alpha_1, \ldots, \alpha_r, \sigma_S^2,$ and σ_N^2. Therefore, as in the case of absence of noise (cf. [24], p. 75 or p. 202) the values of the parameters $\iota_1, \ldots, \iota_q$ here could be expressed in terms of the values of the covariance function $\beta_\Theta(k)$ for $k = r+1, \ldots, r+q$ by means of a solution of a linear system of q equations

$$\beta_\Theta(k) - \iota_1\beta_\Theta(k-1) - \cdots - \iota_q\beta_\Theta(k-q) = 0,$$

(9)

$$k = r+1, \ldots, r+q.$$

Denote by $\iota_1', \ldots, \iota_q'$ the roots of the equations (9). As far as the parameters $\alpha_1, \ldots, \alpha_r, \sigma_S^2,$ and σ_N^2 are concerned, to express their values in terms of $\iota_1', \ldots, \iota_q'$ and $\beta_\Theta(0), \beta_\Theta(1), \ldots, \beta_\Theta(r)$ we have only the first $(r+1)$ equations (8) at our disposal. Since the number of unknowns $(r+2)$ exceeds the number of equations $(r+1)$, the obtained system of $(r+1)$ equations with $(r+2)$ unknowns has no unique solution. Thus the two sets of values

$$\alpha_1', \ldots, \alpha_r', \ (\sigma_S^2)', \ (\sigma_N^2)', \text{ and } \alpha_1'', \ldots, \alpha_r'', (\sigma_S^2)'', (\sigma_N^2)''$$

of the parameters $\alpha_1, \ldots, \alpha_r, \sigma_S^2, \sigma_N^2$ are available, satisfying the first $(r+1)$ equations (8) and also at least one of the inequalities

$$\alpha_1' \neq \alpha_1'' , \ldots, \alpha_r' \neq \alpha_r'', \ (\sigma_S^2)' \neq (\sigma_S^2)'', \ (\sigma_N^2)' \neq (\sigma_N^2)''.$$

It is clear from the arguments presented above that if we denote

$$\Theta_1 = (\iota_1', \ldots, \iota_q', \alpha_1', \ldots, \alpha_r', (\sigma_S^2)', (\sigma_N^2)')$$

and

$$\Theta_2 = (\iota_1', \ldots, \iota_q', \alpha_1'', \ldots, \alpha_r'', (\sigma_S^2)'', (\sigma_N^2)'')$$

then for $\tau = 0,1, \ldots$ the equality (4) will be valid in spite of the fact that $\Theta_1 \neq \Theta_2$. Thus the first proposition of Lemma 1 is verified.

To prove the second proposition we note that for $0 \leqslant r < q$ the first $(q+r+2)$ equations in (8) in view of (7) become

$$\beta_\Theta(0) - \iota_1\beta_\Theta(1) - \cdots - \iota_q\beta_\Theta(q)$$

$$= \sigma_S^2(h_0 - \alpha_1 h_1 - \cdots - \alpha_r h_r) + \sigma_N^2,$$

$$\beta_\Theta(k) - \iota_1\beta_\Theta(k-1) - \cdots - \iota_q\beta_\Theta(k-q)$$

$$= \sigma_S^2(h_{-k} - \alpha_1 h_{1-k} - \cdots - \alpha_r h_{r-k})$$

$$- \iota_k \sigma_N^2, \quad k = 1, \ldots, r,$$

(10) $$\beta_\Theta(k) - \iota_1\beta_\Theta(k-1) - \cdots - \iota_q\beta_\Theta(k-q)$$

$$= \begin{cases} -\iota_k \sigma_N^2, & k = r + 1, ..., q, \\ 0 , & k = q + 1, ..., q + r + 1. \end{cases}$$

It is easy to verify that equation (10) establishes a one-to-one correspondence between $\beta_\Theta(0)$, $\beta_\Theta(1)$, ..., $\beta_\Theta(q+r+1)$ and ι_1, ..., ι_q, α_1, ..., α_r, σ_S^2, σ_N^2 (the proof is analogous to the proof of the related assertion in the case when the noise N_t is absent, i.e., $\sigma_N^2 = 0$; cf. [24], p. 75, and also [128] where the case in which $\alpha_j = 0$, $j = 1, ..., r$ is discussed). From this, and the form of the succeeding equations in (8) for $k > q+r+2$, it follows that $\beta_{\Theta_1}(\tau) \neq \beta_{\Theta_2}(\tau)$ for any $\tau = 0,1, ...,$ provided only that $\Theta_1 \neq \Theta_2$.

Consequently,

$$\int_{-\pi}^{\pi} [f_{\Theta_2}(\lambda) - f_{\Theta_1}(\lambda)]^2 d\lambda \neq 0 \quad \text{for} \quad \Theta_1 \neq \Theta_2. \qquad \square$$

It follows from the first assertion of Lemma 1 that for $0 \leq q \leq r$, the condition (A) stipulated in Section 2 is not fulfilled. Thus, for example for $q = r = 0$ we obtain $f(\lambda) = \sigma_S^2/2\pi + \sigma_N^2/2\pi$ and for $q = 0$ and $r = 1$

$$f(\lambda) = \frac{\sigma_S^2}{2\pi}|1 - \alpha_1 e^{i\lambda}|^2 + \frac{\sigma_N^2}{2\pi} = \frac{\alpha_1 \sigma_S^2}{2\pi c}|1 - c e^{i\lambda}|^2,$$

where c is the root of the equation

$$(1 + c^2)\sigma_S^2 \alpha_1 = c[\sigma_N^2 + \sigma_S^2(1 + \alpha_1^2)].$$

Therefore, here, an infinite set of values of parameters σ_S^2, σ_N^2 (in the first case) and of α_1, σ_S^2, and σ_N^2 (in the second case) corresponds to the very same value of the spectral density f.

Observe that the fact that for $0 \leq q \leq r$ the parameters (ι_1, ..., ι_q, α_1, ..., α_r, σ_S^2, σ_N^2) do not uniquely determine the form of the spectral density $f(\lambda)$ on its own does not seem to be paradoxical; it simply means that in this case one should choose in some different manner the parameters (Θ_1, ..., Θ_p) on which the function $f_\Theta(\lambda)$ is dependent.

In the case when the noise intensity σ_N^2 is known, the problem of estimating the parameters of the spectral density $f^{(S)}$ based on the observed values of the sum $X_t = S_t + N_t$ may often have a sensible solution (also in the case when $0 \leq q \leq r$); however, we shall confine ourselves to a discussion of several examples most frequently encountered in practical applications of the forms of spectral densities of the signal S_t, which depend on the unknown parameters in a rather simple manner. For all the examples presented below condition (A) stated in Section 2 is fulfilled and there exist asymptotic m.l. estimators of unknown parameters. Along with asymptotic likelihood equations for each one of these examples we shall also present the limiting form (as $n \to \infty$) of the corresponding Fisher's information matrix.

Example 1. We shall start with the simplest case when the signal is a Gaussian autoregression of the first order with zero expectation and spectral density

$$(11) \qquad f^{(S)}(\lambda) = \frac{\sigma_S^2}{2\pi} |1 - \iota_1 z|^{-2}, \qquad z = e^{i\lambda}.$$

It is easy to verify that the asymptotic m.l. estimators $\tilde{\iota}_1$, $\tilde{\sigma}_S^2$, and $\tilde{\sigma}_N^2$ of the parameters ι_1, σ_S^2 and σ_N^2 are in this case the roots of a rather complex system of equations.

Note, however, that in the case under consideration it is more convenient to introduce new parameters σ^2 and α_1 related to ι_1, σ_S^2, and σ_N^2 by the equalities

$$(12) \qquad \sigma^2 \alpha_1 = \sigma_N^2 \iota_1, \qquad \sigma^2(1+\alpha_1^2) = \sigma_N^2(1+\iota_1^2) + \sigma_S^2,$$

so that

$$(13) \qquad f(\lambda) = f^{(S)}(\lambda) + \frac{\sigma_N^2}{2\pi} = \frac{\sigma^2}{2\pi}|1-\alpha_1 z|^2 \cdot |1-\iota_1 z|^{-2}$$

(cf. Example 4.3). Next it is convenient to obtain first the asymptotic m.l. estimators $\tilde{\iota}_1$, $\tilde{\alpha}_1$, and $\tilde{\sigma}^2$ of the parameters ι_1, α_1, and σ^2 by means of the relations (4.42)-(4.44) and next to replace in equations (12) the values of ι_1, α_1, and σ^2 by their estimators $\tilde{\iota}_1$, $\tilde{\alpha}_1$, $\tilde{\sigma}^2$ and find the roots $\tilde{\sigma}_S^2$ and $\tilde{\sigma}_N^2$ of the obtained equations with respect to σ_S^2 and σ_N^2. The estimators derived in this manner will be the required asymptotically efficient estimators of parameters ι_1, σ_S^2, and σ_N^2. It is easy to verify that in view of the relations (12), the covariance matrix $D_{\iota_1, \sigma_S^2, \sigma_N^2}$ of the estimators $\tilde{\iota}_1$, $\tilde{\sigma}_S^2$, and $\tilde{\sigma}_N^2$ is of the

form

$$D_{\iota_1, \sigma_S^2, \sigma_N^2} = \Phi' D_{\iota_1, \alpha_1, \sigma^2} \Phi ,$$

where $D_{\iota_1, \alpha_1, \sigma^2}$ is defined by (4.46) and

$$\Phi = \begin{bmatrix} 1 & \sigma^2\alpha_1(1-\iota_1^2)/\iota_1^2 & -\sigma^2\alpha_1/\iota_1^2 \\ 0 & \sigma^2|2\alpha_1-(1+\iota_1^2)/\iota_1 & \sigma^2/\iota_1 \\ 0 & \alpha_1|(1+\alpha_1^2)/\alpha_1-(1+\iota_1^2)/\iota_1 & \alpha_1\iota_1 \end{bmatrix}.$$

Example 2. Let the signal S_t be an autoregressive process of the second order with spectral density of the form

$$f^{(S)}(\lambda) = \sigma_S^2/2\pi\psi(\lambda), \quad \psi(\lambda) = |1-\iota_1 z-\iota_2 z^2|^2, \quad z = e^{i\lambda},$$

so that the spectral density f of the observed process X_t will be of the form

(14) $\qquad f(\lambda) = \phi(\lambda)/2\pi\psi(\lambda), \quad \phi(\lambda) = \sigma_S^2 + \sigma_N^2 \psi(\lambda).$

In this case the asymptotic m.l. estimators $\tilde{\Theta} = (\tilde{\iota}_1, \tilde{\iota}_2, \tilde{\sigma}_S^2, \tilde{\sigma}_N^2)$ of the unknown parameters $\Theta = (\Theta_1, ..., \Theta_4) = (\iota_1, \iota_2, \sigma_S^2, \sigma_N^2)$ are the roots of the system of equations

(15) $\qquad \int_{-\pi}^{\pi} \psi^{-2}(\lambda)(\cos j\lambda - \Theta_j)[I_n(\lambda)-f(\lambda)]d\lambda = 0, \quad j = 1, 2,$

(16) $\qquad \int_{-\pi}^{\pi} \phi^{-2}(\lambda)\psi^j(\lambda)[I_n(\lambda)-f(\lambda)]d\lambda = 0, \quad j = 1,2,$

and the limit as $n \to \infty$ of the Fisher's information matrix is now of the form

(17) $\qquad \Gamma_\Theta = \left[\dfrac{1}{4\pi} \int_{-\pi}^{\pi} \phi^{-2}(\lambda)G_{kj}(\lambda)d\lambda\right]_{k,j=1,\ldots,4}$

where G_{kj} are the elements of the matrix $g(\lambda)g'(\lambda)$ and

$$g(\lambda) = \text{col}(2\sigma_S^2[\cos\lambda(1-\iota_2)-\iota_1]/\psi,$$

$$2\sigma_S^2[\cos\lambda(1-\iota_1)-\iota_2]/\psi, \ 1, \ \psi).$$

Example 3. Consider the case when the signal S_t is a mixture of the autoregressive process of the second order and the moving average of the first order with the spectral density

(18) $f^{(S)}(\lambda) = \dfrac{\sigma_S^2}{2\pi}|1-\alpha_1 z|^2|1-\iota_1 z-\iota_2 z^2|^{-2}.$

The spectral density f of the observed process X_t in this case is again given by the formula (14); however, in this case

$$\Phi(\lambda) = \sigma_S^2|1-\alpha_1 z|^2 + \sigma_N^2\psi(\lambda).$$

In view of the second proposition of Lemma 1 there exist the asymptotic m.l. estimators $\tilde{\iota}_1$, $\tilde{\iota}_2$, $\tilde{\alpha}_1$, $\tilde{\sigma}_S^2$, and $\tilde{\sigma}_N^2$ of the

unknown parameters ι_1, ι_2, α_1, σ_S^2, and σ_N^2 which are the roots with respect to ι_1, ι_2, α_1, σ_S^2, and σ_N^2 of the system of equations

(19) $\displaystyle\int_{-\pi}^{\pi} \boldsymbol{\imath}_j(\lambda)\Phi^{-2}(\lambda)[\Phi(\lambda)-2\pi\psi(\lambda)I_n(\lambda)]d\lambda = 0, \quad j = 1,...,5,$

where

$$\boldsymbol{\imath}_j(\lambda) = 4\pi[(1-\iota_{3-j})\cos\lambda - \iota_j]f^{(S)}(\lambda), \quad j = 1,2,$$

$$\boldsymbol{\imath}_3(\lambda) = (\cos\lambda-\alpha_1)2\sigma_S^2, \qquad \boldsymbol{\imath}_4(\lambda) = |1-\alpha_1 z|^2, \quad \text{and}$$

$$\boldsymbol{\imath}_5(\lambda) = \psi(\lambda).$$

The matrix Γ_Θ where $\Theta = (\iota_1,\iota_2,\alpha_1,\sigma_S^2,\sigma_N^2)$ is also of the form (17) here, however, $k,j = 1, ..., 5$, G_{kj} are the elements of the matrix $g(\lambda)g'(\lambda)$ and $g(\lambda) = (\boldsymbol{\imath}_1(\lambda), ..., \boldsymbol{\imath}_5(\lambda))'$.

Example 4. We shall generalize somewhat the case considered in Example 1, assuming that the spectral density f of the signal S_t is of the form

(20) $f^{(S)}(\lambda) = \dfrac{\sigma_S^2}{2\pi}|1-\iota_q z^q|^{-2}, \quad z = e^{i\lambda},$

where q is a positive integer (the case $q = 1$ was considered in Example 1). In other words, we shall assume that the signal is an autoregressive process of the q-th order with spectral density $f^{(S)}$ of form (20) dependent only on two parameters σ_S^2 and ι_q. In contrast with the case considered here, in the second assertion of Lemma 1 (for $r = 0$ concerning a general autoregressive process of order q) it has been assumed that all parameters ι_1, ..., ι_q, σ_S^2 appearing in the expression for the spectral density f are unknown. Now we are assuming that for $q > 1$ the parameters ι_1, ..., ι_{q-1} are known and equal to zero. Nevertheless, it is easy to verify that also in the case where $\iota_1 = \iota_2 = \cdots = \iota_{q-1} = 0$, $q > 1$, the second proposition of Lemma 1 for $r = 0$ is valid and equation (8) in this case becomes

$$\beta_\Theta(0) - \iota_q \beta_\Theta(q) = \sigma_S^2 + \sigma_N^2,$$

$$\beta_\Theta(k) - \iota_q \beta_\Theta(k-q) = -\iota_q \sigma_N^2 \delta_{qk}, \qquad k > 0,$$

where

$$\Theta = (\iota_q, \sigma_S^2, \sigma_N^2)', \qquad \beta_\Theta(k) = \beta^{(S)}(k) + \sigma_N^2 \delta_{0k}$$

is the covariance function of the observed process $X_t = S_t + N_t$ and

$$\beta^{(S)}(k) = \begin{cases} 0, & k \neq q, 2q, \ldots, \\[2ex] \sigma_S^2 - \dfrac{\iota_q^{k/n}}{1+\iota_q^2}, & k = q, 2q, \ldots, \end{cases}$$

is the covariance function of the signal S_t.

Since condition (A) -- mentioned above on several ocasions -- is fulfilled in this case, the asymptotic m.l. estimators of parameters Θ do exist. However, as in the case $q = 1$ (cf. Example 1), in the case under consideration, it is more appropriate to use the new parameters $\iota_1, \alpha_1, \sigma^2$ obtained by means of equalities (12) where $\iota_1 = \iota_q$ and $\alpha_1 = \alpha_q$. The asymptotic likelihood equations are of the same form as in the case $q = 1$ (cf. (4.42)-(4.44)) the only difference being that in the expressions under the integral in formulas (4.12)-(4.14), $\cos \lambda$ and $z = e^{i\lambda}$ must be replaced by $\cos q\lambda$ and $z^q = e^{i\lambda q}$ respectively (also set $\iota_1 = \iota_q$ and $\alpha_1 = \alpha_q$). The limit as $n \to \infty$ of the Fisher's information matrix Γ_Θ and thus the covariance matrix D_Θ of the asymptotic maximum likelihood estimators $\tilde{\iota}_q, \tilde{\alpha}_q$, and $\tilde{\sigma}^2$ in the case $q > 1$ as well, are given by the formulas (4.45) and (4.46) where, however, $\iota_1 = \iota_q$ and $\alpha_1 = \alpha_q$.

Example 5. Let the signal S_t be a mixture of an autoregressive process of the first order and a moving average process of the first order with the spectral density of the form (3) where, however, $r = q = 1$ and $\alpha_1 = 2\iota_1$. As we have seen above (cf. Lemma 1) in the case when $r = q = 1$, the unknown parameters are the coefficients α_1 and ι_1 of the polynomials $g_1(z)$ and $h_1(z)$ respectively (appearing in formula (3)); together with the parameters σ_S^2 and σ_N^2, the equations (8) do not possess a unique solution with respect to ι_1, α_1, σ_S^2, and σ_N^2 and hence condition (A) presented in Section 2 is not fulfilled. In the case under consideration when $\alpha_1 = 2\iota_1$ equation (8) is of the form

$$\beta_\Theta(0) - \iota_1\beta_\Theta(1) = \sigma_S^2(1+\iota_1^2) + \sigma_N^2,$$

(21) $$\beta_\Theta(1) - \iota_1\beta_\Theta(0) + 2\iota_1\sigma_S^2 + \iota_1\sigma_N^2 = 0,$$

$$\beta_\Theta(k) - \iota_1\beta_\Theta(k-1) = 0, \quad k > 1.$$

From the first three equations (21) we obtain

(22) $$\iota_1 = -\beta_\Theta(2)/\beta_\Theta(1), \quad \sigma_S^2 = -\beta_\Theta(2)/\beta_\Theta^2(1),$$

$$\sigma_N^2 = \beta_\Theta(0) + 2\frac{\beta_\Theta(2)}{\beta_\Theta(1)} - \beta_\Theta(2).$$

Consequently, in this case the assertion of Theorem 4.1 is valid and there exist the asymptotic m.l. estimators $\tilde{\iota}_1$, $\tilde{\sigma}_S^2$, and $\tilde{\sigma}_N^2$ of parameters ι_1, σ_S^2, and σ_N^2. However, as in the Examples 1 and 4 considered above, it is more appropriate here to utilize the new parameters ι_1, α_1, and σ^2 obtained by means of the equations

(23) $$\sigma_S^2(1+4\iota_1^2) + \sigma_N^2(1+\iota_1^2) = \sigma^2(1+\alpha_1^2),$$

$$2\iota_1\sigma_S^2 + \iota_1\sigma_N^2 = \sigma^2\alpha_1.$$

After this the spectral density f of the observed process $X_t = S_t + N_t$ becomes of the form (13) so that the problem of obtaining the estimators of parameters ι_1, α_1, and σ^2 reduces here (analogously to the case in Examples 1 and 4) to the problem of estimating unknown parameters of a mixed first order autoregressive-first order moving average process discussed in Example 4.3.

6. Least-Squares Estimation of Parameters of a Spectrum of a Linear Process

6.1. Above we have always assumed that the process under consideration X_t, $t = ...,$ $-1,0,1,$ $...$ is Gaussian. However, this assumption is often very restrictive in practice. Therefore it is very important to note that much of the statistical inference about the process X_t, obtained under Gaussian assumptions on the distribution of variables X_t, remains valid under much more general assumptions.

In this section we consider as an example the case when the process X_t is linear, i.e., we shall assume that the process X_t can be represented in the form

$$(1) \qquad X_t = \sum_{s=0}^{\infty} g_s \varepsilon_{t-s},$$

where $g_0 = 1$, $\sum_{s=0}^{\infty} g_s^2 < \infty$, and ε_t, $t = ...,$ $-1,0,1,$ $...$ is a sequence of independent identically distributed random variables with

$$E(\varepsilon_t) = 0, \qquad E(\varepsilon_t) = \sigma^2 > 0, \qquad E(\varepsilon_t^4) < \infty$$

and σ^2 is given by formula (1.3). As it is well-known, in this case the spectral density of the process (1) is represented in the form

$$(2) \qquad f(\lambda) = \frac{\sigma^2}{2\pi} g(\lambda)$$

where

$$(3) \qquad g(\lambda) = |\sum_{s=0}^{\infty} g_s z^s|^2, \qquad z = e^{i\lambda}.$$

As usual we shall assume that $\sum_{s=0}^{\infty} g_s z^s$ is an analytic function of a complex variable z nonvanishing in the interior of the unit circle $|z| < 1$. It is known that a Gaussian process X_t is linear provided the integral on the r.h.s. of (1.3) is convergent (cf. [58], Chapter XII), while in the Gaussian case (and only in this case) the distribution of the quantities ε_t will also be Gaussian.

Assume that the spectral density f depends on the unknown parameter $\theta \in \Theta$, i.e., $f = f_\theta$, where Θ is a bounded closed set contained in an open set S of a p-dimensional Euclidean space R_p and that the value of the parameter θ should be estimated by means of n consecutive observations over the random variables $X_1, ..., X_n$.

As in Subsection 2.1 we shall consider the estimator θ determined from condition (2.2) where $U_n(\theta)$ is given by the formula (2.1) assuming that $f_\theta > 0$. Following the terminology used in statistics (cf., e.g., [131,139]) the estimators θ in the case under consideration will be called least squares estimators.

We first show that $\tilde{\theta}$ is a consistent estimator.

Theorem 1. *Let X_t be a linear process possessing spectral density $f = f_\theta$, $\theta \in \Theta$ which satisfies the conditions of Theorem 2.1. Then the least squares estimator $\tilde{\theta}$ is consistent, i.e., $\tilde{\theta} \to \theta$ in $P_n(f_\theta)$ probability as $n \to \infty$.*

Proof. In view of the results pertaining to the linear process X_t presented in Appendix 1 to this chapter, the relations (2.7) and (2.10) are valid (where $H_{\delta,\theta}$ is defined by the formula (2.6)) and moreover

$$\lim_{n\to\infty} nD(H_{\delta,n}) = \left[\frac{\delta}{4\pi} H_{\theta_1,\delta(\theta_1)}\right]^2 \lim_{n\to\infty} nD\left[\int_{-\pi}^{\pi} I_n(\lambda)d\lambda\right]$$

$$= \left[\frac{\delta}{4\pi} H_{\theta_1,\delta(\theta_1)}\right]^2 \left[4\pi \int_{-\pi}^{\pi} f_\theta^2(\lambda)d\lambda\right.$$

$$\left. + \kappa_4 \left[\int_{-\pi}^{\pi} f_\theta(\lambda)d\lambda\right]^2\right] < \infty,$$

where κ_4 is the fourth cumulant of random variables ε_t and

$$\lim_{n\to\infty} D[n^{1/2}(U_n(\theta)-U_n(\theta_1))]$$

$$= \frac{1}{4\pi}\int_{-\pi}^{\pi} \left[1 - \frac{f_\theta(\lambda)}{f_{\theta_1}(\lambda)}\right]^2 d\lambda$$

$$+ \kappa_4 \left[\frac{1}{4\pi} \int_{-\pi}^{\pi} \left[1 - \frac{f_\theta(\lambda)}{f_{\theta_1}(\lambda)}\right]d\lambda\right]^2 < \infty.$$

To complete the proof it remains to repeat the arguments presented above in the course of the proof of Theorem 1 in Section 2.   □

6.2. When proving the asymptotic normality of the estimators $\tilde{\theta}$, one can apply the reasoning utilized in

Subsection 2 of Appendix 2 to the present chapter.

Indeed, tracing the proof of the relation (2.20) we verify that the arguments remain unchanged and only the derivation of the relation (2.23) under the conditions of linearity of X_t is necessary here. More precisely, it is only required to show that the variance of the expression in the l.h.s. of (2.23) converges to zero in the linear case as well. This can be shown easily by using the methods presented in Appendix 1 to this chapter, namely by approximating the continuous function $a = a_\Theta$ in (2.23) by the Féjer sum of its Fourier series $\sigma_n(\lambda, a)$ (cf. (I.2.13)), thus reducing the problem to estimating the covariance between empirical covariances for large values of n (cf. (19)-(21), Appendix 1 to this chapter, and the footnote 10 on page 160).

We now turn to the asymptotic normality of vector $\Delta_{n,\Theta}$. If we apply the part of the assertion of Corollary A1.2 in Appendix 1 dealing with the linear process X_t, then in place of (2.18) we obtain that the vector $\Delta_{n,\Theta}$, as $n \to \infty$, possesses p-dimensional normal distribution $N(0, \Gamma_\Theta + C_{\kappa_4, \Theta})$ where $C_{\kappa_4, \Theta}$

$= \kappa_4 c_\Theta c_\Theta'$, c_Θ being a p-dimensional vector, whose k-th component equals

$$\frac{1}{4\pi}\int_{-\pi}^\pi \frac{\partial}{\partial\Theta_k}\log f_\Theta(\lambda)d\lambda.$$

Thus the following theorem is valid.

Theorem 2. *Let X_t, $t = \dots,-1,0,1,\dots$, be a linear process with spectral density $f = f_\Theta$, $\Theta \in \Theta$ satisfying the conditions of Theorem 2 of Section 2. Then the least squares estimator $\tilde\Theta$ is asymptotically normal, i.e., the random vector $n^{1/2}(\tilde\Theta - \Theta)$, as $n \to \infty$, possesses the normal distribution $N(0, \Gamma_\Theta^{-1} + \Gamma_\Theta^{-1} C_{\kappa_4, \Theta} \Gamma_\Theta^{-1})$.*

6.3. As it was already mentioned in the Introduction, it is almost always assumed in the literature (in particular, in [131]) that the parameter σ^2 (cf. (1.3)) is one of the components of the vector Θ (say the p-th, i.e., $\sigma^2 = \Theta_p$) and that the normalized spectral density $g = 2\pi f/\sigma^2$ (cf. (2) and (3)) depends on the values of other components $(\Theta_1, \dots, \Theta_{p-1}) = \Theta_{(p-1)}$, i.e., $g = g_{\Theta(p-1)}$. This is first and foremost due to the fact that under such an assumption the least squares estimator $\tilde\Theta = (\tilde\Theta_1, \dots, \tilde\Theta_{p-1}, \tilde\sigma^2)$ possesses certain additional "nice" properties. For example, in the next subsection we shall show that $\tilde{\theta}$ enjoys minimal limiting variance (as $n \to \infty$) within a very wide class of estimators. In this subsection we shall deal with the degree of "robustness" of estimators $\tilde{\theta}$, i.e., the extent to which their properties are independent of the assumption on the probability distributions of process X_t.[4]

In view of (1.3), (2.1), and (2) under the conditions that $\theta = (\theta_{(p-1)}, \sigma^2)$

$$U_n(\theta) = \frac{1}{2}\left[\log \frac{\sigma^2}{2\pi} + \frac{1}{\sigma^2}\int_{-\pi}^{\pi} \frac{I_n(\lambda)}{g_{\theta_{(p-1)}}(\lambda)}d\lambda\right],$$

so that the estimators $(\tilde{\theta}_1, \ldots, \tilde{\theta}_{p-1}) = \tilde{\theta}_{(p-1)}$ are determined from the condition

$$(4) \qquad \int_{-\pi}^{\pi} \frac{I_n(\lambda)}{g_{\tilde{\theta}_{(p-1)}}(\lambda)}d\lambda = \max_{\theta} \int_{-\pi}^{\pi} \frac{I_n(\lambda)}{g_{\theta_{(p-1)}}(\lambda)}d\lambda,$$

and

$$(5) \qquad \tilde{\sigma}^2 = \int_{-\pi}^{\pi} \frac{I_n(\lambda)}{g_{\tilde{\theta}_{(p-1)}}(\lambda)}d\lambda.$$

Since in view of (1.3) and (2)

$$\int_{-\pi}^{\pi} \frac{\partial}{\partial \sigma^2}\log f_\theta(\lambda)\frac{\partial}{\partial \theta_k}\log f_\theta(\lambda)d\lambda$$

$$= \frac{1}{\sigma^2}\frac{\partial}{\partial \theta_k}\int_{-\pi}^{\pi}\log f_\theta(\lambda)d\lambda = 0, \qquad k = 1, \ldots, p-1,$$

and

$$\int_{-\pi}^{\pi}\left[\frac{\partial}{\partial \sigma^2}\log f_\theta(\lambda)\right]^2 d\lambda = \frac{2\pi}{\sigma^4},$$

---

[4]The term "robustness" in the sense used here was introduced by Whittle in his paper [125] and since then it has been utilized in the analysis of time series (cf. [165]). However, to avoid any misunderstanding we note that this is in variance with the ordinary meaning of this same term which is common nowadays in statistical literature (cf., for example, [176] as well as [177] where the possibility of extending the ideas of [176] to the case of a linear autoregressive process are discussed); see also Lecture Notes in Statistics 26 (1984), Robust and Nonlinear Time Series Analysis (J. Franke, W. Hardle, and D. Martin, ed.), Springer-Verlag, New York.

in this case the matrix $\Gamma_\Theta = \Gamma_\Theta^{(p)}$ is of the form

(6)
$$\Gamma_\Theta^{(p)} = \begin{bmatrix} \Gamma_\Theta^{(p-1)} & 0 \\ & \\ 0 & 1/2\sigma^4 \end{bmatrix}$$

(here the matrix Γ_Θ is indexed by (p) to indicate that its dimension is p so that the entries of matrix $\Gamma_\Theta^{(p-1)}$ are also given by the formula (2.14) where, however, k and ℓ take the values 1, ..., p-1) and all the entires of vector c_Θ are equal to 0 with the exception of the last one which is equal to $1/2\sigma^2$.

Consequently,

(7)
$$\Gamma_\Theta^{(p)} + C_{\kappa_4,\Theta} = \begin{bmatrix} \Gamma_\Theta^{(p-1)} & 0 \\ & \\ 0 & (2+\kappa_4)/4\sigma^4 \end{bmatrix}$$

and

(8)
$$[\Gamma_\Theta^{(p)}]^{-1}(\Gamma_\Theta^{(p)} + C_{\kappa_4,\Theta})[\Gamma_\Theta^{(p)}]^{-1}$$
$$= \begin{bmatrix} [\Gamma_\Theta^{(p-1)}]^{-1} & 0 \\ & \\ 0 & \sigma^4(2+\kappa_4) \end{bmatrix}.$$

Formula (8) and Theorem 2 yield the following corollary.

Corollary 1. *Under the conditions of Theorem 2, the estimators*

$$(\tilde\Theta_1, ..., \tilde\Theta_{p-1}) = \tilde\Theta_{(p-1)}$$

of the parameters $(\Theta_1, ..., \Theta_{p-1}) = \Theta_{(p-1)}$ *determined from condition* (4) *are asymptotically normal, i.e., the random* $(p-1)$-*dimensional vector* $n^{1/2}(\tilde\Theta_{(p-1)} - \Theta_{(p-1)})$ *as* $n \to \infty$ *possesses the* $(p-1)$-*dimensional normal distribution* $N(0,[\Gamma_\Theta^{(p-1)}]^{-1})$. *The random variable* $n^{1/2}(\tilde\sigma^2 - \sigma^2)$ *is asymptotically independent of the random vector* $n^{1/2}(\tilde\Theta_{(p-1)} - \Theta_{(p-1)})$ *and, as* $n \to \infty$, *it possesses the normal distribution* $N(0,\sigma^4(2+\kappa_4))$.

We thus observe that the limiting distribution of the estimators

$$(\tilde{\theta}_1, ..., \tilde{\theta}_{p-1}) = \tilde{\theta}_{(p-1)}$$

does not depend on the assumption on the distribution of random variables ε_t; in the sense the estimators $\tilde{\theta}_{(p-1)}$ are "robust." As far as $\tilde{\sigma}^2$ is concerned its limiting distribution depends on the fourth cumulant κ_4 of the random variables ε_t so it is not "robust" in the sense in which this term is applied to the estimators $\tilde{\theta}_1, ..., \tilde{\theta}_{p-1}$.

We also note that in view of Corollary 1 for any linear process, estimators $\tilde{\theta}_1, ..., \tilde{\theta}_{p-1}$ of the parameters $\theta_1, ..., \theta_{p-1}$ can be obtained whose limiting dispersion coincides with the dispersion of asymptotically efficient estimators of the same parameters for a Gaussian process. In other words, we see that from the aspect of attainable accuracy of estimators of parameters $\theta_1, ..., \theta_{p-1}$ (but not of σ^2) the Gaussian processes are "the worst ones" among all linear processes.[5] It is however essential that for all such processes the least square estimators $\tilde{\theta}_1, ..., \tilde{\theta}_{p-1}$ possess a "universal" property in the sense that their properties remain unaltered as the distribution of random variables ε_t in formula (1) changes. As far as estimators optimal for a certain (non-Gaussian) linear process X_t are concerned, these do not possess such a "universality" and for some distributions of ε_t they may be worse than the least squares estimators (cf. [125] and also [165]).

6.4.   In Section 2, when studying the special case of normally distributed ε's in (6.1), we presented conditions under which the estimator $\tilde{\theta}$ of parameter θ of the spectral density $f_\theta > 0$ determined by condition (2.2) is asymptotically efficient in Fisher's sense (cf. Theorem 2.1). It was also

---

[5]For a given distribution of ε's in (6.1) one can, in principle, determine the limit of Fisher's information matrix relative to $\theta_{(p-1)}$. Indeed, informal considerations (analogous to those applied in [125] to the particular case of linear autoregression) result in the following limiting expression: $\gamma\Gamma_\theta^{(p-1)}$ where γ is the Fisher's information quantity relative to the scale parameter in the general distribution of variables ε_t/σ, i.e., $\gamma = \int_{-\pi}^{\pi} [g'(x)]^2/g(x)\,dx$ where g is the density of the distribution. Moreover, $\gamma \geqslant 1$ with equality only if g is the standard Gaussian density (see Appendix 3 of this chapter).

indicated that (cf. footnote on page 108) since the related family of Gaussian distributions $\{P_n(f_\Theta), \Theta \in \Theta\}$ is locally asymptotically normal (LAN), in the sense of the definition presented in the paragraph between formulas (2.17) and (2.18), one could, utilizing methods of general asymptotic theory, further trace some more refined properties of the estimators $\tilde\Theta$. This would, however, entail a number of notions and results from the general theory which we are purposely avoiding in order not to exceed the predesigned framework of this book. Nevertheless, in Appendix 2 of this chapter, the most basic definitions and conclusions of the general theory are presented which actually lead to establishing further properties of the estimators $\tilde\Theta$. Namely, under the LAN conditions on general sequences of families of distributions $\{P_{n,\Theta}, \Theta \in \Theta\}$, $n = 1, 2, \ldots$ (Definition A2.1) Hájek's [175] definition of "regularity of an estimator of parameter Θ" is given (Definition A2.2) which allows us inter alia to formulate the following assertion.

For any regular estimator $\hat\Theta$ in Hájek's sense of the parameter Θ the matrix inequality[6]

$$\lim_{n\to\infty} nE\{(\hat\Theta-\Theta)^{\otimes 2} | P_{n,\Theta}\} \geqslant \Gamma_\Theta^{-1}$$

is valid where Γ_Θ is a positive definite matrix appearing in the definition of LAN of the family $\{P_{n,\Theta}, \Theta \in \Theta\}$ (which is the limit as $n \to \infty$ of the Fisher's information matrix per unit of observations). In other words the limiting dispersion of any regular estimator $\hat\Theta$ is bounded below by an ellipse generated by matrix Γ_Θ^{-1}.

Of course, under the conditions of Section 2, the last assertion is applicable to the case studied therein. As a result, under these conditions the estimator $\tilde\Theta$ defined by condition (2.2) is itself regular in Hájek's sense (cf. the remark immediately following the statement of Theorem A2.1), it possesses the limiting covariance matrix Γ_Θ^{-1} (cf. (2.14)), and hence its limiting dispersion is not "worse" than the limiting dispersion of any other regular estimator.

---

[6]For convenience of notation here and below we use the symbol $\otimes$ of the Kronecker product of matrices; in particular, for any vector x we have $xx' = x \otimes x = x^{\otimes 2}$.

When we deviate from the Gaussian assumption, the estimator $\tilde{\Theta}$ loses this useful property. Nevertheless, it turns out that it retains the smallest limiting dispersion albeit in a much narrower (but still practically important) class of regular estimators. The remainder of this section is devoted to the proof of this assertion.

Before describing precisely the above mentioned narrow class of regular estimators, we note that this class actually contains all the estimators of parameters of a spectrum which are mentioned in this book. All of them are constructed solely by means of "statistics of the second order"; empirical covariances and smoothed periodograms (cf. Section 2 and Section 3 of the next chapter). In the context of this book, such a restriction, which is self-evident in the Gaussian case, seems to be natural also in the case of general linear processes in which the hypothesis of belonging of the spectral density to a particular parametric family is supplemented only by a general assumption concerning the linear structure of the process under consideration (6.1). (This is because any attempt to include statistics of a higher order when deducing certain statistical conclusions would require additional specification of the model which would result in a deviation from the direction and scope of this volume. For this reason a brief and informal discussion of this topic is relegated to the Appendix 3 to this chapter).

Recall that under the conditions of Theorem 2 the least squares estimator $\tilde{\Theta}$ satisfies relation (2.20), which in the notation

$$(9) \qquad A_\Theta(\lambda) = \Gamma_\Theta^{-1}\dot{\phi}_\Theta(\lambda)$$

(as usual here $\dot{\phi}_\Theta$ is the gradient vector $(\partial/\partial\Theta)\log f_\Theta$ so that $\Gamma_\Theta = (1/4\pi)\int_{-\pi}^{\pi}\dot{\phi}_\Theta(\lambda)^{\otimes 2}$ by (2.14)),

$$(10) \qquad I_n(h) = n^{1/2}\int_{-\pi}^{\pi} h(\lambda)[I_n(\lambda) - f(\lambda)]d\lambda$$

and under the assumption that

$$(11) \qquad \tilde{\Theta} = \hat{\Theta}_A \quad \text{if} \quad A_\Theta = \Gamma_\Theta^{-1}\dot{\phi}_\Theta$$

can be expressed in the following form:

(12) $|n^{1/2}(\hat{\theta}_A - \theta) - I_n(A_\theta/4\pi f_\theta)| \rightarrow 0$

in $P_n(f_\theta)$ probability as $n \rightarrow \infty$.   Moreover, in view of the same Theorem 2

$$\lim_{n \to \infty} E_{n,\theta}(\hat{\theta}_A - \theta)^{\otimes 2} = \frac{1}{4\pi} \int_{-\pi}^{\pi} A_\theta(\lambda)^{\otimes 2} d\lambda$$

(13)

$$+ \kappa_4 \left[\frac{1}{4\pi} \int_{-\pi}^{\pi} A_\theta(\lambda) d\lambda\right]^{\otimes 2}$$

$$= \Gamma_\theta^{-1} + \Gamma_\theta^{-1} C_{\kappa_4,\theta} \Gamma_\theta^{-1}, \quad \text{if} \quad A_\theta = \Gamma_\theta^{-1} \dot{\phi}_\theta.$$

Evidently the asymptotic relation (12) determines the estimator $\hat{\theta}_A$ up to asymptotic equivalence.[7] In particular, in (11) the estimator $\tilde{\sigma}$ can be replaced by the estimator $\vec{\sigma}$ which will be introduced in the next chapter (by means of formula (IV.1.6) with $\tau_n = n^{-1/2}$) since $\vec{\sigma}$ satisfies (III.1.15) (with $\tau_n = n^{-1/2}$).

We now present a couple of other examples of estimators for the parameter θ of spectral density f_θ satisfying (12) but with A_θ different from (9).   This is taken from Section 2 of the next chapter.

We define an estimator of an appropriate root with respect to θ of a system of equations $I_n(\mathbf{h}) = 0$ with some p-vector-valued function $\mathbf{h}$ (which like f may depend on θ). Following arguments in the spirit of those presented on page 109 based essentially on the utilization of the mean value theorem (cf. also Subsection 1.3 of the next chapter), we can assure the usual asymptotic properties of such an estimator denoted here by $\hat{\theta}_A$, provided only that the smoothness of function $\mathbf{h}_\theta$ and f_θ allows us to represent $\hat{\theta}_A$ in the form (12) with

(14) $A_\theta(\lambda) = \left[\frac{1}{4\pi} \int_{-\pi}^{\pi} \mathbf{h}_\theta(\mu) \otimes \dot{\phi}(\mu) d\mu\right]^{-1} \mathbf{h}_\theta(\lambda).$

---

[7] Actually one should talk about a family $\{\hat{\theta}_A\}$ of asymptotically equivalent estimators corresponding to a fixed A_θ in (12).

The asymptotic value of the covariance matrix of this estimator is determined from the first row of formula (13) (with the obvious modification in the case of a complex valued h, in particular in the widely used practical case

$$h(\lambda) = \text{col}\{e^{i\lambda\tau_1}, ..., e^{i\lambda\tau_p}\},$$

in which the initial system of equations becomes $\beta_n^*(\tau_j) = \beta(\tau_j)$, $0 \leqslant \tau_j \leqslant n-1$, $j = 1, ..., p$, where $\beta_n^*(\tau)$ is the empirical covariance function (4.5)).

As another example we present the estimator $\hat{\theta}_A$, defined as the appropriate root of the system of equations $\rho_n^*(\tau_j) = \rho(\tau_j)$, $j = 1, ..., p$, where $\rho(\tau) = \beta(\tau)/\beta(0)$ and $\rho^*(\tau) = \beta_n^*(\tau)/\beta_n^*(0)$ (correlation function and empirical correlation function respectively). It is clear that the same considerations based on the utilization of the mean value theorem yield the representation (12) where A satisfies (14) with a particular h of the form

$$h(\lambda) = \text{col}\left\{\int_{-\pi}^{\pi} e^{i\tau_j(\lambda-\mu)} f(\mu)d\mu, \quad j = 1, ..., p\right\}.$$

Let us turn back now to the general representation (12), the special cases of which have just been presented. It is shown in Appendix 2 to this chapter (Theorem A2.1) that if we confine ourselves to the Gaussian case, then the estimator $\hat{\theta}_A$ satisfying (12) is regular in Hajek's sense [175] iff the following condition on A holds:

(15) $\qquad \dfrac{1}{4\pi} \int_{-\pi}^{\pi} A_\theta(\lambda) \otimes \dot{\phi}_\theta(\lambda)d\lambda = I_p$

(cf. Definition A2.2 of the Hájek regularity and the subsequent discussion which emphasizes the importance of this property). In Appendix 2 the role of the condition (15) -- when the Gaussian hypothesis is deleted -- is also clarified. In view of these considerations, we shall, for brevity, refer to the estimator $\hat{\theta}_A$ represented in form (12) as a regular one provided only A satisfies (15).

We now formulate the basic result of this subsection concerning the lower bound for asymptotic dispersion of regular estimators represented in form (12).

Theorem 3. *Let the conditions of Corollary 1 be fulfilled. Then for any regular estimator* $\hat{\theta}_A$, *satisfying (12) and*

*possessing the asymptotic covariance matrix defined by the first
row in formula* (13), *the following matrix inequality is valid:*

(16) $\lim_{n \to \infty} n E_{n, \Theta}(\hat{\Theta}_A - \Theta)^{\otimes 2} \geqslant \Gamma_\Theta^{-1} + \Gamma_\Theta^{-1} C_{K_4, \Theta} \Gamma_\Theta^{-1}.$

The proof, which is very simple, is relegated to Appendix 2.
We note that this proof uses to a large extent, the
assumptions that the parameter σ^2 (cf. (1.3)) is one of the
unknowns so that the covariance matrix on the r.h.s. of (16)
has the structure (8). Since the latter matrix is the limit of
the covariance matrix for the least squares estimator $\tilde{\Theta}$, its
limiting dispersion does not exceed the limiting dispersion of
any regular estimator which is represented in the form (12).

7. Estimation by Means of the Whittle Method of Spectrum Parameters of General Processes Satisfying the Strong Mixing Condition

7.1. In the preceding section we studied the properties of
estimators $\tilde{\Theta}$ obtained using Whittle's method (i.e., least
squares estimators) under the assumption that X_t, $t = ...,-1,0,1,$
... belongs to a relatively general class of linear processes
(which includes Gaussian ones as a very special case).
However, such an assumption may be inapplicable in certain
situations.

Taking this into account, we shall discuss in the current
section the case when X_t is a stationary random process (in
the narrow sense) such that $E(|X|^{2\beta}) < \infty$ for some $\beta > 2$ while
$E(X_t) = 0$. Then for any choice of random variables
$X_{t_1}, X_{t_2}, X_{t_3}, X_{t_4}$ there exists a mixed cumulant of the fourth

order

$$c_4(t_1, ..., t_4) = E(X_{t_1} ... X_{t_4}) - E(X_{t_1} X_{t_2}) E(X_{t_3} X_{t_4})$$

$$- E(X_{t_1} X_{t_3}) E(X_{t_2} X_{t_4}) - E(X_{t_1} X_{t_4}) E(X_{t_2} X_{t_3}).$$

The spectral density of the fourth order $f_4(\lambda_1, ..., \lambda_4)$ is
determined from the equality

$$c_4(t_1, ..., t_4) = \int_{-\pi}^{\pi} \cdots \int_{-\pi}^{\pi} f_4(\lambda_1, ..., \lambda_4)$$

$$\times \exp\left\{i \sum_{k=1}^{4} t_k \lambda_k\right\} \tilde{\delta}(\lambda_1 + \cdots + \lambda_4) d\lambda_1 \ldots d\lambda_4,$$

where

$$\tilde{\delta}(\lambda) = \sum_{k=-\infty}^{\infty} \delta(\lambda + 2k\pi)$$

and $\delta(\lambda)$ is the Dirac's δ-function (cf. Appendix 1, the concluding part of Subsection 2).

Assume furthermore that process X_t satisfies Rosenblatt's mixing condition, i.e., that

$$\alpha(\tau) = \sup\{|P(B_1 B_2) - P(B_1) \cdot P(B_2)|;$$

$$B_1 \in A_{-\infty}^{t}, \ B_2 \in A_{t+\tau}^{\infty}\} \to 0$$

as $\tau \to \infty$, where A_s^t, $s \leqslant t$ is a σ^2-algebra generated by random variables $X_s, \ldots, X_t$. Moreover, we shall also assume that for $\beta > 2$ such that $E(|X_t|^{2\beta}) < \infty$ the following inequality holds:

$$\sum_{\tau=1}^{\infty} [\alpha(\tau)]^{1-2/\beta} < \infty$$

In view of the results of Appendix 1 to this chapter dealing with such processes the following theorem is valid.

Theorem 1. *Let a random process X_t be such as it is stated above, with the spectral density $f = f_\Theta$ satisfying the conditions of Theorem 2.1. Then the estimator $\tilde{\Theta}$ determined from the condition (2.2) (where $U_n(\Theta)$ is given by the formula (2.1)) is a consistent estimator, i.e., $\tilde{\Theta} \to \Theta$ in probability $P_n(f_\Theta)$.*

The proof is the same as in the case of the related Theorems 2.1 and 6.1 taking into account that in this case

$$\lim_{n\to\infty} D\left[\int_{-\pi}^{\pi} I_n(\lambda) d\lambda\right] = 4\pi \int_{-\pi}^{\pi} f^2(\lambda) d\lambda$$

$$+ 2\pi \int_{-\pi}^{\pi} \int f_4(\lambda_1, -\lambda_1, \lambda_2, -\lambda_2) d\lambda_1 d\lambda_2$$

and

$$\lim_{n\to\infty} D[n^{1/2}(U_n(\Theta) - U_n(\Theta_1))]$$

$$= \frac{1}{4\pi} \int_{-\pi}^{\pi} \left[1 - \frac{f_\Theta(\lambda)}{f_{\Theta_1}(\lambda)} \right]^2 d\lambda$$

$$+ \frac{1}{8\pi} \int_{-\pi}^{\pi} \int [f_\Theta^{-1}(\lambda_1) - f_{\Theta_1}^{-1}(\lambda_1)][f_\Theta^{-1}(\lambda_2) - f_{\Theta_1}^{-1}(\lambda_2)]$$

$$\cdot f_4(\lambda_1, -\lambda_1, \lambda_2, -\lambda_2) d\lambda_1 d\lambda_2.$$

7.2. Assume now that f_Θ satisfies the conditions of Theorem 2.2 and let us follow once again the argument presented in Subsection 6.2. It will then be clear that the processes which satisfy the conditions presented in the preceding subsection lead to the same conclusions as in the linear case, the only difference being that in view of the corresponding conclusions of Corollary A1.3, as $n \to \infty$, $\Delta_{n,\Theta}$ possesses a normal distribution with zero expectation, but with a different variance which is equal to $\Gamma_\Theta + C_{f_4,\Theta}$ where

$C_{f_4,\Theta}$ is a $(p \times p)$-matrix whose (k,ℓ)-th term is of the form

$$\frac{1}{8\pi} \int\int_{-\pi}^{\pi} f_4(\lambda_1, -\lambda_1, \lambda_2, -\lambda_2) \frac{\partial}{\partial \Theta_k} \frac{1}{f_\Theta(\lambda_1)}$$

$$\cdot \frac{\partial}{\partial \Theta_\ell} \frac{1}{f_\Theta(\lambda_2)} d\lambda_1 d\lambda_2.$$

From here and the relation (2.20) (which is valid in this case) follows the theorem.

Theorem 2. *Let the random process X_t be as it is stated in Subsection 1 and let its spectral density $f = f_\Theta$ satisfy the conditions of Theorem 2.2. Then Whittle's estimator $\hat{\Theta}$ of parameter Θ is asymptotically normal and the distribution of the random vector $n^{1/2}(\hat{\Theta}-\Theta)$ as $n \to \infty$ converges to the normal distribution $N(0, \Gamma_\Theta^{-1} + \Gamma_\Theta^{-1} C_{f_4,\Theta} \Gamma_\Theta^{-1})$.*

Appendix 1

1. In this appendix, the necessary results about the asymptotic normality of statistics commonly appearing in this book, which are quadratic forms in observations, are collected. In this subsection the case of Gaussian observations is considered while in Subsections 2 and 3 the observed

process is assumed to be linear and strongly mixing, respectively.

Let $\mathbf{X}_N$ be a random vector possessing an N-dimensional Gaussian distribution with zero mean and positive definite covariance matrix $E\mathbf{X}_N\mathbf{X}_N' = B_N$.

Consider a positive definite quadratic form $\mathbf{X}_N'A_N\mathbf{X}_N$. As it is known (cf., e.g., [37], p. 273 or [66]), the characteristic function of the normalized random variable

(1) $\xi_N = \{D(\mathbf{X}_N'A_N\mathbf{X}_N)\}^{-1/2}\{\mathbf{X}_N'A_N\mathbf{X}_N - E(\mathbf{X}_N'A_N\mathbf{X}_N)\}$

where

(2)
$$E(\mathbf{X}_N'A_N\mathbf{X}_N) = \mathrm{tr}(A_NB_N),$$
$$D(\mathbf{X}_N'A_N\mathbf{X}_N) = 2\mathrm{tr}[(A_NB_N)^2],$$

is of the form

$$\psi_N(x,\xi_N) = E[\exp(ix\xi_N)]$$
$$= \exp[-iy\ \mathrm{tr}(A_NB_N)][\det(I_N-2iyA_NB_N)]^{-1/2}$$

where $y = x\{2\mathrm{tr}[(A_NB_N)^2]\}^{-1/2}$.

Consequently,

$$\log\psi_N(x,\xi_N) + \frac{x^2}{2} = -\frac{1}{2}\ U_N(-2iyA_NB_N) \to 0$$

provided only

1) $\{\mathrm{tr}(A_NB_N)^2\}^{-1/2}|A_NB_N|$ is bounded,

2) $\{\mathrm{tr}(A_NB_N)^2\}^{-1/2}\|A_NB_N\| \to 0$ as $N \to \infty$

(cf. Lemma A1.1 presented in Appendix 1 to Chapter I). This fact can be stated in the form of the following proposition.

Proposition A1.1. *Under conditions 1) and 2) the sequence of distributions* $L(\xi_N)$, $N = 1,2, \dots$ *converges to the standard normal distribution* $N(0,1)$.

We apply these rather simple arguments to the particular case under consideration when the components X_N are the observations $(X_1, ..., X_n) = \mathbf{X}_n$ generated by a stationary Gaussian process X_t with zero mean and covariance function $\beta(\tau)$. (Here evidently, the dimensionality N of vector $\mathbf{X}$ is the "sample size" n, and $B_N = B_n$ is a Toeplitz matrix $[\beta(\tau-s), \tau,s = 1, ..., n]$ of dimension n). Assume that the covariance function is square summable

(3) $\qquad \sum_{\tau=-\infty}^{\infty} |\beta(\tau)|^2 < \infty.$

First we shall verify the validity of 1) and 2) for the case of special matrix

$$A_N = A_n = \sum_{k,j=1}^{m} a_k a_j \, I_{|t_k-t_j|},$$

where $a_1, ..., a_m$ are fixed real numbers, $t_1, ..., t_m$ are integers such that $0 \leqslant |t_k-t_j| < n$, and

$$I_t = \frac{1}{2}\left(\begin{bmatrix} 0 & I_{n-t} \\ 0 & 0 \end{bmatrix} + \begin{bmatrix} 0 & 0 \\ I_{n-t} & 0 \end{bmatrix}\right).$$

We are interested in this particular case because here

$$\mathbf{X}_N' A_N \mathbf{X}_N = \mathbf{X}_n' A_n \mathbf{X}_n = n \sum_{k,j=1}^{m} a_k a_j \beta_n^*(t_k-t_j),$$

where

$$\beta_n^*(\tau) = \beta_n^*(-\tau) = \begin{cases} \dfrac{1}{n} \sum_{j=1}^{n-\tau} X_j X_{j+\tau}, & t = 0,1, ..., n-1 \\[2mm] 0, & \tau = n,n+1, ... \end{cases}$$

is the empirical covariance function so that the assertion about the asymptotic normality of this quadratic form is equivalent to the similar assertion concerning empirical covariances (cf. Theorem A1.1 below).

Clearly for any $\mathbf{x} = \mathrm{col}(x_1, ..., x_n)$ with $|\mathbf{x}| = 1$,

$$|\mathbf{x}' I_t \mathbf{x}| = |\sum_{j=1}^{n-t} x_j x_{j+t}| \leqslant \left[\sum_{j=1}^{n-t} x_j^2 \sum_{j=1}^{n-t} x_{j+t}^2\right]^{1/2} \leqslant 1,$$

so that $\|I_t\| \leqslant 1$, $\|A_n\| \leqslant (\sum_{k=1}^{m}|a_k|)^2$. Moreover, in view of

(3), $n^{-1/2}\|B_n\| \to 0$ and $n^{-1/2}|B_n| \to \gamma$ so that

$$n^{-1/2}|A_n B_n| \leqslant n^{-1/2}\|A_n\| \cdot |B_n| < \infty$$

and

$$n^{-1/2}\|A_n B_n\| \leqslant n^{-1/2}\|A_n\| \cdot \|B_n\| \to 0.$$

Thus the conditions 1) and 2) are verified by means of the following relation

$$2n^{-1}\mathrm{tr}[(A_n B_n)^2] = nD\left[\sum_{k,\,j=1}^{m} a_k a_j \beta_n^*(t_k - t_j)\right]$$

$$\to \sum_{t=-\infty}^{\infty}[b_t^2 + b_t b_{-t}],$$

where

$$b_t = \sum_{k,\,j=1}^{m} a_k a_j \beta(t + |t_k - t_j|).$$

Hence we have the following corollary of Proposition A1.1.

Theorem A1.1. *Let X_t be a standard Gaussian process with zero mean and covariance function $\beta(\tau)$ satisfying condition (3).*
   Then the distribution of the vector

(4) $n^{1/2}\mathrm{col}\{\beta_n^*(t) - \beta(t), \quad t = t_1, ..., t_m\}$

($0 \leqslant t_j \leqslant n-1$) converges as $n \to \infty$ to the m-dimensional normal distribution with zero mean and covariance structure given by the relation

(5)
$$n\,\mathrm{cov}\{\beta_n^*(k), \beta_n^*(\ell)\} \to \sum_{t=-\infty}^{\infty}[\beta(t+|k|)\beta(t+|\ell|)$$
$$+ \beta(t-|k|)\beta(t+|\ell|)].$$

The assertion of this theorem and its generalizations for the cases without the Gaussian restriction (cf. succeeding sections) is of special interest in applications -- indeed in the parametric models which are usually considered (these are discussed in detail in succeeding chapters) the empirical covariance function is the basis of the construction of almost all the estimators and tests which are important in applications.

We now turn to the derivation of the corollaries of Proposition A1.1, which are especially important in the present context.

Theorem A1.2. *Let X_t be a Gaussian process with zero mean generating the observations $(X_1, ..., X_n) = \mathbf{X}_n$ with a positive definite covariance matrix B_n. Let a be, as usual, square integrable function such that*

(6) $$\gamma^2 = \frac{1}{2\pi}\int_{-\pi}^{\pi} a^2(\lambda)d\lambda > 0$$

Then

1)   *as $n \to \infty$*

(7) $$\mathfrak{G}_n = \frac{1}{2}n^{-1/2}\left\{\mathbf{X}_n' B_n^{-1} B_{a/2\pi}\mathbf{X}_n - \frac{n}{2\pi}\int_{-\pi}^{\pi} a(\lambda)d\lambda\right\}$$

   is normally distributed $N(0, \gamma^2/2)$.

   Let the additional conditions of Lemma I.3.1 be valid also. Then

2)   *assertion 1) is valid also for*

(8) $$\mathfrak{G}_n = \frac{1}{2}n^{-1/2}\{\mathbf{X}_n' B_{a/(2\pi)^2 f}\mathbf{X}_n - \frac{n}{2\pi}\int_{-\pi}^{\pi} a(\lambda)d\lambda\}$$

Proof. Assertion 1) follows from conditions 1) and 2) and assertions 3) and 4) of Lemma A1.2 in Appendix 1 to Chapter I. Next, applying Lemma I.3.1 we arrive also at assertion 2). □

We shall use the notation and conditions presented in Section 3 of Chapter I. We say that a sequence of Gaussian distributions $P_n(f)$, $n = 1,2, ...$, possesses the local asymptotic normality (LAN) if for any square integrable a the following relation holds

(9) $\Lambda(f,g) - \mathfrak{G}_n + \gamma^2/4 \to 0$ as $n \to \infty$

in $P_n(f)$ probability. (Recall that $\Lambda(f,g)$ is the logarithm of the likelihood ratio of measure $P_n(g) = P_n(f + n^{-1/2}fa)$ with respect to $P_n(f)$). Here $\mathfrak{G}_n$ is a sequence of random variables such that

(10) $\{\delta_n|P_n(f)\} \rightarrow N(0,\gamma^2/2).$

It should be noted, however, that the term introduced above is usually used in connection with a sequence of parametric families of distributions (cf., e.g., [70,80,110] and also Section 1 of the next Appendix) such as, for example, the sequence of Gaussian distributions $\{P_n(f_\Theta), \Theta \in \Theta\}$, $n = 1,2, \dots$. In the usual sense of this term, the last family is called locally asymptotically normal (at a fixed point Θ) provided there exist a p-dimensional random variable $\Delta_{n,\Theta}$ and a positive definite $(p{\times}p)$-matrix Γ_Θ such that (9) and (10) are valid with $\delta_n = \mathbf{h}'\Delta_{n,\Theta}$ and $\gamma^2/2 = \mathbf{h}'\Gamma_\Theta\mathbf{h}$ for any p-dimensional vector $\mathbf{h}$.

Taking into account the results of Section 3 of Chapter I we obtain from Theorem A1.2 the following corollary.

Corollary A1.1. *Let the conditions presented at the beginning of Section 3 of Chapter* I *prior to the statement of Theorem 1 be valid. Then*

1)   *The sequence of Gaussian distributions $P_n(f)$, $n = 1,2, \dots$, possesses the* LAN *property with δ_n given by the relation* (7), *i.e., for such a δ_n* (9) *and* (10) *are satisfied.*

   Let the conditions of Lemma I.3.1 *be fulfilled also. Then*

2)   *Assertion* 1) *is valid for δ_n of the form* (8).

   Next, let the spectral density of the observed Gaussian process belong to the family $\{f_\Theta, \Theta \in \Theta\}$ and, moreover, let the conditions presented in the statement of Corollary I.3.1 *before assertions* 1) *and* 2) *be fulfilled. Then*

3)   *The family $\{P_n(f_\Theta), \Theta \in \Theta\}$ possesses the* LAN *property with*

(11)
$$\Delta_{n,\Theta} = \mathrm{col}\left\{\frac{1}{2}n^{-1/2}(\mathbf{X}_n'B_{f_\Theta}^{-1}\dot{B}_{\Phi_{k,\Theta}}/2\pi\, \mathbf{X}_n \right.$$
$$\left. -\frac{n}{2\pi}\int_{-\pi}^{\pi}\dot{\Phi}_{k,\Theta}(\lambda)d\lambda), \quad k = 1, \dots, p\right\}$$

   and

$$(12) \qquad \Gamma_\Theta = \left[\frac{1}{4\pi} \int_{-\pi}^{\pi} \dot{\Phi}_{k,\Theta}(\lambda) \dot{\Phi}_{\ell,\Theta}(\lambda) d\lambda, \quad k,\ell = 1, ..., p \right].$$

(Recall that $\dot{\Phi}_{k,\Theta}$ is the logarithmic derivative of f_Θ with respect to the k-th component of Θ in the L_2 sense.)

Finally, let the conditions of Lemma I.3.1 for $a = h'\dot{\Phi}_\Theta$ be fulfilled also. Then

4) Assertion 3) is valid for $\mathbf{\Delta}_{n,\Theta}$ of the form (I.3.23).

To conclude this subsection we shall briefly discuss the possibility of extending the above stated considerations to the case -- which is permissible in Section 3 (as well as in the corresponding subsections of the preceding chapter) -- of complex-valued observations.

First, we note that Proposition A1.1 remains valid when applied to

$$\xi_N = \{D(\mathbf{X}_N^* A_N \mathbf{X}_N)\}^{-1/2} \{\mathbf{X}_N^* A_N \mathbf{X}_N - E(\mathbf{X}_N^* A_N \mathbf{X}_N)\},$$

where $\mathbf{X}_N$ is an N-dimensional complex-valued Gaussian random variable with zero mean and Hermitian covariance matrix $B_N = E\mathbf{X}_N \mathbf{X}_N^*$ (with the additional convention that $E\mathbf{X}_N \mathbf{X}_N' = 0$), A_N is a Hermitian matrix and

$$E(\mathbf{X}_N^* A_N \mathbf{X}_N) = \mathrm{tr}(A_N B_N), \quad D(\mathbf{X}_N^* A_N \mathbf{X}_N) = \mathrm{tr}[(A_N B_N)^2]$$

(cf. [166]).

Therefore, the assertion 1) of Theorem A1.2 has a natural generalization to the case of a complex-valued Gaussian process X_t generating observations $\mathbf{X}_n = (X_1, ..., X_n)$ with Hermitian covariance matrix $B_n = E\mathbf{X}_n \mathbf{X}_n^*$.

As far as an assertion analogous to 2) of the same theorem is concerned then, under the condition and notation given in Section 4 of Chapter I for

$$(13) \qquad \delta_n = \frac{1}{2} n^{-1/2} \{ \tilde{\mathbf{Y}}_n^* B_{a/(2\pi)^2 f_0} \tilde{\mathbf{Y}}_n - \frac{n}{2\pi} \int_{-\pi}^{\pi} a(\lambda) d\lambda \}$$

we have (10) (where obviously $f = f_q$). More precisely, the following theorem with the corresponding corollary are valid.

Theorem A1.3. Let the observations $\mathbf{X}_n = (X_1, ..., X_n)$ generated by a complex-valued Gaussian process X_t with zero mean possess *positive definite covariance matrix* $EX_nX_n^* = B_n$. *Then*

1) *for*

(14) $$\delta_n = \frac{1}{2} n^{-1/2} \left\{ X_n^* B_n^{-1} B_{a/2\pi} X_n - \frac{n}{2\pi} \int_{-\pi}^{\pi} a(\lambda) d\lambda \right\}$$

assertion 1) *of Theorem A1.2 is valid;*

2) *under the validity of assertion 4) of Theorem I.4.1 the same conclusion is valid for* δ_n *of the form* (13).

Corollary A1.2.

1) *Under the condition and notation given in the beginning of Section 4 of Chapter I prior to the statement of Theorem 1, a sequence of Gaussian distributions* $P_n(f)$, $n = 1,2,...$, *possesses the LAN property with* δ_n *given by the relation* (14).

2) *If moreover, additional conditions preceding assertion 4) in the statement of Theorem I.4.1 are fulfilled, then the concluding part of the previous assertion is valid also for* δ_n *of the form* (13).

3) *Let the spectral density of the observed process depend on the parameter* $f = f_\Theta$, $\Theta \in \Theta$ *in the manner indicated in the beginning of Subsection 4.2 of Chapter I. Furthermore, let conditions presented in Corollary I.4.1 prior to assertions 1) and (3) be fulfilled. Then the family of distributions* $\{P_n(f_\Theta), \Theta \in \Theta\}$ *possesses the LAN property with* $\Delta_{n,\Theta}$ *and* Γ_Θ *of the form* (11) *(with the obvious replacement of the sign* ' *by* *)* *and* (12) *respectively.*

4) *If, in addition, the conditions of assertion 3) of Corollary I.4.1 are also valid then in the concluding part of the preceding proposition,* $\Delta_{n,\Theta}$ *may have components given by the formula* (I.4.22).

2. By eliminating the confining hypothesis of the Gaussian nature of the observed series we follow, in Section 6, the long established tradition in analysis of time series and assume that X_t is a linear process of the form (6.1) with a sequence of independent identically distributed random variables $\{\varepsilon_t, t = ...,-1,0,1, ...\}$ such that $E(\varepsilon_t) = 0$, $E(\varepsilon_t^2) = \sigma^2 > 0$, and $E(\varepsilon_t^4) < \infty$. This is due, first of all, to the fact that many of the conclusions arrived at under Gaussian assumptions are either independent of the type of distribution of ε's or are simply

generalized to the case of linearity. Moreover, a number of actual problems which occur in the case of such a generalization (such as the problem of the asymptotic efficiency and robustness of estimators or that of the feasibility of adaptive estimation) have a simple, easily interpreted solution (cf. [125,172,177,178]).

One should, however, bear in mind that the assumption of independence of the variables ε_t is difficult to verify in practice. Therefore the fact that this assumption can also be relaxed in the case of investigating certain problems is indeed of some importance. Namely, it can be assumed that $\{\varepsilon_t, t = ...-1,0,1, ...\}$ is a sequence of martingale differences

(15) $E(\varepsilon_t \mid \mathfrak{U}_{t-1}^{(\varepsilon)}) = 0$ a.s. for all t

where $\mathfrak{U}_t^{(\varepsilon)}$ is the σ-algebra generated by ε_s, $s \leqslant t$.

Indeed, the last assumption lies somewhere between the assumption that ε's are uncorrelated (which as stipulated in the representation of X_t in the form of moving averages) and the assumption of independence which is valid in Section 6.[8]

Following Hannan [142] (see also [39], remark on page 492), we shall assume in addition to (15) the validity of the following conditions

(16) $E(\varepsilon_t^2 \mid \mathfrak{U}_{t-1}^{(\varepsilon)}) = \sigma^2 > 0$ a.s.,

(17) $E(\varepsilon_t^3 \mid \mathfrak{U}_{t-1}^{(\varepsilon)}) = \text{const}$ a.s.

---

[8]Actually the assumption that ε_t are uncorrelated is expressed as $E(\varepsilon_t \varepsilon_s) = E(\varepsilon_t)E(\varepsilon_s)$, $t \neq s$, while the condition (15) can be restated as follows: if $\phi(\varepsilon_{t-1}, \varepsilon_{t-1}, ...)$ is a $\mathfrak{U}_{t-1}^{(\varepsilon)}$-measurable function of its variables then

$$E[\varepsilon_t \phi(\varepsilon_{t-1}, \varepsilon_{t-2}, ...)] = E(\varepsilon_t)E[\phi(\varepsilon_{t-1}, \varepsilon_{t-2}, ...)] = 0.$$

This is stronger than the absence of correlation. On the other hand, the indepenence of ε's is equivalent to the even stronger condition that for any ϕ (the same as above) and any measurable functions $\psi(\varepsilon_t)$ of ε_t the equality

$$E[\psi(\varepsilon_t)\phi(\varepsilon_{t-1}, \varepsilon_{t-2}, ...)] = E[\psi(\varepsilon_t)]E[\phi(\varepsilon_{t-1}, \varepsilon_{t-2}, ...)]$$

is valid.

and

(18) $E(\varepsilon_t^4) < \infty.$

Denote by κ_4 the coefficient of excess $E(\varepsilon_t^4)\sigma^{-4}$ - 3 of the variables ε_t. Obviously, if ε_t are Gaussian, $\kappa_4 = 0$.

Under these conditions the following theorem is valid.

Theorem A1.4. *Let a process X_t, stationary in the narrow sense, be represented in form (6.1) with the same coefficients as in Section 6, but with the sequence $\{\varepsilon_t, t = \ldots -1,0,1, \ldots\}$ satisfying the conditions (15)-(18) now. Then condition (3) is necessary and sufficient for the assertion stated in the conclusion of Theorem A1.1 with the new summands $\kappa_4\beta(k)\beta(\mathbf{l})$ appearing on the r.h.s. of (5).*

The proof presented in [142] essentially reduces to obtaining the limiting distribution of a linear combintion of a finite number of expressions of the form

$$S_n = \sum_{t=1}^{n} [\varepsilon_t \varepsilon_{t-t_k} - \sigma^2 \delta_{0t_k}]$$

(with corresponding obvious normalization by the factor $n^{-1/2}$). Since $\{S_n, \, \mathfrak{U}_n^{(\varepsilon)}; \, n = 1,2, \ldots\}$ is a martingale this distribution turns out to be Gaussian.[9]

As far as the limiting covariance structure is concerned, it is easily verified by direct calculations: under condition (3), up to asymptotic terms which can be neglected,[10] we have

(19)
$$n \, \text{cov} \, \{\beta_n^*(k),\beta_n^*(\mathbf{l})\} = \sum_{t=-n}^{n} \left[1 - \frac{|t|}{n}\right][\beta(t)\beta(t + |k-\mathbf{l}|) + \beta(t + |k|)\beta(t - |\mathbf{l}|) + c_4(0,k,t,t+\mathbf{l})],$$

where

---

[9] The martingale central limit theorem due to Billingsley [173] is used to establish this fact in [142]. This theorem assumes ergodicity of the sequence $\{\varepsilon_t, t = \ldots -1, 0, 1, \ldots\}$. Actually one can utilize here the theorem given in [174] which allows us to eliminate the last requirement (cf. also [162], p. 193).

[10] The exact expression is given for example on page 464 of the book [4]; it is cumbersome and of no special interest.

$$c_4(t_1, ..., t_4) = E(X_{t_1} \cdots X_{t_4}) - E(X_{t_1} X_{t_2}) \cdot E(X_{t_3} X_{t_4})$$

$$- E(X_{t_1} X_{t_3}) \cdot E(X_{t_2} X_{t_4}) - E(X_{t_1} X_{t_4}) \cdot E(X_{t_2} X_{t_3})$$

is the cumulant of the fourth order of the quantities involved (see the general definition just below). Thus, when in the r.h.s. of (19) one can pass to the limit as $n \to \infty$ (cf. [4], Theorem 8.3.3), we then have

(20)
$$n \, \text{cov}\{\beta_n^*(k), \beta_n^*(\ell)\} \to \sum_{t=-\infty}^{\infty} \{\beta(t+|k|)\beta(t+|\ell|)$$

$$+ \, \beta(t+|k|)\beta(t-|\ell|) + c_4(0,k,t,t+\ell)\}$$

$$= 2\pi \int_{-\pi}^{\pi} (e^{i\lambda(k-\ell)} + e^{i\lambda(k+\ell)}) f^2(\lambda) d\lambda$$

$$+ 2\pi \int \int_{-\pi}^{\pi} e^{i\lambda_1 - i\lambda_2 \ell} f_4(\lambda_1, -\lambda_1, \lambda_2, -\lambda_2) d\lambda_1 d\lambda_2,$$

where f_4 is the spectral density of the 4th order of the process $\dot{X}_t$. It remains only to verify that (cf., e.g., [17])

(21)
$$f_4(\lambda_1, -\lambda_1, \lambda_2, -\lambda_2) = \frac{\kappa_4}{2\pi} f(\lambda_1) \cdot f(\lambda_2).$$

To conclude this subsection we recall that the cumulant of the m-th order of the variables $X_{t_1}, ..., X_{t_m}$, $\text{cum}\{X_{t_1}, ..., X_{t_m}\} = c_m(t_1, ..., t_m)$ is by definition the coefficient at $(i)^m x_1 \cdots x_m$ in the Taylor expansion of the logarithm of characteristic function $E[\exp(ix_1 X_{t_1} + \cdots + ix_m X_{t_m})]$. A periodic function f_m (the spectral density of the m-th order) is associated with it.[11] This function is concentrated on the manifold $\lambda_1 + \cdots + \lambda_m = 0$ and is such that

---

[11] Evidently $c_1(t_1) = E(X_{t_1})$ and $c_2(t_1, t_2) = \text{cov}(X_{t_1}, X_{t_2})$ so that for our process X_t, $c_1(t_1) = 0$ and $c_2(t_1, t_2) = \beta(t_1 - t_2)$; under the additional Gaussian assumption on X_t all the higher order cumulants equal zero. Thus in this sense the higher order cumulants characterize the "degree of deviation from Gaussian assumption." Obviously $f_2(\lambda_1, \lambda_2)$ as a function of a single variable on the line $\lambda_1 + \lambda_2 = 0$ is the ordinary spectral density of the process X_t.

$$c_m(t_1, ..., t_m) = \int_{-\pi}^{\pi} \cdots \int exp\left[i \sum_{j=1}^{m} \lambda_j t_j\right) \cdot$$

$$\cdot \delta(\lambda_1 + \cdots + \lambda_m) f_m(\lambda_1, ..., \lambda_m) d\lambda_1 ... d\lambda_m,$$

where $\delta(\lambda)$ is the Dirac's δ-function. A sufficient condition for the existence of f_m

$$\sum_{t_1, ..., t_{m-1}} |c_m(0, t_1, ..., t_{m-1})| < \infty$$

guarantees its boundedness and uniform continuity on the manifold $\lambda_1 + \cdots + \lambda_m = 0$.

3. We shall continue to assume that the process X_t is stationary in the narrow sense but in this subsection we shall assume analogously to Section 7 that the interdependence among the terms of the series decreases as the distance between them increases. More precisely, we shall assume that the observed process X_t is strongly mixing in the sense of the following definition [109].

Definition A1.1. A process ξ_t, stationary in the narrow sense is strongly mixing if

(22) $\alpha_\tau^\xi = \sup\{|P(AB)-P(A)P(B)|; A \in A_{-\infty}^t, B \in A_{t+\tau}^\infty\} \to 0$

as $\tau \to \infty$, where A_s^t, $s \leqslant t$ is the σ-algebra generated by the variables $\xi_s, ..., \xi_t$. The quantity α_τ^ξ characterizing the "speed of mixing" is called the mixing coefficient.

The following central limit theorem due to Ibragimov [167] is the basis for derivations in this subsection (see also [169], pp. 346-351).

Proposition A1.2. *Let a process, stationary in the narrow sense with zero mean, ξ_t be strongly mixing in the sense of the definition given above. Let $E|\xi_t|^{2+\delta} < \infty$ for some $\delta > 0$. If*

$$\sum_{\tau=1}^{\infty} [\alpha_\tau^\xi]^{\delta/(2+\delta)} < \infty,$$

then

$$\sigma^2 = E(\xi_0^2) + 2 \sum_{j=1}^{\infty} E(\xi_0 \xi_j) < \infty$$

and if $\sigma^2 \neq 0$ then $(\xi_1 + \cdots + \xi_n)/\sigma n^{1/2}$ as $n \to \infty$ possesses the standard normal distribution $N(0,1)$.

Our aim is to apply this theorem to the case in which

$$(23) \qquad \xi_t = \sum_{k=1}^{m} a_k [X_t X_{t+t_k} - \beta(t_k)],$$

where $a_1, ..., a_m$ are arbitrary real numbers, $t_1, ..., t_m$ are the same as in (4), and X_t is the observed process strongly mixing as indicated above with correlation function $\beta(\tau)$. If α_t^X is the coefficient of mixing of X_t, then α_t^ξ for the process (23) satisfies the inequality $\alpha_t^\xi \leqslant \alpha_{t+N}^X$ for a fixed N determined by the values of $t_1, ..., t_m$. Therefore the process (23) is also strongly mixing. Moreover, let

(i) $E|X_t|^{2\beta} < \infty$ for some $\beta > 2$ and

(ii) $\sum_{\tau=1}^{\infty} [\alpha_\tau^X]^{1-2/\beta} < \infty.$

Then the process (23) satisfies the corresponding conditions of Proposition A1.2 so that by applying the Cramér-Wold device (cf. [170], p. 48) we arrive at the following theorem.

Theorem A1.5. *Let a process X_t, stationary in the narrow sense, be strongly mixing and satisfy the conditions* (i) *and* (ii) *stated above. Then as $n \to \infty$ the vector* (4) *possesses a normal distribution with covariance structure defined by the relation* (20).

4. Here we present a corollary (important in the context of this book) of the two last theorems and of Theorem A1.2 (cf. Corollary A1.2 below). Several preliminary facts are stated in the form of the following two lemmas:

Lemma A1.1.

1) *Let the Fourier coefficients $\beta(\tau)$ and $\rho(\tau)$ of function f and h respectively, satisfy the conditions*

$$\sum_{j=1}^{\infty} j|\beta(j)|^2 < \infty, \qquad \sum_{j=1}^{\infty} j|\rho(j)|^2 < \infty.$$

Then if we denote

$$I_n(h) = n^{-1/2}[X'B_{h/2\pi}X - tr(B_{hf})]$$

(24)

$$= n^{1/2}\int_{-\pi}^{\pi} h(\lambda)[I_n(\lambda) - f(\lambda)]d\lambda$$

(where $X = \text{col}(X_1, ..., X_n)$ is as usual a vector of observations over process X_t with spectrum f), then $E[I_n(h)] = O(n^{-1/2})$.

2) If $f \in L_2$ and h is a function of bounded variation, then $E[I_n(h)] = o(1)$.

3) If $h \in L_1$ and f satisfy the Lipschitz condition of order $1/2 + \alpha$, $\alpha > 0$, then $E[I_n(h)] = O(n^{-\alpha})$.

Proof. Evidently

$$E[I_n(h)] = n^{-1/2}tr(B_f B_{h/2\pi} - B_{hf})$$

(25)

$$= n^{1/2}\int_{-\pi}^{\pi} h(\lambda)\{E[I_n(\lambda)] - f(\lambda)\}d\lambda$$

$$= n^{1/2}\int_{-\pi}^{\pi} h(\lambda)[\sigma_n(\lambda;f) - f(\lambda)]d\lambda$$

(cf. (I.2.13)).

Assertion 1) is a corollary of the arguments utilized in the course of the proof of the first of the assertions of Lemma A1.4 presented in Chapter I.

If in the condition of assertion 2) var$h$ is the total variation of function h, then $|\rho(j)| \leqslant$ var $h/|j|$, $j \neq 0$ (cf. [65], p. 48) so that the absolute value of the expression

$$tr[B_{2\pi h} - B_f B_h] = \sum_{j=1}^{\infty} \min(n,j)\beta(j)\rho(j)$$

does not exceed

$$2 \text{ var } h \left[\sum_{0 \leqslant j \leqslant \sqrt{n}} |\beta(j)| + \sum_{\sqrt{n} < j \leqslant n} |\beta(j)| + n \sum_{j>n} |\beta(j)|/j \right]$$

$$\leqslant 2 \text{ var}h \left[n^{1/4} \left[\sum_{j=1}^{\infty} |\beta(j)|^2 \right]^{1/2} \right]$$

$$+ n^{1/2} \left[\sum_{j>\sqrt{n}} |\beta(j)|^2 \right]^{1/2} + n^{1/2} \left[\sum_{j>n} |\beta(j)|^2 \right]^{1/2} \right]$$

$$= o(n^{1/2}).$$

Assertion 2) is thus proved.

If f satisfies the conditions of assertion 3) then by Theorem 3.15 on page 31 of the book [65] $n^{1/2}[\sigma_n(\lambda;f) - f(\lambda)] = O(n^{-\alpha})$ uniformly in λ so that from the last equality in (25) $E[I_n(h)] = O(n^{-\alpha})$ for $h \in L_1$. Assertion 3) and thus Lemma A1.1 are proved.   □

Lemma A1.2. *Let time series X_t be such that the vector (4) constructed from it possesses as $n \to \infty$ a normal distribution with zero mean and covariance structure given by formula (20). Then the sequence $I_n(h) = I_n(h) - E[I_n(h)]$, $n = 1, 2, \ldots$ (cf. (24)), where h is a continuous function, converges to a random variable possesing a normal distribution with zero mean and variance*

$$(26) \quad \begin{aligned} & 2\pi \int_{-\pi}^{\pi} h^2(\lambda) f^2(\lambda) d\lambda + 2\pi \int_{-\pi}^{\pi} h(\lambda) h(-\lambda) f^2(\lambda) d\lambda \\ & + 2\pi \int \int_{-\pi}^{\pi} h(\lambda_1) h(\lambda_2) f_4(\lambda_1, -\lambda_1, \lambda_2, -\lambda_2) d\lambda_1 d\lambda_2. \end{aligned}$$

The proof of this lemma actually appears in a number of papers (cf., e.g., [39,165,171]); therefore we shall indicate very briefly the course of the arguments utilized therein.

First, we observe that by Féjer's theorem ([65], p. 89) $\sigma_n(\lambda,h)$ uniformly converges to h (cf. (I.2.13)). Thus given an ε one can choose N so large that $|\delta_N(\lambda)| \leqslant \varepsilon$, where $\delta_N(\lambda) = h(\lambda) - \sigma_N(\lambda,h)$. Now the asymptotic normality of vector (4) easily implies the asymptotic normality of $I_n(\sigma_N)$ for a fixed N. The proof is then completed by verifying the bound $E[I_n(\delta_N)]^2 < C\varepsilon^2$ where the constant C does not depend on n and N (for this purpose we utilize (19) or more precisely the formula for $\text{cov}\{\beta_n^*(k), \beta_n^*(\ell)\}$ presented, for example on page 464 of the book [4]).

Corollary A1.3. *Let the process X_t which is stationary in the narrow sense be either (1) Gaussian[12] satisfying the conditions of the Theorem A1.1, (2) linear satisfying the conditions of the Theorem A1.4, or (3) strongly mixing satisfying the conditions of the Theorem A1.5.*

Let the spectral density f of the process X_t and a certain continuous function h satisfy one of the conditions 1)-3) of

---

[12] A uniform version of the assertion of this corollary concerning the Gaussian case can be found in the recent paper by R. Z. Hasminskii and I. A. Ibragimov (1985), Asymptotically efficient nonparametric estimation of functionals of a spectral density function (submitted to Z. Wahr. theorie und verw. Gebiete).

Lemma A1.1.
Then the sequence of random variables $I_n(h)$, $n = 1, 2, ..., (cf.$
(24)) is weakly convergent to the random variable possessing the
Gaussian distribution with zero mean and variance given by
expression (26). Moreover, in the case of the Gaussian process X_t
the last summand in this expression equals 0 *since* $f_4 = 0$. *In*
the case of a linear process X_t, *this summand equals*

$$\kappa_4 \left[\int_{-\pi}^{\pi} h(\lambda) f(\lambda) d\lambda \right]^2$$

in view of relation (21).

Appendix 2

1. In this subsection we shall briefly present the most
essential facts about general asymptotic theory (selected
mainly from the book [70] where additional details and
appropriate references can be found; cf. also [110]) dealing
with a given sequence of "experiments"

(1) $E_n = \{X_n, \mathfrak{U}_n, P_{n, \Theta}\}$, $\Theta \in \Theta$, $n = 1, 2, ... ,$

where X_n is the space of all possible outcomes of the n-th
experiment E_n, $\mathfrak{U}_n$ is the σ-algebra defined on X_n, and $\{P_{n, \Theta}, \Theta$
$\in \Theta\}$ is a parametric family of distributions on $\mathfrak{U}_n$ possessing
the LAN property in the sense of the following definition.[13]

Definition A2.1. A parametric family of distributions $\{P_{n, \Theta}, \Theta \in \Theta\}$, $\Theta \subset R_p$ is called locally asymptotically normal (at a
fixed point Θ) if there exists a sequence of p-dimensional
random vectors $\Delta_{n, \Theta}$, $n = 1, 2, ...,$ and a positive definite
matrix Γ_{Θ} of dimension p such that

1)   for some increasing sequence of positive numbers τ_n, $n =$

---

[13] It is useful to trace the connection between the LAN and the "asymptotic
differentiability" of the family of distribuitons $\{P_{n, \Theta}, \Theta \in \Theta\}$ defined on page
20 of the Introduction (conditions (D1)-(D4)). Condition 1) in Definition
A2.1 coincides with (D2) and conditions 1) and 2) imply (D1) in view of
Proposition 6.1 in the book [110] Chapter II. Conversely, under conditions
(D1)-(D4), Lemma 4 presented in Subsection 1.4 of the next chapter is
valid.

1,2, ...

$$\Lambda(\Theta, \Theta+h/\tau_n) - h'\Delta_{n,\Theta} + (1/2)h'\Gamma_\Theta h \to 0$$

in $P_{n,\Theta}$ probability as $n \to \infty$, for any p-dimensional vector vector h (recall that $\Lambda(\Theta_1,\Theta_2) = \log[dP_{n,\Theta_2}/dP_{n,\Theta_1}]$);

2) a sequence of distributions of vectors $\Delta_{n,\Theta}$, $n = 1,2, ...,$ converges to a normal distribution with mean zero and covariance matrix Γ_Θ:

$$L\{\Delta_{n,\Theta} \mid P_{n,\Theta}\} \to N(0,\Gamma_\Theta).$$

It is particularly important in the context of this book that in view of the corresponding assertions of the Corollary A1.1, the arguments presented in the present subsection are directly applicable to the particular case in which the n-th experiment E_n is generated by the observations $X_1, ..., X_n$ over a stationary Gaussian process X_t, $t = ...-1,0,1, ...,$ with zero mathematical expectation and spectral density belonging to the parametric family $\{P_{n,\Theta}, \Theta \in \Theta\}$, i.e., to the case in which X_n is the set of sample values of the vector $X_n = \text{col}(X_1, ..., X_n)$ and $P_{n,\Theta} = P_n(f_\Theta)$ is the corresponding Gaussian distribution. We shall discuss this application in a separate subsection below; here we shall only note that in the last particular case $\tau_n = n^{-1/2}$, so that for the sake of simplicity, we shall assume that $\tau_n = n^{-1/2}$ also in the general Definition A2.1.

We now return to the general scheme of a sequence (1) of experiments E_n, $n = 1,2,...$ with the family of distributions $\{P_{n,\Theta}, \Theta \in \Theta\}$ possessing the LAN property. We shall pose the problem of estimating the unknown value of the parameter Θ and start from a natural supposition that only those estimators, whose limiting distribution (as $n \to \infty$) is stable relative to small variations of the "true" value of the parameter, will be of any practical interest. In view of this we shall confine our attention to "regular" estimators in the sense of the following definition due to Hájek [175] (cf. also [70], p. 151).

Definition A2.2. An estimator $\hat{\Theta}_n$ of the parameter Θ is called regular (at a fixed point Θ) if for some nonsingular distribution function F the weak convergence takes place

(2) $\qquad L\{n^{1/2}(\hat{\Theta}_n - (\Theta+n^{-1/2}h)) \mid P_{n,\Theta+n^{-1/2}h}\} \to F$

as $n \to \infty$ for any p-dimensional vector $\mathbf{h}$ and this convergence is uniform in $|\mathbf{h}| < b$ for any $b > 0$.

The following characterization of the limiting distribution in (2) given by Hájek [175] is quite remarkable (cf. also [70], Chapter II, Theorem 9.1, or [110], Chapter V, Theorem 3.1).

Proposition A2.1. *Let the family of distributions* $\{P_{n,\Theta}, \Theta \in \Theta\}$ *in the sequence of experiments* (1) *satisfy the LAN condition and let* $\hat{\Theta}_n$ *be a regular estimator of the parameter* Θ. *Then*

1)   *the limiting distribution* F *of a random vector* $\zeta_n = n^{1/2}(\hat{\Theta}_n - \Theta)$ *is a convolution of* $N(0, \Gamma_\Theta^{-1})$ *and some other distribution* G:

$$F = N(0, \Gamma_\Theta^{-1}) * G$$

(here $N(0, \Gamma_\Theta^{-1})$ *denotes a normal distribution with zero mean and covariance matrix* Γ_Θ^{-1});

2) G *is the limiting distribution of the difference* $\zeta_n - \Gamma_\Theta^{-1}\Delta_{n,\Theta}$.

It follows from 1) and 2) that in particular $\zeta_n - \Gamma_\Theta^{-1}\Delta_{n,\Theta}$ and $\Gamma_\Theta^{-1}\Delta_{n,\Theta}$ are asymptotically independent: the $P_{n,\Theta}$ probability of the event $\{\zeta_n - \Gamma_\Theta^{-1}\Delta_{n,\Theta} < x, \Gamma_\Theta^{-1}\Delta_{n,\Theta} < y\}$ converges to the product of the distribution functions G and $N(0, \Gamma_\Theta^{-1})$ evaluated at points x and y respectively (cf. [70], Chapter II, Theorem 9.2).

We are not presenting proofs of these assertions; we note only that they substantially rely on the approximation of the measure $P_{n,\Theta+n^{-1/2}\mathbf{h}}$ for a fixed Θ by an exponential measure $Q_{n,\Theta,\mathbf{h}}$ constructed by the formula

$$Q_{n,\Theta,\mathbf{h}}(A) = \int_A \exp(\mathbf{h}'\hat{\Delta}_{n,\Theta})dP_{n,\Theta}/E_{n,\Theta}\{\exp(\mathbf{h}'\hat{\Delta}_{n,\Theta})\}$$

(3)

$$A \in \mathfrak{U}_n$$

and satisfying the condition: for any $b > 0$

(4) $\displaystyle\sup_{|\mathbf{h}| < b} \sup_{A \in \mathfrak{U}_n} |Q_{n,\Theta,\mathbf{h}}(A) - P_{n,\Theta+n^{-1/2}\mathbf{h}}(A)| \to 0$ as $n \to \infty$

(cf. [70], Chapter II, Theorem 8.1, or [110], Chapter III, Theorem 1.1). In (3) the vector-valued random variable $\hat{\Delta}_{n,\Theta}$ is a truncation of $\Delta_{n,\Theta}$ such that the difference between them disappears in $P_{n,\Theta}$ probability as $n \to \infty$, and also for any $b > 0$

(4')
$$\sup_{|h|<b} |E_{n,\Theta} \exp\{h'\hat{\Delta}_{n,\Theta} - \frac{1}{2} h'\Gamma_\Theta h\} - 1| \to 0$$

(for more details cf. pp. 148 and 149 of [70]).

Proposition A2.1 leads to the following idea: since a convolution "spreads the mass," a regular estimator $\hat{\Theta}_n$ can be called asymptotically efficient if the corresponding distribution G is degenerated at the point $\{0\}$. In this sense a regular estimator $\hat{\Theta}_n$ satisfying the relation

$$n^{-1/2}(\hat{\Theta}_n - \Theta) - \Gamma_\Theta^{-1} \Delta_{n,\Theta} \to 0$$

in $P_{n,\Theta}$ probability as $n \to \infty$, is asymptotically efficient. Proposition A2.1 allows us to obtain an asymptotic bound from below for risks of regular estimators ([70], pp. 160-161).

Corollary A2.1. *Let $w(x) \geqslant 0$, $x \in R_p$ be a continuous even function such that the set $\{x: w(x) < b\}$ is convex for any $b > 0$. Then under the conditions of Proposition A2.1*

$$\lim_{n\to\infty} E_{n,\Theta} w(\zeta_n) \geqslant Ew(\zeta),$$

where ζ is a random vector distributed according to the normal distribution $N(0,\Gamma_\Theta^{-1})$ and $\zeta_n = n^{1/2}(\hat{\Theta}_n - \Theta)$.

In particular the matrix inequality

$$\varliminf_{n\to\infty} n \, E_{n,\Theta}(\hat{\Theta}_n - \Theta)^{\otimes 2} \geqslant \Gamma_\Theta^{-1}$$

is valid. Thus as $n \to \infty$ the ellipsoid of concentration of a regular estimator cannot be smaller than the ellipsoid generated by the matrix Γ_Θ^{-1} (cf. p. 144).

2. In this subsection we again return to the basic problem of this Chapter -- the problem of estimating unknown parameters of the spectral density f_Θ. We shall confine ourselves only to the particular case where X_t is a Gaussian process such that the corresponding Gaussian family of distributions $\{P_{n,\Theta} = P_n(f_\Theta), \Theta \in \Theta\}$ possesses the LAN property in the sense of Definition A2.1 with $\tau_n = n^{-1/2}$ and $\Delta_{n,\Theta}$ and Γ_Θ are defined by the relations (23) and (24) respectively in Section 3 of Chapter I (cf. assertion 4) of the Corollary A1.1).

In addition we shall assume that the assertion of corollary
A1.3 dealing with Gaussian processes is applicable to the
vector-valued function $H_\Theta = (A_\Theta, \dot\phi_\Theta)$ (where for simplicity A
is assumed to be real and even, while $\dot\phi$ is as usual the
gradient vector of logarithmic derivatives of f_Θ). This leads
to

(5) $$L\{I_n(H_\Theta/4\pi f_\Theta) \mid P_n(f_\Theta)\} \rightarrow N(0, C_\Theta)$$

with

(6) $$C_\Theta = \frac{1}{4\pi} \int_{-\pi}^{\pi} H_\Theta(\lambda) H_\Theta'(\lambda) d\lambda = \frac{1}{4\pi} \int_{-\pi}^{\pi} H_\Theta(\lambda)^{\otimes 2} d\lambda$$

and

(7) $$I_n(H_\Theta/4\pi f_\Theta) = \frac{n^{1/2}}{4\pi} \int_{-\pi}^{\pi} H_\Theta(\lambda) \frac{I_n(\lambda) - f_\Theta(\lambda)}{f_\Theta(\lambda)} d\lambda;$$

(cf. (24) of Appendix 1). Evidently, in accordance with
(I.3.23),

(8) $$I_n(\dot\phi_\Theta/4\pi f_\Theta) = \Delta_{n,\Theta}.$$

As in Subsection 6.4 of this Chapter we associate with each
p vector-valued function A a class of asymptotically
equivalent estimators $\{\hat\Theta_A\}$ satisfying the relation

(9) $$|n^{1/2}(\hat\Theta_A - \Theta) - I_n(A_\Theta/4\pi f_\Theta)| \rightarrow 0$$

in $P_n(f_\Theta)$ probability as $n \rightarrow \infty$ and possessing therefore the
following limiting covariance matrix:

(10) $$\lim_{n\to\infty} n E_{n,\Theta}(\hat\Theta_A - \Theta)^{\otimes 2} = \frac{1}{4\pi} \int_{-\pi}^{\pi} A_\Theta(\lambda)^{\otimes 2} d\lambda$$

(cf. (6.12) and (6.13) with $\kappa_4 = 0$ due to the Gaussian
assumption).

In view of the above (cf. Subsection 1) it seems important
to determine the conditions on A which would assure the
regularity of estimators $\{\hat\Theta_A\}$ in the sense of Definition A2.2.
However, by virtue of (2), it is not sufficient to know for
this purpose that (5) is valid; the knowledge of the limiting
distribution of the corresponding statistic is required under
the condition that the observations $(X_1, ..., X_n) = X$ possess the

distribution $P_n(f_{\Theta+n^{-1/2}h})$. This is given in the following proposition.[14]

Proposition A2.2. *Let the above stated conditions be satisfied. Then for any* $h \in R_p$

(11) $L\{I_n(H_\Theta/4\pi f_\Theta)|P_n(f_{\Theta+n^{-1/2}h})\} \rightarrow N\left[C_\Theta\begin{bmatrix} 0 \\ h \end{bmatrix}, C_\Theta\right].$

Proof.   In view of (8) condition (5) can be written in the form

$$E\{\exp I_n(ix'H_\Theta/4\pi f_\Theta) \mid P_n(f_\Theta)\}$$

(12) $$= E\{\exp I_n(ix_1'A_\Theta/4\pi f_\Theta)\exp ix_2'\Delta_{n,\Theta} \mid P_n(f_\Theta)\}$$

$$\rightarrow \exp\{-\frac{1}{2}x'C_\Theta x\},$$

for any $x = (x_1, x_2)$, with x_1 and x_2 being of the appropriate dimensionality.

Let $\hat{\Delta}_{n,\Theta}$ be the truncation of the vector $\Delta_{n,\Theta}$ with the properties stipulated in the preceding subsection, and $Q_{n,\Theta,h}$ be an approximation of $P_n(f_{\Theta+n^{-1/2}h})$ constructed by means of formula (3) and possessing property (4). Then as $n \rightarrow \infty$ the relation

(13) $$\sup_{|\xi|<1}\left|\int \xi \, dP_n(f_{\Theta+n^{-1/2}h}) - \int \xi \, \exp\{h'\hat{\Delta}_{n,\Theta}\right.$$
$$\left. - \frac{1}{2}h'\Gamma h\}dP_n(f_\Theta)\right| \rightarrow 0$$

is valid (cf. [70], Chapter II, formula (9.1)). From (12) and (13) we obtain for any x_1 and x_2 that

$$E\{\exp I_n(ix_1'A_\Theta/4\pi f_\Theta)\exp(ix_2'\hat{\Delta}_{n,\Theta})\exp(-h'\hat{\Delta}_{n,\Theta}$$
$$+ \frac{1}{2}h'\Gamma_\Theta h)|P_n(f_{\Theta+n^{-1/2}h})\} \rightarrow \exp\{-\frac{1}{2}x'C_\Theta x\},$$

---

[14]In the scheme of a general sequence of experiments (1), the related assertions which are, however, of a greater generality, are due to Le Cam [80] (cf. also [31], Section VI. 1.4, the so-called "third Le Cam's lemma").

so that

(14)

$$E\{\exp I_n(ix_1' A_\Theta/4\pi f_\Theta)\exp((ix_2-h)'\hat{\Delta}_{n,\Theta})|P_n(f_{\Theta+n^{-1/2}h})\}$$
$$\to \exp\left\{-\frac{1}{8\pi}\int_{-\pi}^{\pi}\left\{[x_1' A_\Theta(\lambda)]^2 + 2[x_1' A_\Theta(\lambda)][x_2'\dot{\phi}_\Theta(\lambda)]\right.\right.$$
$$\left.\left. + [x_2'\dot{\phi}_\Theta(\lambda)]^2 + [h'\dot{\phi}_\Theta(\lambda)]^2\right\}d\lambda.\right.$$

Actually the last relation is valid also for complex-valued x_2. This can be established by using the arguments analogous to those in [70] p. 154 (or [110], Lemma V.3.2) which lead to similar conclusions concerning the relation (9.5). Thus in (14) x_2 can be replaced by x_2-ih which results in

(15)

$$E\{\exp I_n(ix_1' A_\Theta/4\pi f_\Theta)\exp(ix_2'\hat{\Delta}_{n,\Theta})|P_n(f_{\Theta+n^{-1/2}h})\}$$
$$\to \exp\left\{ix'C_\Theta\begin{bmatrix}0\\h\end{bmatrix} - \frac{1}{2}x'C_\Theta x\right\}.$$

However, since the difference between $\Delta_{n,\Theta}$ and its truncation $\hat{\Delta}_{n,\Theta}$ vanishes in $P_n(f_\Theta)$ probability (by contiguity considerations also in $P_n(f_{\Theta+n^{-1/2}h})$ probability), $\hat{\Delta}_{n,\Theta}$ in (15) can be replaced by $\Delta_{n,\Theta}$. In summary, in view of (8) we have

$$E\{\exp I_n(ix'H_\Theta/4\pi f_\Theta) \mid P_n(f_{\Theta+n^{-1/2}h})\}$$

$$\to \exp\left\{ix'\ C_\Theta\begin{bmatrix}0\\h\end{bmatrix} - \frac{1}{2}x'C_\Theta x\right\}$$

and this is equivalent to the required relation (11). Proposition A2.2 is thus proved.   □

From (9) and (11) we have in particular that

(16)

$$L\{n^{1/2}(\hat{\Theta}_A-(\Theta+n^{-1/2}h)) \mid P_n(f_{\Theta+n^{-1/2}h})\}$$
$$\to N\left\{\left[\frac{1}{4\pi}\int_{-\pi}^{\pi}A_\Theta \otimes \dot{\phi}_\Theta(\lambda)d\lambda - I_p\right]h, \frac{1}{4\pi}\int_{-\pi}^{\pi}A_\Theta(\lambda)^{\otimes 2}d\lambda\right\}$$

Thus the asymptotic distribution is invariant here with respect to h if and only if the condition

(17) $\dfrac{1}{4\pi} \int_{-\pi}^{\pi} A_\Theta \otimes \dot{\phi}_\Theta(\lambda) d\lambda = I_p$

is fulfilled.

Now we are in the position to state the conditions for the regularity of estimators $\{\hat{\Theta}_A\}$ representable in the form (9).

Theorem A2.1. *For a Gaussian random process X_t, let the above-stated conditions be fulfilled. Then an estimator $\hat{\Theta}_A$ possessing representation (9) is regular in Hájek's sense if and only if A_Θ appearing in representation (9) satisfies the condtition (17). Moreover, for any* **h**

(18)
$$L\{n^{1/2}(\hat{\Theta}_A - (\Theta + n^{-1/2}h)) \mid P_n(f_{\Theta + n^{-1/2}h}))\}$$
$$\rightarrow N\left[0, \dfrac{1}{4\pi} \int_{-\pi}^{\pi} A_\Theta(\lambda)^{\otimes 2} d\lambda\right].$$

Under the condition of validity of representation (2.20) the asymptotic maximum likelihood estimator $\tilde{\Theta}$ is regular since A_Θ is of the form (6.11) satisfies condition (17).

Obviously A_Θ of the form (6.14) also satisfies condition (17). This allows us to determine the regularity of the estimators defined as an appropriate root of some reasonably constructed system of equations $I_n(h) = 0$ (natural examples of such a system are mentioned in Section 6).

Note that (16) implies in paticular the relation

(19) $n^{1/2}E_{n,\,\Theta + n^{-1/2}h}(\hat{\Theta}_A - (\Theta + n^{-1/2}h)) \rightarrow 0$

provided only that (17) is valid and that property (19) of the estimator $\hat{\Theta}_A$ can be interpreted as a weakened version of "uniform in a neighborhood of Θ asymptotic unbiasedness."

It is important also to note that for any estimator $\hat{\Theta}_A$ representable in the form (9) we have in view of (5)-(8)

(20)
$$L\left\{\begin{bmatrix} n^{1/2}(\hat{\Theta}_A - \Theta) - \Gamma_\Theta^{-1}\Delta_{n,\,\Theta} \\[2mm] \Gamma_\Theta^{-1}\Delta_{n,\,\Theta} \end{bmatrix} \mid P_n(f_\Theta)\right\}$$
$$\rightarrow N\left[0, \dfrac{1}{4\pi} \int_{-\pi}^{\pi} \begin{bmatrix} A_\Theta(\lambda) - \Gamma_\Theta^{-1}\dot{\phi}_\Theta(\lambda) \\[2mm] \Gamma_\Theta^{-1}\dot{\phi}_\Theta(\lambda) \end{bmatrix}^{\otimes 2} d\lambda\right].$$

If $\hat{\theta}_A$ is a regular estimator we thus have from condition (17) and formula (20) that

$$L\left\{\left[\begin{array}{c} n^{1/2}(\hat{\theta}_A-\theta)-\Gamma_\theta^{-1}\Delta_{n,\,\theta} \\ \Gamma_\theta^{-1}\Delta_{n,\,\theta} \end{array}\right]\Big| P_n(f_\theta)\right\}$$

(21)

$$\Rightarrow N\left[0,\; \left[\begin{array}{cc} \frac{1}{4\pi}\int_{-\pi}^\pi A_\theta(\lambda)^{\otimes 2}d\lambda - \Gamma_\theta^{-1} & 0 \\ 0 & \Gamma_\theta^{-1} \end{array}\right]\right];$$

in particular the p-vector-valued random variables $n^{1/2}(\hat{\theta}_A-\theta)$ - $\Gamma_\theta^{-1}\Delta_{n,\,\theta}$ and $\Gamma_\theta^{-1}\Delta_{n,\,\theta}$ are found to be asymptotically independent (cf. the statement following immediately after the formulation of Proposition A2.1). Moreover, under condition (17) we have[15]

(22) $$\frac{1}{4\pi}\int_{-\pi}^\pi A_\theta(\lambda)^{\otimes 2}d\lambda \geq \Gamma_\theta^{-1}.$$

Since in the l.h.s. of this matrix inequality the limiting covariance matrix of the estimator $\hat{\theta}_A$ appears (cf. (10)), the inequality (22) can be interpreted as follows: *Among regular estimators $\{\hat{\theta}_A\}$ representable in form (9) for some A_θ the estimators which correspond to A_θ of the form (6.11) possess the smallest asymptotic variance; under the conditions which assure the representation (2.20), the estimator $\tilde{\theta}$ determined from condition (2.2) is one of such optimal estimators.*

Actually, however, under the conditions of this subsection, i.e., under the conditions imposed above on the observed Gaussian process X_t with associated LAN family of distributions $P_n(f_\theta)$, $\theta \in \Theta$, the last estimator is "optimal" among all possible regular estimators (and not only those which possess the representation (9)); moreover, they are "optimal" in a wider sense of the word. To show this it is

---

[15]Note that the implication (17) $\Rightarrow$ (22) is an elementary corollary of the matrix Cauchy-Schwarz inequality

$$\int_{-\pi}^\pi A^{\otimes 2}d\lambda \geq \int_{-\pi}^\pi \dot{\phi}\otimes A\; d\lambda \left[\int_{-\pi}^\pi \dot{\phi}^{\otimes 2}d\lambda\right]^{-1}\int_{-\pi}^\pi A\otimes\dot{\phi}\; d\lambda.$$

required only to verify the applicability to the present case of the general derivation carried out in the preceding subsection. We shall, however, postpone any further discussion of the Gaussian case to the final subsection.

In the next subsection we shall drop the Gaussian assumption and will consider the process X_t to be linear in the sense of Section 6; thus the methods of Subsection 1 and more generally the methods of the book [70] will not be applicable herein, for in general statements attainable by applying these methods to the special Guassian case, specifically those concerning the "optimality" of the estimator $\hat{\theta}$ among all regular estimators, will lose their meaning. None the less, if the class of estimators under consideration is narrowed down to the class of estimators which are discussed above in the italicized statement then, as it will be shown below this statement can be extended to certain non-Gaussian situations (namely, to the case studied in Subsections 6.3 and 6.4. In these situations one can characterize more or less distinctly the notion of regularity of estimators belonging to the narrowed down class which is described in the italicized statement.)

3. Assume that the conditions of the validity of Corollary A1.3 are fulfilled under which the relation (5) holds but with C_Θ of the form (6) only in the particular case of a Gaussian process, while in general

(23)
$$C_\Theta = \frac{1}{4\pi} \int_{-\pi}^{\pi} H_\Theta(\lambda)^{\otimes 2} d\lambda + \frac{1}{4\pi} \iint_{-\pi}^{\pi} H_\Theta(\lambda_1)$$
$$\otimes H_\Theta(\lambda_2)\psi(\lambda_1,\lambda_2) d\lambda_1 d\lambda_2$$

where

(24) $\psi(\lambda_1,\lambda_2) = f_4(\lambda_1,-\lambda_1,\lambda_2,-\lambda_2)/f(\lambda_1)f(\lambda_2),$

so that in the particular case of a linear process we have

(25) $\psi(\lambda_1,\lambda_2) = \kappa_4/2\pi$

(cf. Appendix 1, formulas (21) and (26)).

In the present subsection we shall deal mainly with the latter particular case - namely we shall give the proof of Theorem 6.3 which has been formulated in Subsection 6.4 without proof.

We shall thus assume that X_t is a linear process of the form (6.1) with ε's having an undefined common distribution.[15] As in Subsections 6.3 and 6.4 suppose that $\Theta = (\Theta_1, ..., \Theta_{p-1}, \sigma^2)$. Consider the estimator $\hat{\Theta}_A$ representable in the form (9) and call it regular if the corresponding function A satisfies (17).[16] Before proceeding with the proof of Theorem 6.3 let us make the following remark.

In accordance with the statement following directly from the formulation of Proposition A2.1 one should note that the asymptotic independence of $n^{1/2}(\hat{\Theta}_A-\Theta) - \Gamma_\Theta^{-1}\Delta_{n,\Theta}$ and $\Gamma_\Theta^{-1}\Delta_{n,\Theta}$ does not follow in general from (17) and the representability of the estimator $\hat{\Theta}_A$ in the form (9). This independence is however, assured under the additional stipulation $\Theta = (\Theta_1, ..., \Theta_{p-1}, \sigma^2)$ which is adopted in Subsections 6.3 and 6.4. Indeed, it is easy to verify in this case the validity of a formula of the form (21) but with a different limiting covariance matrix -- in the right hand side lower corner Γ_Θ^{-1} should be replaced by the matrix $\Gamma_\Theta^{-1} + \Gamma_\Theta^{-1}C_{K_4,\Theta}\Gamma_\Theta^{-1}$ of the form (6.8). This

remark seems to indicate that the definition of regularity of estimators $\hat{\Theta}_A$ presented above makes complete sense only after the above stipulation is accepted.

It is also easy to verify the important corollary of the modified (in the manner indicated above) formula (21) -- the matrix inequality (6.16) (obviously the matrix (6.8) appears in its right-hand-side). Thus the Theorem 6.3 is valid.

---

[15] If one views the distribution of ε's as an "abstract parameter' additional to Θ it is then possible in principle to extend the ideas and notions of Subsection 1 to this "semiparametric" model in the spirit of paper [179], say. We note a recent work in this direction [178]. See also the informal discussion of Remarks 3 and 4 to Section 6 in the next appendix.

[16] If one limits the discussion to the estimators $\hat{\Theta}_A$ representable in the form (9) then, under the conditions stipulated below in Subsection 1.2 of Chapter V, one can arrive at a formula of the form (16) and then, assuming the validity of (17), a formula of the form (18) (with the corresponding modification due to the presence of an additional term in the limiting expression for the covariance matrix in accordance with (23)-(25)) where $P_n(f_{\Theta+n^{-1/2}h})$ corresponds to the alternative hypothesis H_1 for $g_n = f_{\Theta+n^{-1/2}h}$ which is introduced therein. Such an analysis does not however seem to be of special interest: only the above-mentioned role of condition (17) is important at least in the present context, which allows us to name the estimator $\hat{\Theta}_A$ representable in the form (9) as a regular estimator provided (17) is valid.

In conclusion, it may be desirable to summarize below the above stated results. Before doing this, however, we shall observe the following obvious general fact: the estimator $\tilde{\Theta}$ of the parameter Θ cannot be worse than any estimator $\hat{\Theta}$ satisfying the relation[17]

$$(26) \qquad \lim_{n \to \infty} n \, E_{n, \Theta}(\hat{\Theta}-\tilde{\Theta}) \otimes (\tilde{\Theta}-\Theta) = 0,$$

in the sense that (26) implies[18]

$$\lim_{n \to \infty} \{n \, E_{n, \Theta}(\tilde{\Theta}-\hat{\Theta})^{\otimes 2} - n \, E_{n, \Theta}(\tilde{\Theta}-\Theta)^{\otimes 2}\} \geqslant 0.$$

Specifically, let Θ be an unknown parameter of the spectrum f_Θ and $\tilde{\Theta}$ be its estimator determined from condition (2.2) and possessing representation (2.20). Then under the conditions stated in the beginning of this subsection for any estimator $\hat{\Theta}_A$ satisfying (19) we have

$$
\begin{aligned}
(27) \qquad & \lim_{n \to \infty} n E_{n, \Theta}(\hat{\Theta}_A - \tilde{\Theta}) \otimes (\tilde{\Theta}-\Theta) \\
& = \left\{ \frac{1}{4\pi} \int_{-\pi}^{\pi} A_\Theta(\lambda) \otimes \dot{\phi}_\Theta(\lambda) d\lambda + I_p \right\} \Gamma_\Theta^{-1} \\
& \quad - \frac{1}{4\pi} \iint_{-\pi}^{\pi} [A_\Theta(\lambda_1) - \Gamma_\Theta^{-1}\dot{\phi}_\Theta(\lambda_1)] \\
& \quad \otimes \Gamma_\Theta^{-1}\dot{\phi}_\Theta(\lambda_2)\psi(\lambda_1,\lambda_2)d\lambda_1 d\lambda_2.
\end{aligned}
$$

Thus the estimator Θ_A representable in the form (9) satisfies the condition (26) provided only (17) is valid, and moreover, the last term in the r.h.s. of (27) vanishes. Aside from the simple case of a Gaussian process which is of no interest to us at present, the last condition is verified as indicated above in the linear case also when ψ is a constant (cf. (25)) and $\Theta = (\Theta_1, ..., \Theta_{p-1}, \sigma^2)$.

---

[17]Roughly speaking, the estimator $\hat{\Theta}$ cannot be "improved" by adding to it a statistic which is asymptotically independent of $\sqrt{n}(\tilde{\Theta}-\Theta)$.

[18]Indeed we simply have

$$0 \leqslant n \, E_{n, \Theta}(\hat{\Theta}-\tilde{\Theta})^{\otimes 2} = n \, E_{n, \Theta}(\hat{\Theta}-\Theta)^{\otimes 2} - n \, E_{n, \Theta}(\tilde{\Theta}-\Theta)^{\otimes 2}$$

$$- n \, E_{n, \Theta}(\hat{\Theta}-\tilde{\Theta}) \otimes (\tilde{\Theta}-\Theta) - n \, E_{n, \Theta}(\tilde{\Theta}-\Theta) \otimes (\hat{\Theta}-\tilde{\Theta}).$$

and this yields the desired implication.

4. We now return to the Gaussian case. In the beginning of this Chapter we introduced an MLE estimator for the parameter θ of spectral density f_θ (it was denoted there by $\bar{\theta}$ and it is determined from condition (1.1)). Later however, we ignored this estimator and considered a much simpler estimator $\tilde{\theta}$ justifying this replacement with their asymptotic equivalence as $n \to \infty$. Without making this last property precise we were motivated by the simple consideration that it is possible to state very general conditions under which $\tilde{\theta}$ is consistent, asymptotically normal, and asymptotically efficient (in a certain sense); in other words, $\tilde{\theta}$ possesses asymptotic properties shared by the MLE.

In this connection, the following question arises naturally. What can be specifically asserted about the asymptotic properties of the MLE itself? The question is definitely relevant[19] and we shall discuss it now utilizing the method presented in [70].

Specifically we shall base our discussion on the material presented in Section 3.1 of the book which deals with the determination of the asymptotic properties of the MLE in the general framework of sequences of experiments (1) satisfying the assumptions N1-N4 ([70], pp. 173-174).

One should clearly start with a stipulation of the conditions to be imposed on the family of spectral densities $\{f_\theta, \theta \in \Theta\}$ under which these assumptions N1-N4 are verified (adopted of course, to our specific problem).

Let the spectral density f_θ of a Gaussian process X_t be such that $m \leqslant f_\theta \leqslant M, -\pi < \lambda \leqslant \pi, \theta \in \Theta$, where m and M are positive numbers. Let the vector-valued function ϕ_θ -- the gradient vector of continuous derivatives of the logarithm of spectral density -- satisfy the conditions presented in the beginning of Subsection 2.2 so that the matrix Γ_θ (cf. (2.14)) will be nondegenerate and possess entries continuous in θ for $\theta \in \Theta$. Then the assumptions N1 and N2 are satisfied ([70], pp. 173 and 174) since the Assertion 3) of Corollary A1.1 of the preceding Appendix is valid (which states the LAN

---

[19]The importance of this question is enhanced by taking into consideration the fact that under the usual conditions of asymptotic equivalence of estimators $\bar{\theta}$ and $\tilde{\theta}$ one can, under scrutiny, observe the superiority of $\bar{\theta}$ over $\tilde{\theta}$ provided such a comparison is possible (an informal discussion of this problem is given in the beginning of the next Appendix).

property of the family of distributions $\{P_n(f_\Theta), \Theta \in \Theta\}$ in a certain sense) and since in our case, the matrix $(n\Gamma_\Theta)^{-1/2}$ plays the role of a normalizing matrix in N1 and N2 (this normalization also satisfies the condition, additional to N1-N4 of the Corollary 1.1 presented in [70], p. 175).

As far as the conditions N3 and N4 are concerned, they are satisfied if one shows that

(A)   for any integer s there exists $n_0 > 0$ and $C > 0$ such that for $n > n_0$ and any $\Theta \in \Theta$, $h_1, h_2 \in R_p$,

$$E_{n,\Theta} \mid Z_{n,\Theta}^{1/2s}(h_1) - Z_{n,\Theta}^{1/2s}(h_2)\mid^{2s} \ll C\mid h_1 - h_2\mid^{2s},$$

where

$$Z_{n,\Theta}(h) = dP_n(f_{\Theta+(n\Gamma_\Theta)^{-1/2}h})/dP_n(f_\Theta);$$

(B)   there exist values $c > 0$ and $n_0 > 0$ such that for $n > n_0$ and any $\Theta \in \Theta$, $h \in R_p$,

$$E_{n,\Theta} Z_{n,\Theta}^{1/2}(h) \ll \exp\{-c\mid h\mid^2\}.$$

To get the idea of the possibilities, consider the following lemma.

Lemma A2.1. *Let f be a spectral density such that $m \ll f \ll M$ and $g_i = f(1+n^{-1/2}a_i)$, $i = 1,2$, be two spectral densities with square integrable a_i. Then for $0 \ll \beta \ll 1$*

(28)
$$\lim_{n \to \infty} E_{n,g_1}\{dP_n(g_2)/dP_n(g_1)\}^\beta$$
$$= \exp\left\{-\frac{\beta(1-\beta)}{4}\mid a_1 - a_2\mid^2\right\},$$

where

$$\mid a_1 - a_2\mid^2 = \frac{1}{2\pi}\int_{-\pi}^{\pi}[a_1(\lambda) - a_2(\lambda)]^2 d\lambda.$$

Proof.   Observe first of all, that using the methods of Appendix 1 to Chapter I (Lemma A1.2) it is easy to show that

(29) $n^{1/2}\|B_{f(a_1-a_2)} B_{g_2}^{-1}\| \to 0,$

(30) $n^{-1/2}|B_{f(a_1-a_2)}B_{g_2}^{-1}| \to |a_1-a_2|$

and

(31) $n^{-1}\mathrm{tr}[(B_{f(a_1-a_2)}B_{g_2}^{-1})^2] \to |a_1-a_2|^2.$

From (29), (30), and Lemma A1.1 of Chapter I we have

(32) $|U_n(n^{-1/2}B_{f(a_1-a_2)}B_{g_2}^{-1})| \to 0.$

Since for $0 \leqslant \beta \leqslant 1$

$$(2\pi)^{-n/2}\int_{x\in R_p} \exp\left\{-\frac{1}{2}\, x'[\beta B_{g_2}^{-1}+(1-\beta)B_{g_1}^{-1}]x\right\} dx$$

$$= \{\det[\beta B_{g_2}^{-1} + (1-\beta)B_{g_1}^{-1}]\}^{-1/2}$$

(cf. for example, [76], p. 120) taking (I.3.1)-(I.3.6) into account we obtain

$$E_{n,\,g_1}\{dP_n(g_2)/dP_n(g_1)\}^\beta$$

$$= [\det(I_n + n^{-1/2}B_{f(a_1-a_2)}B_{g_2}^{-1}]^{\beta/2}$$

$$\times\, [\det(I_n + \beta_n^{-1/2}B_{f(a_1-a_2)}B_{g_2}^{-1}]^{-1/2}$$

$$= \exp\left\{\frac{1}{2}\beta\, U_n(n^{-1/2}B_{f(a_1-a_2)}B_{g_2}^{-1})\right.$$

$$-\frac{1}{2}U_n(\beta n^{-1/2}B_{f(a_1-a_2)}B_{g_2}^{-1})$$

$$\left.-\frac{\beta(1-\beta)}{4n}\mathrm{tr}[(B_{f(a_1-a_2)}B_{g_2}^{-1})^2]\right\}.$$

From here the relations (31) and (32) imply (28).   □

Assertion A2.2. *Let the conditions of [156] be fulfilled.   Then for any integer s there exist $n_0 > 0$ and $C > 0$ such that*

(33)
$$E_{n,\,f}|\{dP_n(g_1)/dP_n(f)\}^{1/2s}$$

$$- \{dP_n(g_2)/dP_n(f)\}^{1/2s}|^{2s} \leqslant C|a_1-a_2|^{2s}.$$

For the proof of this result we refer the reader to [156].

Here we note only that in view of Lemma A2.1 the left hand side of the inequality (33) which is equal to

$$E_{n,\,g_1}|1 - \{dP_n(g_2)/dP_n(g_1)\}^{1/2s}|^{2s}$$

$$= \sum_{k=0}^{2s} \binom{2s}{k}(-1)^k E_{n,\,g_1}\{dP_n(g_2)/dP_n(g_1)\}^{k/2s}$$

converges to

$$\sum_{k=0}^{2s} \binom{2s}{k}(-1)^k \exp\left\{\frac{k}{2s}\left[\frac{k}{2s} - 1\right]\frac{|a_1 - a_2|^2}{4}\right\}.$$

Thus for the derivation of (33) we utilize here the arguments presented on page 202 of the book [70] for the proof of Lemma 5.2 (truly, as (28) is only an asymptotic result, these arguments ought to be extended as indicated in [156]).

Assertion A2.3. *If under the conditions indicated above in Lemma A2.1 we set* $a_1 = 0$, $a_2 = a$, *and* $g = f(1 + n^{-1/2}a)$ *then there exist* $c > 0$ *and* $n_0 > 0$ *such that for* $n > n_0$

$$E_{n,\,f}\{dP_n(g)/dP_n(f)\}^{1/2} \leqslant \exp\{-cy^2\},$$

where, as usual $y^2 = (1/2\pi)\int_{-\pi}^{\pi}a^2(\lambda)d\lambda$.

For proof, again see [156].

Remark 1. These properties are retained also for $\exp\{-n[U_n(\Theta + n^{-1/2}\Gamma_\Theta h) - U_n(\Theta)\}$ in place of $Z_{n,\,\Theta}(h)$ where $U_n(\Theta)$ is given by the expression (2.1) provided only such a replacement is justified from asymptotic considerations of Subsection 2.1 in Chapter I.

Properties (A) and (B) allow us to utilize the Theorem 5.1 from [70], p. 42 in order to arrive at the following conclusion: there exist values $C_0 > 0$ and $c_0 > 0$ such that for n sufficiently large and H sufficiently large

(34) $$P_{n,\,\Theta}\{|n\Gamma_\Theta)^{1/2}(\overline{\Theta} - \Theta)| > H\} \leqslant C_0\exp\{-c_0 H^2\}.$$

From here we obtain in particular that for any loss function w possessing a polynomial majorant (and satisfying the conditions (1)-(4) on page 18 of the book [70]),

$$\overline{\lim_{n\to\infty}} \, E_{n,\,\Theta} w((n\Gamma_\Theta)^{1/2}(\overline{\Theta}-\Theta)) < \infty$$

([70], Corollary 5.2, p. 44).

Next, since the conditions N1-N4 which precede the Theorem 1.1 on page 174 of the book [70] are fulfilled, it follows that the assertions 1)-3) of this theorem as well as that of the next Theorem 1.2 are valid for the MLE $\overline{\Theta}$ (however, with a fixed $t = \Theta$, $\phi(\varepsilon,t) = (n\Gamma_\Theta)^{-1}$ and $\Delta_{\varepsilon,t} = \Delta_{n,\,\Theta}$ of the form (11) presented in the preceding Appendix).

Finally it is important that one can utilize here the Theorem 1.3 presented on page 176 of the book [70] and its Corollary 1.1 which states the asymptotic efficiency of $\overline{\Theta}$ in the sense of the definition on page 162 of the book [70].

In conclusion we especially emphasize the fact that based on the Remark 1 presented above and the Theorem 1.3 ([70], p. 176) of a general nature, one can, in principle, extend the asymptotic conclusion about MLE $\overline{\Theta}$ as given in this subsection to the case of an asymptotic MLE $\tilde{\Theta}$. For this, however, it is necessary to require additional conditions which assure the validity of assertion 4) of Corollary A1.1 (instead of assertion 3)), and then carry out the corresponding modifications in the arguments. In spite of the indisputable importance of these results concerning $\tilde{\Theta}$ we shall not dwell upon them herein (limiting ourself to Remark 2 to Section 2 in the next appendix; see also Remark 3 ibid).

Appendix 3. Remarks and Bibliography

Section 1

1. A brief discussion of the properties of MLE (in the spirit of [70]) appears in the last subsection of the preceding Appendix. See [156] for the continuous time case.

2. As it is noted in [143] one can, in general, reveal the superiority of MLE over the estimator $\tilde{\Theta}$ if, in the expansion in the powers of $1/n$ of expressions for their covariance

matrices, one compares not only the corresponding factors at $1/n$ but also the factors at $1/n^2$.

We shall consider a very simple_ example which demonstrates completely the superiority of $\bar{\Theta}$ over $\tilde{\Theta}$.

Example. Let the spectral density f_Θ of a Gaussian process X_t depend multiplicatively on an unknown scalar parameter $\Theta > 0$: $f_\Theta(\lambda) = \Theta f(\lambda)$. The Fisher's information quantity (2.15) is then equal to $1/2\Theta^2$.

For all n the MLE $\bar{\Theta} = X'B_f^{-1}X/n$ is efficient since $n\bar{\Theta}/\Theta$ possesses the χ^2-distribution with n degrees of freedom, i.e., $E(\bar{\Theta}) = \Theta$ and $D(\bar{\Theta}) = 2\Theta^2/n$. On the other hand, the estimator $\tilde{\Theta} = X'B_{1/4\pi^2 f}X/n$ is biased, $E(\tilde{\Theta}) = \Theta \, \mathrm{tr}\{B_f B_{1/4\pi^2 f}\}/n$ and is

inefficient $D(\tilde{\Theta}) = 2\Theta^2 \mathrm{tr}\{(B_f B_{1/4\pi^2 f})^2\}/n^2$. In fact, $\tilde{\Theta} \geqslant \bar{\Theta}$ for any observations since $B_{1/4\pi^2 f} - B_f^{-1} \geqslant 0$ (cf. [181]).

In accordance with Remark 3 in Section 3 presented in Appendix 3 to Chapter I, the bias of $\tilde{\Theta}$ is computed as follows:

$$E(\tilde{\Theta})-\Theta = \Theta\left\{1 - \frac{1}{2\pi}\iint_{-\pi}^{\pi} K_n(\lambda-\mu)\frac{f(\mu)}{f(\lambda)}d\lambda d\mu\right\}.$$

Sections 2-3

1. In [123], [32,33,159], and [48] the definition of an asymptotic maximum likelihood estimator $\tilde{\Theta}$ maximizing the "principal part" of the logarithm of the likelihood function is naturally carried over to the case of a multidimensional process X_t, $t = 0,\pm1, ...$, a random field X_t, $t = (t_1, ..., t_d)'$, and a continuous time process X_t, $-\infty < t < \infty$ respectively. Moreover, the consistency, asymptotic normality, and asymptotic efficiency of this estimator are proved (in the same spirit as in these sections).

2. The results of Sections 2-3 can be refined by extending the method of Ibragimov and Has'minskii [70], Section 1.5 (namely, the method of estimating the probabilities of large deviations of the MLE from the true value of the parameter, via inequalities of type (34), Appendix 2), as it is done by A. Sieders (1985), Research Report, TH Delft.

The basic result of this report can be informally described as follows:

Consider a sequence of experiments (1) of Appendix 2, and let $\hat{\Theta}_n$ be an estimator for Θ defined by maximizing with respect to Θ a certain functional of observations (e.g., the likelihood function, or, in our special case, the functional (2.1) taken with the opposite sign). If this functional satisfies certain conditions similar to the conditions imposed on the likelihood function in [70], Section 1.5, then the estimator $\hat{\Theta}_n$ is not only consistent (in $P_{n, \Theta}$ probability), but also the inequality (34) of Appendix 2 holds with $\hat{\Theta}_n$ in place of $\overline{\Theta}$.

Among various applications of this rather general result, in the above-mentioned report one can find the application to the situation of Section 2: the conditions are sought under which the estimator $\overline{\Theta}$ defined by (2.2) satisfies the inequality

$$P_{n, \Theta}\{|(n\Gamma_\Theta)^{1/2}(\overline{\Theta}-\Theta)| > H\} \leqslant C_0 \exp\{-c_0 H^2\}$$

with the constants specified as in (34), Appendix 2.

As a consequence of the last inequality, one can refine the property (2.19) of asymptotic normality of the estimator $\overline{\Theta}$ by stating the convergence in all moments instead of the weak convergence in (2.19).

3.   Again, consider a sequence of experiments (1) of Appendix 2, and suppose that the conditions of "asymptotic differentiability" (D1)-(D4) given on page 21 of the Introduction are satisfied. Moreover, suppose that the condition 2) of Definition A2.1 is also satisfied, so that actually the family of distributions $\{P_{n, \Theta}, \ \Theta \in \Theta\}$ under consideration is LAN in the sense of Definition A2.1.[20]   For simplicity, restrict the considerations to the case in which $\tau_n = \sqrt{n}$.

Here we make use of the following general result:

Proposition ([203], Theorems 2.6 *and* 6.3). *Let* w *be a loss function on* R_p *in the sense of* [203], *Definition* 2.3. *Then under the above conditions the following two statements hold:*

---

[20]Cf. [203], Definition 2.2:   the family of distributions $\{P_{n, \Theta}, \ \Theta \in \Theta\}$ satisfying the conditions 1) and 2) of our Definition A2.1 is called there HLAN (Hajek's LAN), whereas the term LAN in [203] is reserved for the stronger condition, namely HLAN plus our condition (D3).

(i)   *For any estimator Θ_n^* of Θ*

$$\lim_{K\to\infty} \lim_{n\to\infty} \sup_{|(n\Gamma_t)^{1/2}(t-\Theta)|\leqslant K} E_{n,t} w((n\Gamma_t)^{1/2}(\Theta_n^*-t))$$

$$\geqslant \frac{1}{(2\pi)^{1/2}} \int_{x\in R_p} w(x) e^{-\frac{1}{2}|x|^2} dx$$

(ii)   *Any estimator $\hat{\Theta}_n$ of Θ satisfying the asymptotic relation*

$$\Gamma_\Theta n^{1/2}(\hat{\Theta}_n-\Theta)-\Delta_{n,\Theta} \to 0$$

in $P_{n,\Theta}$ probability as $n \to \infty$ (cf. (2.20)) is Hájek's regular by Definition A2.2, with $F = N(0,\Gamma_\Theta^{-1})$; the latter property, in turn, implies that $\hat{\Theta}_n$ is locally asymptotically minimax (LAM) in the sense that it attains the lower bound given in (i), namely

$$\lim_{K\to\infty} \lim_{n\to\infty} \sup_{|(n\Gamma_t)^{1/2}(t-\Theta)|\leqslant K} E_{n,t} w((n\Gamma_t)^{1/2}(\hat{\Theta}_n-\Theta))$$

$$= \frac{1}{(2\pi)^{1/2}} \int_{x\in R_p} w(x) e^{-\frac{1}{2}|x|^2} dx.$$

Evidently, the conditions of the present remark cover the special situation discussed in Sections 2 and 3, therefore, as a corollary of this proposition and the Theorems 2.2 and 3.2 we arrive at the following important conclusion.

Corollary. *Under the conditions of Section 2 (Section 3), the estimator $\tilde{\Theta}$ defined by (2.2) is asymptotically efficient not only in Fisher's sense, as it is stated in Section 2 (Section 3), but also in the sense that it is LAM: the assertion (ii) of the above proposition holds with $\hat{\Theta}_n = \tilde{\Theta}$ and $\Delta_{n,\Theta}, \Gamma_\Theta$ defined in Section 2 (Section 3).*

Section 4

1. Let X_t satisfy (1) where ε_t are such that $E\varepsilon_t = 0$ and $E\varepsilon_t\varepsilon_s$ $= \sigma^2\delta_{ts}$ (ε_t are not necessarily Gaussian as in Subsection 4.1). In order that the stochastic difference equation (1) possess a

stationary solution (expressed only in terms of the "past" of ε's) it is necessary and sufficient that all the roots of the characteristic equation (1) do not exceed one in their absolute value [182]. Violation of the last condition obviously results in substantial complications (cf., e.g., [183,184] and the references therein; the latter paper is devoted to the special case for which some of the roots are equal to 1 in their absolute value; cf. also Remark 5 to Section 6).

2. In Section 3, when studying the case of a Gaussian process with a degenerate spectral density on certain frequencies λ_1, ..., λ_q, we purposely did not mention the problem which can naturally arise in applications: the problem of estimating these frequencies λ_1, ..., λ_q in the case when they are also unknown along with the usual parameters appearing in the expression for f_0. The basic reason for this was the quite different (as far as our book is concerned) nature of this problem: it is remarkable that when estimating the frequencies λ_1, ..., λ_q the "correct" normalizing factor is n and not $\sqrt{n}$ (as is the situation in all the cases studied in this text). What we mean here is that if one follows the established route of studying properties of MLE for λ_i (determined in the same manner as in Section 1 and denoted by $\bar{\lambda}_i$) which will require a preliminary determination of the LAN property in the sense of Definition A2.1 with $p = q$ and $\theta = \text{col}\{\lambda_1, ..., \lambda_q\}$, one then encounters the necessity to specify τ_n (in condition 1) of Definition A2.1) as n rather than as $\sqrt{n}$. Consequently, the limiting properties of the vector $n(\bar{\theta}-\theta)$ where $\bar{\theta} = \text{col}\{\bar{\lambda}_1, ..., \bar{\lambda}_q\}$ must be discussed here. We shall illustrate this by means of the simplest possible example confining ourselves solely to the determination of the LAN property.

Example. Let a complex-valued Gaussian process X_t satisfy the relation $X_t = \varepsilon_t - e^{i\theta}\varepsilon_{t-1}$, where ε_t is a sequence such that

$E\varepsilon_t = 0$, $E\varepsilon_t\bar{\varepsilon}_s = \sigma^2\delta_{ts}$, $\sigma^2 > 0$ with the usual convention that $E\varepsilon_t\varepsilon_s = 0$. Then X_t possesses the spectral density $f(\lambda) = \sigma^2|e^{i\lambda} - e^{i\theta}|^2/2\pi$ which degenerates at frequency θ. We assume that θ is an unknown parameter to be estimated based on the observations $\mathbf{X} = \text{col}\{X_1, ..., X_n\}$.

In order to apply the maximum likelihood method to this, we derive the expression for the logarithm of the likelihood function L_n. Taking into account that

$$B_f = \text{diag}\{e^{i\Theta}, ..., e^{in\Theta}\}S^{-1}HS^{-1}\text{diag}\{e^{-i\Theta}, ..., e^{-in\Theta}\},$$

where $H = \sigma^2(I_n + vv')$, $v = \text{col}(1, ..., 1)$ (cf. Example 4 presented in Subsection 4.5), and that S is a lower triangular $(n \times n)$-matrix with ones on the intersection of the k-th row and ℓ-th column for $k \geqslant \ell$, we obtain

$$L_n = \{-n \log \pi + \log \det B_f + X^*B_f^{-1}X\}$$

$$= -\{n \log \pi\sigma^2 + \log(n+1) + Y_\Theta^* H^{-1} Y_\Theta\};$$

here

$$Y_\Theta = S \text{ diag}\{e^{-i\Theta}, ..., e^{-in\Theta}\}X$$

$$= \text{col}\left\{Y_k = \sum_{j=1}^{k} e^{-i\Theta j}X_j, \ k = 1, ..., n\right\}.$$

Consequently,[21]

$$(\partial/\partial\Theta)L_n = -X^*(\partial/\partial\Theta)B_f^{-1}X = 2\text{Im } \tilde{Y}_\Theta^* S\tilde{Y}_\Theta/\sigma^2,$$

where

$$\tilde{Y}_\Theta = \sigma^2 H^{-1} Y_\Theta = \text{col}\left\{Y_t - \frac{1}{n+1}\sum_{s=1}^{n} Y_s, \ t = 1, ..., n\right\}$$

possesses the covariance matrix

$$E\tilde{Y}\tilde{Y}^* = \sigma^4 H^{-1} = \sigma^2\left[I_n - \frac{1}{n+1}vv'\right]$$

(and $E\tilde{Y}\tilde{Y}' = 0$). It is easy to compute that

---

[21] The second of these equations follows from the relation

$$H^{-1}SNS^{-1} - S^{-1'}NS'H^{-1} = \sigma^2H^{-1}(S'-S)H^{-1}$$

with $N = \text{diag}\{1,...,n\}$ which can easily be established by using the equation $SNS^{-1} = I+N-S$.

$$E(\partial/\partial\theta)L_n = i\mathrm{tr}\left\{(S'-S)\left[I_n - \frac{1}{n+1}vv'\right]\right\} = 0$$

and

$$D(\partial/\partial\theta)L_n = -\mathrm{tr}\left\{(S'-S)\left[I_n - \frac{1}{n+1}vv'\right]\right\}^2 = \frac{n(n-1)}{3}.$$

To prove the asymptotic normality $N(0,1)$ of the statistic $\Delta_{n,\theta} = \{n(n-1)/3\}^{-1/2}(\partial/\partial\theta)L_n$ we shall apply Proposition A1.1 presented in Appendix 1 to this Chapter. For this purpose it is necessary to verify that[22]

$$\left\|\begin{bmatrix} 0 & -S \\ S & 0 \end{bmatrix}\begin{bmatrix} (1/2)\sigma^2 H^{-1} & 0 \\ 0 & (1/2)\sigma^2 H^{-1} \end{bmatrix}\right\|$$

$$= |S\sigma^2 H^{-1}| = 0(n)$$

and

$$\left\|\begin{bmatrix} 0 & -S \\ S & 0 \end{bmatrix}\begin{bmatrix} (1/2)\sigma^2 H^{-1} & 0 \\ 0 & (1/2)\sigma^2 H^{-1} \end{bmatrix}\right\|$$

$$= \frac{1}{2}\sup_{|x|^2+|y|^2=1}|y'(S\sigma^2 H^{-1}-\sigma^2 H^{-1}S')x| \leqslant Cn^{1/2},$$

where C is a positive constant independent of n. To prove the last inequality, the Cauchy-Schwarz inequality is used, as well as the fact that a bilinear form whose absolute value is to be minimized can be written in terms of the components x_j and y_j, $j = 1, ..., n$, of the vectors x and y respectively in the following manner

$$\sum_{k-1}^{n-1} y_k\left[\sum_{j=k+1}^{n}\left(1-\frac{j}{n+1}\right)x_j - \frac{1}{n+1}\sum_{j=1}^{k}jx_j\right] - \frac{y_n}{n+1}\sum_{j=1}^{n}jx_j$$

$$-\sum_{k-1}^{n-1} x_k\left[\sum_{j=k+1}^{n}\left(1-\frac{j}{n+1}\right)y_j - \frac{1}{n+1}\sum_{j=1}^{k}jy_j\right] - \frac{x_n}{n+1}\sum_{j=1}^{n}jy_j.$$

[22]Here the relation

$$\mathrm{Im}\ \bar{Y}^*(S'-S)\bar{Y} = \begin{pmatrix}\mathrm{Re}\ \bar{Y} \\ \mathrm{Im}\ \bar{Y}\end{pmatrix}'\begin{pmatrix} 0 & -S \\ S & 0 \end{pmatrix}\begin{pmatrix}\mathrm{Re}\ Y \\ \mathrm{Im}\ Y\end{pmatrix}$$

is taken into account.

Now to establish the desired LAN property the Taylor expanison is applied:

$$\Lambda(f_\Theta, f_{\Theta_n}) = h\Delta_{n,\Theta} - \frac{1}{2}h^2 + o_P(1),$$

where $\Theta_n = \Theta + \{n(n-1)/3\}^{-1/2}h$.

Section 5

1. Some additional aspects of the problem discussed herein are presented in [39] and [165].

2. In [51] the problem of estimating the unknown spectral parameters is considered in relation to "signal" and "noise" with discrete as well as continuous times. In the latter case, it is assumed that the "signal" and the "noise" are both Gaussian processes with a rational and constant spectral density respectively (cf. also [152]).

In [156] the method of the book [70] is applied to the study of properties of an MLE for the parameter of spectral density of a continuous time Gaussian process observed on a white noise background.

Section 6

1. Since the characteristic functional of a linear process (1) can be expressed in terms of the characteristic function ψ_ε of variables ε_t as

$$E \exp\left\{i \sum_{-\infty}^{\infty} \lambda_t X_t\right\} = \prod_{-\infty}^{\infty} \psi_\varepsilon(\mu_t), \quad \mu_t = \sum_0^{\infty} \lambda_{s+t} g_s,$$

the n-dimensional probability density $p_n(x_1, ..., x_n)$ of the random variables $X_1, ..., X_n$ (assuming that it exists) is given by the following n-fold integral

$$p_n(x_1, ..., x_n) = \int_{-\infty}^{\infty} \cdots \int \exp\left\{i \sum_1^n \lambda_t x_t\right\} \prod_{-\infty}^{\infty} \psi_\varepsilon(\mu_t) d\lambda_1 \cdots d\lambda_n$$

$$\mu_t = \sum_{-\min(0, t-1)}^{n-t} \lambda_{s+t} g_s,$$

In the particular case of an autoregressive process of q-th order (4.1) with ε's possessing probability density p_ε (obviously not necessarily Gaussian), this integral is reduced to the following expression

$$p_n(x_1, ..., x_n) = p_n(x_1, ..., x_q) \prod_{q+1}^{n} p_\varepsilon(x_t - \iota_1 x_{t-1} - \cdots - \iota_1 x_{t-q})$$

namely a linear autoregression is a Markov process.

2. Arguing analogously it is easy to obtain also the density of the conditional distribution of variables $X_1, X_2, ...$ (under the condition $X_0 = 0$, $X_{-1} = 0$, ...):

$$p(X_1, X_2, ... \mid X_0 = 0, \ X_{-1} = 0, \ ...)$$

$$= \prod_{t-1}^{\infty} p_\varepsilon(x_t + d_1 x_{t-1} + \cdots + d_{t-1} x_1),$$

where d_j is the coefficient at z^j in the expansion of $(\sum_{s=0}^{\infty} g_s z^s)^{-1}$ in the powers of z (cf. Appendix 3 to the preceding Chapter, Remark 7 in Section 2).

3. We now turn to the case of a linear autoregression (cf. Remark 1). Assume that the density p_ε of the distribution of ε's determines a positive and bounded quantity of Fisher's information with respect to the location parameter

$$I_\iota = \int_{-\infty}^{\infty} \iota^2(x) p_\varepsilon(x) dx, \quad \iota = (\partial/\partial x)\log p_\varepsilon$$

We shall assume that, along with $\iota_1, ..., \iota_q$, the density p_ε is also an unknown parameter. We thus have here a case of the so-called semi-parametric (or parametric-nonparametric) model in the spirit of the paper [179], for example.

Let $P_n(p_\varepsilon, \iota)$ be a probability distribution induced on the sample space with the density indicated at the end of Remark 1. Since we are interested here in the asymptotic results, the first factor in the expression for this density can be neglected.

Let for some $\iota_1^{(1)}, ..., \iota_q^{(1)}$ and functions a with

$$\int_{-\infty}^{\infty} x^i a(x) p_\varepsilon(x) dx = 0, \quad i = 0, 1,$$

$$\int_{-\infty}^{\infty} a^2(x) p_\varepsilon(x) dx = I_a < \infty.$$

we have

$$\iota_{in} = \iota_i + n^{-1/2}\iota_i^{(1)}, \quad i = 1, ..., q,$$

$$p_\varepsilon^{(n)} = p_\varepsilon(1 + n^{-1/2}a)$$

where $p_\varepsilon^{(n)}$ is a probability density. Along with $\varepsilon_t = X_t - \iota_1 X_{t-1} - \cdots - \iota_q X_{t-q}$ denote $\varepsilon_t^{(n)} = X_t - \iota_{1n}X_{t-1} - \cdots - \iota_{qn}X_{t-q}$.

Consider the logarithm of the likelihood ratio

$$\Lambda(p_\varepsilon, \iota; p_\varepsilon^{(n)}, \iota_n) = \log[dP_n(p_\varepsilon^{(n)}, \iota_n)/dP_n(p_\varepsilon, \iota)]$$

of measures corresponding to the values of the parameters p_ε, ι, and p_ε, ι_n respectively.[23]

In accordance with what is stated above, that the first factor in the expression for the n-dimensional density is immaterial (presented at the end of Remark 1), we have up to asymptotically negligible summands

$$\Lambda(p_\varepsilon, \iota; p_\varepsilon^{(n)}, \iota^{(n)}) = \sum_{t=q+1}^{n} \{\log[p_\varepsilon(\varepsilon_t^{(n)})/p_\varepsilon(\varepsilon_t)]$$

$$+ \log[1 + n^{-1/2}a(\varepsilon_t^{(n)})]\}.$$

Using standard arguments based on Taylor's expansion, this representation of the logarithm of the likelihood ratio leads to an asymptotic formula for the form (9) in Appendix 1 of the present Chapter, i.e.,

$$\Lambda(p_\varepsilon, \iota; p_\varepsilon^{(n)}, \iota^{(n)}) - \delta_n + \frac{1}{2}(I_a + \gamma^2) \to 0 \quad \text{as } n \to \infty$$

in $P_n(p_\varepsilon, \iota)$ probability, where

$$\delta_n = n^{-1/2} \sum_{t=q+1}^{n} \{a(\varepsilon_t) - \ell(\varepsilon_t) \sum_{s=1}^{q} \iota_s^{(1)} X_{t-s}\},$$

and

$$\gamma^2 = I_\ell \sigma^2 \cdot \iota^{(1)'} \Gamma_\iota^{(q)} \iota^{(1)}.$$

---

[23] Here $\iota = \mathrm{col}(\iota_1, \ldots, \iota_q)$ and $\iota_n = \mathrm{col}(\iota_{1n}, \ldots, \iota_{qn})$. Below we also introduce $\iota^{(1)} = \mathrm{col}(\iota_1^{(1)}, \ldots, \iota_q^{(1)}) = n^{1/2}(\iota_n - \iota)$.

Evidently, in the particular case of a Gaussian density p_ε with parameters $(0,\sigma^2)$ we have $\mathit{l}(x) = -x/\sigma^2$ and $I_{\mathit{l}}\sigma^2 = 1$, in full agreement with what is known about this special case. Observe that, in a non-Gaussian case, $I_{\mathit{l}}\sigma^2 > 1$ since from the Cauchy-Schwarz inequality $\{I_{\mathit{l}}\sigma^2\}^{1/2} \geqslant \int x\mathit{l}(x)p_\varepsilon(x)dx = 1$ (with the equality only in the case when $\mathit{l}(x)$ is proportional to x).

Next, the independence and identical distribution of ε's leads to the assumption that

$$\{\delta_n \mid P_n(p_\varepsilon, \iota)\} \rightarrow N(0, I_a + \gamma^2).$$

This means that in the general linear case also, one can substantiate the LAN property of distribution induced on the sample space.

As far as the problem of estimating the parameters $\iota_1, ..., \iota_q$ is concerned, under the conditions discussed herein, the following results are generally valid:

(i)   If an observer knows the distribution p_ε, he (she) can determine (the asymptotic) MLE from the condition of maximizing $\Pi_{t=q+1}^{n} p_\varepsilon(\varepsilon_t)$ with respect to the parameters $\iota_1, ..., \iota_q$ hoping to attain here the lower bound for possible asymptotic values of covariances matrices of "regular" estimators which is in accordance with the Cramér-Rao inequality, the inverse of Fisher's information matrix $I_{\mathit{l}}\sigma^2 \Gamma_{\iota}^{(q)}$. The inequality $I_{\mathit{l}}\sigma^2 \geqslant 1$ (with the equality only in the Gaussian case) explains the assertion about the "worst case" -- the Gaussian case presented in the last paragraph of Subsection 6.3.

(ii) In principle, one can construct estimators of the parameters $\iota_1, ..., \iota_q$ with the above-mentioned optimal asymptotic variance since here an adaptation is admissible (cf. [179], Corollary 3.1). Examples of such adaptive estimators are given in [172] and [178]. (In the latter paper, general ARMA models are considered.)

(iii) The quantity $\{I_{\mathit{l}}\sigma^2\}^{-1} \leqslant 1$ may serve as the measure of asymptotic efficiency of the least squares estimators $\tilde{\iota}_1, ..., \tilde{\iota}_q$ (cf. [189]).

4.   One can assume that the results about autoregressive parameters presented above can be extended also to the case of a general linear process (6.1) with parameter θ (only the

coefficients g_s depend on this parameter). However, the modifications in the arguments required here are far from being always easily substantiated.

In accordance with Remark 2, the asymptotic MLE will now be determined from the condition of maximization with respect to Θ of the expression $\prod_{t=1}^{n} p_{\mathcal{E}}(\eta_t)$, $\eta_t = X_t + d_1 X_{t-1} + \cdots + d_{t-1} X_1$ which depends on Θ via the coefficients d_j.

If we now consider $\Lambda(p_{\mathcal{E}}, \Theta; p_{\mathcal{E}}^{(n)}, \Theta^{(n)})$ where $\Theta^{(n)} = \Theta + n^{-1/2} \mathbf{h}$ then the above-stated remains in general valid with

$$\delta_n = n^{-1/2} \sum_{t=1}^{n} [a(\eta_t) + \mathbf{1}(\eta_t) \cdot \mathbf{h}'(\partial/\partial\Theta)\eta_t]$$

and

$$\gamma^2 = I_{\mathbf{1}} \sigma^2 \cdot \mathbf{h}' \Gamma_{\Theta} \mathbf{h}$$

(cf. (2.14)).

5. In the autoregressive representation (4.1) we shall assume that the innovations ε_t are martingale differences satisfying the conditions (15) and (16) of Appendix 1. The predictable ($A_{t-1}^{(\mathcal{E})}$-measurable) regressor $a_t = a_{t,\iota} = \iota_1 X_{t-1} + \cdots + \iota_q X_{t-q}$ of the martingale difference ε_t is linearly dependent on the unknown parameters $\iota = \mathrm{col}\{\iota_1, \ldots, \iota_q\}$ so that the autoregressive scheme can be considered as a special case of the general linear regression model admitting stochastic regressors also ([190,191]).[24]

In the framework of the approach described herein, the transition from an autoregressive scheme to a more general ARMA model is not a simple step (cf. [194]). Readers interested in learning about the degree of expected difficulty associated with the nonlinearity of the regressor with respect to the estimated parameters can find the relevant information in paper [195] devoted to nonlinear regression.

---

[24]The nice example of applying martingale methods is given in [192], Section VII.5. Some useful discussion on related subjects can be found in [193].

Section 7

1. If the time series X_t is allowed to be d-vector-valued with an everywhere nondegenerate spectral density matrix $f = f_\Theta$, $\Theta \in \Theta$, then the formulas (2.17) and (2.14) are extended as follows:

$$\Delta_{n,\Theta} = -\frac{1}{4\pi} \mathrm{col}\left\{ I_n\left[\frac{\partial}{\partial\Theta_1}f_\Theta^{-1}\right], ..., I_n\left[\frac{\partial}{\partial\Theta_p}f_\Theta^{-1}\right] \right\}$$

and

$$\Gamma_\Theta = \left\| \frac{1}{4\pi}\int_{-\pi}^{\pi} \mathrm{tr}\left\{ f_\Theta^{-1}(\lambda)\,\frac{\partial}{\partial\Theta_i}f_\Theta(\lambda)f_\Theta^{-1}\,\frac{\partial}{\partial\Theta_j}f_\Theta(\lambda)\right\}d\lambda \right\|_{i,j=1,...,p}.$$

Here (and below) $I_n(h)$ is a "smoothed periodogram" (cf. Appendix 1, formula (24)):

$$I_n(h) = n^{1/2}\int_{-\pi}^{\pi} \mathrm{tr}\{h(\lambda)[I_n(\lambda) - f(\lambda)]\}d\lambda,$$

$I_n(\lambda) = d_n(\lambda)d_n^*(\lambda)$ being the d-dimensional periodogram (see, e.g., [61]),

$$d_n(\lambda) = \frac{1}{2\pi n^{1/2}}\sum_{t=1}^{n} X_t e^{i\lambda t}.$$

As it is well-known (see [26, 165; also Dahlhaus (1981), Schwache konvergenz bei einer klasse von spectralschatzern. Dissertation zur Erlangung des Grades eines Dr.rer.nat., Universitat Essen]), under appropriate regularity conditions on a fourth-order stationary time series X_t the variables $I_n(h_1), ..., I_n(h_k)$ are asymptotically normal for a wide class of $(d \times d)$-matrices $h_1, ..., h_k$ and the covariance structure of the limiting normal distribution is given by (cf. Appendix 1, formula (26))

$$\frac{1}{(4\pi)^2}\mathrm{cov}\{I_n(h_i), I_n(h_j)\}$$
$$\to \frac{1}{4\pi}\int_{-\pi}^{\pi} \mathrm{tr}\{h_i(\lambda)f(\lambda)h_j(\lambda)f(\lambda)\}d\lambda$$
$$+ \frac{1}{8\pi}\int\int_{-\pi}^{\pi} \mathrm{tr}\{[h_i(\lambda) \otimes h_j(\mu)]f_4(\lambda,-\lambda,\mu,-\mu)\}d\lambda d\mu$$

Here $f_4(\lambda_1,\lambda_2,\lambda_3,\lambda_4)$ is a $(d^2 \times d^2)$-matrix with the fourth-order cospectral density $f_{ijk\ell}(\lambda_1, \ldots, \lambda_4)$ between i, j, k, and ℓ-th components of X_t (defined in the symmetric form; see [26], p. 26) as an entry at the $((i-1)d+k)$-th row and the $((j-1)d+\ell)$-th column; in the Gaussian case $f_4 = 0$, of course. But in the general (not necessarily Gaussian) case we have

$$\Delta_{n,\Theta} \sim N(0, \Gamma_\Theta + C_\Theta)$$

where C_Θ is the $(p \times p)$-matrix with entry (cf. Subsection 7.2, the expression for $C_{f_4,\Theta}$)

$$\frac{1}{8\pi} \iint_{-\pi}^{\pi} \mathrm{tr}\left\{ \left[\frac{\partial}{\partial\Theta_i} f_\Theta^{-1}(\lambda) \otimes \frac{\partial}{\partial\Theta_j} f_\Theta^{-1}(\mu) \right] f_4(\lambda,-\lambda,\mu,-\mu) \right\} d\lambda\, d\mu$$

$$= \frac{1}{8\pi} \iint_{-\pi}^{\pi} \mathrm{tr}\left\{ \left[f_\Theta^{-1}(\lambda) \otimes f_\Theta^{-1}(\mu) \right] \right.$$

$$\left. \cdot \left[\frac{\partial}{\partial\Theta_i} f_\Theta(\lambda) \otimes \frac{\partial}{\partial\Theta_j} f_\Theta(\mu) \right] \psi(\lambda,\mu) \right\} d\lambda\, d\mu$$

at the i-th row and j-th column. Obviously (cf., Appendix 2, formula (24))

$$\psi(\lambda,\mu) = [f_\Theta(\lambda) \otimes f_\Theta(\mu)]^{-1} f_4(\lambda,-\lambda,\mu,-\mu).$$

Thus for the Whittle estimator $\tilde\Theta$ we have (cf., Section 7, Theorem 2)

$$n^{1/2}(\tilde\Theta-\Theta) \sim N(0, \Gamma_\Theta^{-1}(\Gamma_\Theta + C_\Theta)\Gamma_\Theta^{-1}).$$

2. As in Subsection 6.4 consider the class of estimators $\hat\Theta_h$ that satisfy the asymptotic relation

$$\left| n^{1/2}(\hat\Theta_h-\Theta) - \frac{1}{4\pi}\mathrm{col}\{I_n(h_1), \ldots, I_n(h_p)\} \right| \to 0$$

in $P_n(f_\Theta)$ probability for any $(d \times d)$ matrix-valued functions $h_1, \ldots, h_p$ such that

$$(*) \qquad \frac{1}{4\pi} \int_{-\pi}^{\pi} \mathrm{tr}\left[h_i(\lambda)\frac{\partial}{\partial\Theta_j} f(\lambda) \right] d\lambda = \delta_{ij}, \qquad i,j = 1, \ldots, p,$$

and

$$\frac{1}{4\pi} \iint_{-\pi}^{\pi} \text{tr}\left\{ \left[h_i(\lambda) + \sum_k y^{ik} \frac{\partial}{\partial\theta_k} f^{-1}(\lambda) \right] \right.$$

(**)

$$\left. \otimes \sum_l y^{jl} \frac{\partial}{\partial\theta_l} f^{-1}(\mu) f_4(\lambda,-\lambda,\mu,-\mu) d\lambda d\mu = 0, \right.$$

$$i,j = 1, ..., p.$$

where y^{kl} are the entries of Γ_θ^{-1}.

In the important particular case of a (multidimensional) linear process X_t the last highly cumbersome condition follows from the preceding one. Indeed representing X_t in the form (6.1) we shall assume that the innovations ε_t are vector-valued with a nondegenerate covariance matrix Σ and that the entries σ_{ij} of this matrix are unknown. (As in the scalar case, Subsections 6.3 and 6.4, we shall assume that the matrix-valued coefficients g_s do not depend on σ_{ij} but that they may of course depend on some other unknown parameters.) Then (cf. [165])

$$f_4(\lambda,-\lambda,\mu,-\mu) = \frac{1}{2\pi} \sum_{i,j,k,l} \kappa_{ijkl} \frac{\partial}{\partial\sigma_{ij}} f(\lambda) \otimes \frac{\partial}{\partial\sigma_{kl}} f(\mu),$$

where κ_{ijkl} is the fourth cumulant among the corresponding components of ε_t. Thus

$$\psi(\lambda,\mu) = \frac{1}{2\pi} \sum_{i,j,k,l} k_{ijkl} \Phi_{ij}(\lambda) \otimes \Phi_{kl}(\mu)$$

where

$$k_{ijkl} = \sum_{a,b} \sigma^{ia} \sigma^{kb} \kappa_{ajbl}, \qquad [\sigma^{ij}] = \Sigma^{-1},$$

while

$$\Phi_{ij}(\lambda) = \sum_k \sigma_{ik} f^{-1}(\lambda) \frac{\partial}{\partial\sigma_{kj}} f(\lambda).$$

It is easily seen that

$$\frac{1}{2\pi} \int_{-\pi}^{\pi} \text{tr}[\Phi_{ij}(\lambda)\Phi_{kl}(\lambda)] d\lambda = \delta_{il}\delta_{kj}.$$

Taking into acount that σ_{ij} are some of the components of vector θ these formulas allow us to verify that (*) implies (**).

We now turn to the class of estimators $\hat{\theta}_h$. This class contains Whittle's estimator $\tilde{\theta}$ with a special choice of $h_i = -\Sigma_k y^{ik}(\partial/\partial\theta_k)f^{-1}$. The asymptotic variance of the estimators $\hat{\theta}_h$ is determined by the asymptotic relation

$$nE(\hat{\theta}_h-\theta)^{\otimes 2} \to G_h^{-1}[\Gamma_h+C_h]G_h^{-1}{}',$$

where the $(p \times p)$-matrices G_h, Γ_h, and C_h have entries

$$\frac{1}{4\pi}\int_{-\pi}^{\pi} \text{tr}\left[h_i(\lambda)f^{-1}(\lambda)\frac{\partial}{\partial\theta_j} f(\lambda)\right]d\lambda,$$

$$\frac{1}{4\pi}\int_{-\pi}^{\pi} \text{tr}[h_i(\lambda)h_j(\lambda)]d\lambda,$$

$$\frac{1}{8\pi}\iint_{-\pi}^{\pi} \text{tr}\{[h_i(\lambda) \otimes h_j(\mu)]\ \psi(\lambda,\mu)\}d\lambda d\mu$$

at the i-th row and j-th column.

Now it is easy to verify the following generalization of formula (6.16) (cf. Appendix 2, Subsection 3)

$$\lim_{n\to\infty} nE(\hat{\theta}_h-\theta)^{\otimes 2} \geqslant \Gamma_\theta^{-1}(\Gamma_\theta + C_\theta)\Gamma_\theta^{-1},$$

provided only that h satisfies (*) and (**) (moreover in the linear case described above (**) follows from (*)).

Chapter III
SIMPLIFIED ESTIMATORS POSSESSING "NICE" ASYMPTOTIC PROPERTIES

1. Asymptotic Properties of Simplified Estimators

1.1. The examples considered in Sections 4 and 5 of the preceding Chapter indicate that the asymptotic m.l. estimators $\tilde{\theta}$ of the parameters θ appearing in the expression for spectral density f_θ of a Gaussian random process X_t, $t = ...,-1,0,1, ...$ while they are simpler than the exact m.l. estimators $\bar{\theta}$, they are nevertheless most often roots of rather complex nonlinear equations so that their determination also requires a substantial amount of time and effort. Only the problem of estimating the parameters ι_1, ..., ι_q and σ^2 in the autoregressive process with spectral density (II.4.3) was an exception. In Subsection 4.1 of the preceding Chapter it was shown that for this problem the asymptotic m.l. estimators $\tilde{\iota}_1$, ..., $\tilde{\iota}_q$ are roots of a simple system of linear equations (II.4.6) with respect to the variables ι_1, ..., ι_q and that the estimator $\tilde{\sigma}^2$ of the parameter σ^2 is given by a relatively simple formula (II.4.7). However, already in the case of a moving average process of the first order (not to mention more complex moving average processes of higher orders or mixed autoregressive-moving average processes) the determination of an asymptotic m.l. estimator turns out to be a complicated problem.

In view of this complexity, substantial attention is being devoted in scientific literature to construct various simplified estimators which possess for large n sufficiently high accuracy. The first relatively crude results of this kind dealing with the estimation of coefficients of polynomials in expressions for rational (with respect to $z = e^{i\lambda}$) spectral density $f(\lambda)$ of the observed process X_t were obtained by Durbin [59,60] and Walker [129,130]. These results are described in detail, for example in Anderson's book [4] (Sections 5.7 and 5.8), and we shall not dwell on them here (cf. also [140] Chapter VI, Section 4 or [71] Chapter 50). Recently, however, substantially better results were obtained in relation to this problem in papers by Hannan [141], Clevenson [74], Parzen [100], and Anderson [5]. These results allow us, starting with the practically arbitrary consistent estimators Θ_*, to construct improved estimators $\tilde{\Theta}$ which in the case of Gaussian processes, are asymptotically efficient (while under more general conditions possess the same asymptotic properties as those introduced in the preceding Chapters for the estimator $\tilde{\Theta}$). The proof of asymptotic efficiency in [140,141] and [74] was based on direct (and rather cumbersome) calculations of the covariance matrix D_Θ of the proposed estimators and of verification that $\lim_{n\to\infty} nD_\Theta$ indeed coincides with the matrix Γ_Θ^{-1}, while in the paper [100] a proof for asymptotic efficiency is not given, although it seems that the author had in mind the same approach. Later, however, it was shown (cf. [2,53,54]) that the results of all the papers mentioned above can be obtained by means of a simple general argument which has been utilized -- albeit in a somewhat different form -- in applied mathematics since the 17th Century and which was utilized in mathematical statistics in relation to other problems. Below we shall discuss this general approach to the determination of simplified estimators.

1.2. We shall begin -- following the author's paper [42] -- by considering the general problem of estimating the unknown value of some p-dimensional vector-valued parameter $\Theta = (\Theta_1, ..., \Theta_p)'$ which appears in the expression for the n-dimensional distribution $P_{n,\Theta}$ of the observations $X_1, ..., X_n$ (it is possible that this parameter does not completely specify the distribution). The results related to this problem will be applied below to the case when Θ is a parameter of the spectrum of a stationary process X_t.

Definition. Let τ_n^*, $n = 1,2, ...,$ be an unboundedly increasing sequence of positive numbers. The estimator

$$\Theta_* = \Theta_{n*}(X) = \{\Theta_{1*}(X), ..., \Theta_{p*}(X)\}, \quad X = (X_1, ..., X_n)'$$

of parameter Θ is called τ_n^*-consistent if the p-dimensional random vector $\tau_n^*(\Theta_*-\Theta)$ is bounded in probability $P_{n,\Theta}$, i.e., if for any $\varepsilon > 0$ there exist positive numbers $c_0(\varepsilon)$ and $n_0(\varepsilon)$ such that

$$P\{\tau_n^* |\Theta_{j*}-\Theta_j| > c_0(\varepsilon), \quad j = 1, ..., p\} < \varepsilon$$

for all $n > n_0(\varepsilon)$.

Assume that the following assumptions are satisfied.

Assumption 1. There exists an τ_n^*-consistent estimator Θ_* and bounded in probability p-dimensional random column-vector $\Phi_{n,\Theta} = \Phi_{n,\Theta}(X)$ depending on the values of the parameter Θ such that

(1) $$\Phi_{n,\Theta_*} - \Phi_{n,\Theta} + J_\Theta \tau_n(\Theta_*-\Theta) \to 0$$

in $P_{n,\Theta}$ probability, where J_Θ is a nonsingular $(p\times p)$-matrix with non random entries (depending in general on the values of Θ) and τ_n, $n = 1,2, ...,$ is a sequence of positive numbers unboundedly increasing at least as fast as τ_n^*, but such that

(2) $$\tau_n^{1/2}/\tau_n^* \to 0$$

as $n \to \infty$.

Assumption 2. There exist $(p\times p)$-matrix $D_* = D_{n*}(X)$ with random elements which is a τ_n^*-consistent estimator of the matrix $D_\Theta = J_\Theta^{-1}$ in the sense that all the entries of the matrix $\tau_n^*(D_*-D_\Theta)$ are bounded in probability $P_{n,\Theta}$.

Under the above stated assumptions the following lemma is valid.

Lemma 1. *Let*

(3) $$\vec{\Theta}_n(X) = \vec{\Theta} = \Theta_* + \tau_n^{-1}D_*\Phi_{n,\Theta_*}$$

where

(4) $\Phi_{n,\Theta} - J_\Theta \tau_n(\vec{\Theta}-\Theta) \to 0$

in probability $P_{n,\Theta}$.

Proof. Since, in view of (1) and (3)

$$\tau_n(\vec{\Theta}-\Theta) = \tau_n(\Theta_*-\Theta) + D_*[\Phi_{n,\Theta} - J_\Theta \tau_n(\Theta_*-\Theta)+\xi_n]$$

(5) $$= D_\Theta \Phi_{n,\Theta} + \tau_n^*(D_*-D_\Theta)[(\tau_n^*)^{-1}\Phi_{n,\Theta}$$

$$- \tau_n(\tau_n^*)^{-2}J_\Theta \tau_n^*(\Theta_*-\Theta)] + D_*\xi_n,$$

where ξ_n is a random p-dimensional column-vector all of whose entries converge to zero in probability; the assertion of Lemma 1 follows from condition (2), the τ_n^*-consistency of the estimator D_* of the matrix D, and boundedness in probability of the vectors $\Phi_{n,\Theta}$ and $\tau_n^*(\Theta_*-\Theta)$.   □

Suppose now that, in addition, the following assumption is valid.

Assumption 3. The distribution of a random vector $\Phi_{n,\Theta}$ approaches, as $n \to \infty$, a p-dimensional normal distribution $N(0,\mathbf{J})$ with a zero expected value and a fixed covariance matrix $\mathbf{J}$ (whose entries in general in addition to the value of Θ may also depend on the values of other parameters appearing in the expressions of finite dimensional probability densities of the process X_t).

Under the assumptions 1-3 the following lemma is valid.

Lemma 2. *The random vector* $\tau_n(\vec{\Theta}-\Theta)$, *as* $n \to \infty$, *possesses p-dimensional normal distribution* $N(0,D_\Theta \mathbf{J} D_\Theta')$.

The proof of Lemma 2 easily follows from (4) and Assumption 3.

It follows from the assertion of Lemma 2 that the random vector $\vec{\Theta}$ is a τ_n-consistent asymptotically normal estimator of the parameter Θ.

1.3.   As it is well-known, the usual methods for constructing an estimator $\tilde{\theta}$ of an unknown parameter θ utilized in mathematical statistics (such as the method of moments, least squares method, or the maximum likelihood method) are usually reduced to a determination of a root $\tilde{\theta}$ of some system of equations in unknown θ of the form

(6) $\Phi_{n, \theta}(X) = 0$,

where $\Phi_{n, \theta}$ is a random p-dimensional column-vector whose elements are functionals in X dependent on θ. Here, in many cases, one succeeds in proving that for sufficiently large values of n with probability close to 1 there exists a root $\tilde{\theta}$ of the system (6), belonging to the convex region $\theta \in R_p$ of admissible values, which is a consistent estimator of θ. It is known that classical estimation methods utilized in mathematical statistics often lead to asymptotically normal τ_n-consistent estimators with "nice" asymptotic properties. In such cases $\tilde{\theta}$ usually satisfies relation (1) for $\theta_* = \tilde{\theta}$ and hence, in view of (6),

(7) $J_\theta \tau_n(\tilde{\theta}-\theta) - \Phi_{n, \tilde{\theta}} \to 0$

in probability $P_{n, \theta}$ as $n \to \infty$, while $\Phi_{n, \theta}$ satisfies Assumption 3, so that as $n \to \infty$ the random vector $\tau_n(\tilde{\theta}-\theta)$ possesses p-dimensional normal distribution $N(0, D_\theta J D_\theta')$. However, the problem of determining the root $\tilde{\theta}$ of the system of equations (6) turns out often to be computationally extremely complex. Moreover, in those cases where there exist several distinct roots of equation (6), the problem acquires additional difficulty to single out the root which actually estimates θ. Therefore, the problem of determining a more easily constructed estimator which possesses the same asymptotic properties as the estimator $\tilde{\theta}$ is of substantial interest.

Now let the random vector $\Phi_{n, \theta}$ appearing in formula (3) coincide with the random vector appearing in the l.h.s. of (6) and let D_* be a τ_n^*-consistent estimator of the matrix D_θ satisfying relation (1). Then, in view of (4) and (7)

(8) $\tau_n(\tilde{\theta}-\hat{\theta}) \to 0$

in $P_{n, \theta}$ probability as $n \to \infty$, and thus the estimator $\hat{\theta}$ determined by formula (3) may just serve as an example of the desired simplified estimator. Indeed, in view of (8) the distributions of the probabilities of estimators $\tilde{\theta}$ and $\vec{\theta}$, as $n \to \infty$, behave in the same manner, while the structure of the estimator $\vec{\theta}$ is far simpler than that of $\tilde{\theta}$. Indeed, to determine $\vec{\theta}$ one must only possess certain τ_n^*-consistent estimators θ_* and D_* of the parameter θ and the matrix D_θ respectively, and then solve a linear system of equations obtained as a result of multiplying both sides of (3) by J_*.

Observe that in the case when the entries of the vector $\Phi_{n,\theta}$ satisfy certain regularity condition in θ by applying the mean value theorem we obtain

$$(9) \qquad \Phi_{n,\theta_*} - \Phi_{n,\theta} = J_{\theta'}\tau_n(\theta_*-\theta),$$

where J_θ is the Jacobian with the multiplier τ_n^{-1} and $\theta' \in [\theta^*,\theta]$. From (1), (9), and the boundedness of the vector $\tau_n^*(\theta_*-\theta)$ in $P_{n,\theta}$ probability we have

$$\frac{\tau_n}{\tau_n^*}[J_{n,\theta'} + J_\theta] \to 0$$

in $P_{n,\theta}$ probability, so that, in view of (2), the matrix $J_{n,\theta'}$ may serve in this case as a τ_n^*-consistent estimator of the matrix $-J_\theta$ (under the above mentioned regularity conditions on $\Phi_{n,\theta}$ the matrix J_{n,θ_*} is also such an estimator). Therefore setting in (3) $D_* = -J_{n,\theta_*}^{-1}$ we obtain that the random vector

$$(10) \qquad \vec{\theta}^{(1)} = \theta_* - (\tau_n J_{n,\theta_*})^{-1}\Phi_{n,\theta_*}$$

will also be an estimator of θ asymptotically equivalent to the root $\tilde{\theta}$ of the system of equations (6).

On the other hand, it is well-known that if the roots of the system of equations (6) are obtained using the approximate iterative Newton-Raphson's method, the first iterative cycle results in the estimator $\vec{\theta}^{(1)}$ (provided a certain τ_n^*-consistent estimator θ_* is chosen as the initial value). Recall that Newton-Raphson's method is the iterative process

$$(11) \qquad \theta^{(k+1)} = \theta^{(k)} - (\tau_n J_{n,\theta^{(k)}})^{-1}\Phi_{n,\theta^{(k)}}, \quad k = 0,1,...,$$

which, under certain natural restrictions on $\Phi_{n,\theta}$, is convergent to the root $\tilde{\theta}$ of the equation (6) (cf., e.g., [94]). Now if as the initial approximate solution $\theta^{(0)}$ of equation

(9) a certain τ_n^*-consistent estimator Θ_* is chosen, then already the first approximation $\Theta^{(1)}$ of the root $\tilde{\Theta}$ of system (6) -- which in view of (10) and (11) for $k = 1$ coincides with the estimator $\tilde{\Theta}^{(1)}$ -- will have, as $n \to \infty$, the same distribution as the estimator $\tilde{\Theta}$. Obviously, all the succeeding approximations $\Theta^{(k)}$, $k = 2,3, ...$, will be asymptotically equivalent (as $n \to \infty$) to the estimator $\tilde{\Theta}$ (in the sense described in [165], Section 5).

It was the Newton-Raphson method that we had in mind in Section 1.1 when we referred to a simple argument widely utilized in applied mathematics. The reasoning of the method presented here as applied to the classical problem of determining the maximum likelihood estimators of the parameter Θ of probability density $p(x,\Theta)$ for independent and identically distributed random variables X_1, ..., X_n is well-known. In this case the role of the k-th entry of the vector $\Phi_{n,\Theta}$ is played by the partial derivative with respect to the k-th component Θ_k of the vector Θ of the logarithm of the usual likelihood function $\sum_{i=1}^{n}\log p(X_i,\Theta)$ multiplied by $n^{-1/2}$ (here $\tau_n = \sqrt{n}$); the (k,ℓ)-entry of the matrix $J_{n,\Theta}$ is the random variable (cf., e.g., [64] Section 5.2 and Section 5.5 or [77] Section 18.21)

$$\frac{1}{n}\sum_{j=1}^{n} \frac{\partial^2}{\partial\Theta_k\partial\Theta_\ell}\log p(X_j,\Theta).$$

We also note that in the case when all the entries of a nonrandom matrix J_Θ are continuous in Θ, one can choose in particular the matrix J_{Θ_*} as a τ_n^*-consistent estimator J_* of the matrix J_Θ. Thus

(12) $\tilde{\Theta}^{(2)} = \Theta_* + (\tau_n J_{\Theta_*})^{-1}\Phi_{n,\Theta_*}$

is an important example of an estimator of parameter Θ, asymptotically equivalent to the root $\tilde{\Theta}$ of the system (6).

The estimator $\tilde{\Theta}^{(2)}$ defined by (12) where $\Phi_{n,\Theta}$ is as above while J_Θ the Fisher's information matrix (cf. formula (21) of the Introduction) was introduced by Fisher [135] in 1935 under the name "method of scoring for parameters." Fisher applied this to the problem of estimating the parameter Θ appearing in the expression for the probability density $p(x,\Theta)$. The word scoring signified that in the cases when the "improved" estimator $\tilde{\Theta}^{(2)}$ may seem to be not sufficiently adequate, we can resort to the next iteration and again repeat the construction where $\vec{\theta}^{(2)}$ plays the role of the initial estimator θ_*; operation (12) may thus be repeated many times, successively accumulating corrections. Nowadays Fisher's method of scoring is described in many well-known textbooks on mathematical statistics (cf., e.g., [106] Section 5.2, [64], Section 5.2, [71], paragraph 18.21). Fisher applied the scoring method to a sample of a fixed (albeit large) size n in order to proceed from an arbitrary consistent estimator θ_* to a new estimator possibly close to a maximum likelihood estimator (which is actually of no special advantage for finite values of n); the motivation for using J_{θ_*} rather than J_{n,θ_*} as the estimator of J_θ is that J_{θ_*} is of a simple form and is relatively close to J_{n,θ_*} for large n. Later in 1956 LeCam [79] observed that if in formula (12) (again applied only to estimators of parameters of probability density constructed from independent observations) one sets $n \to \infty$, then a family of estimators $\vec{\theta}^{(2)}$ is obtained which is asymptotically equivalent to the maximum likelihood estimator $\tilde{\theta}$, and hence are asymptotically efficient under general conditions. (Note that it was mentioned in [79] that the conditions for the asymptotic efficiency of $\vec{\theta}^{(2)}$ do not in general assure the existence of maximum likelihood estimators $\tilde{\theta}$; moreover, even if a maximum likelihood estimator $\tilde{\theta}$ exists and is uniquely determined, these conditions do not imply its consistency.)

1.4. Later LeCam generalized the results of his paper [79] to the case of an arbitrary sequence of experiments

$$E_n = \{X_n, \mathfrak{U}_n, P_{n,\theta}\}, \quad \theta \in \Theta, \quad n = 1, 2, \dots ,$$

where X_n is a set of possible outcomes of the n-th experiment (i.e., for example, the set of possible values of the random vector $\mathbf{X} = (X_1, \dots, X_n)'$), $\mathfrak{U}_n$ is a σ-algebra defined on X_n, and $P_{n,\theta}, \theta \in \Theta$ is a family of distributions on $\mathfrak{U}_n$ asymptotically differentiable in the sense of the definition presented on the Introduction on page 21 (cf. [80-82]).

The following two well-known facts which are formulated for convenience below in the form of two lemmas serve as the basis for all of the arguments utilized by LeCam for constructing asymptotically efficient estimators of an unknown parameter θ.

Lemma 4. *Let* $P_{n,\Theta}$, $\Theta \in \Theta$ *be a family of asymptotically differentiable distributions. Then the following assertions are valid.*

1) *For all* $\Theta \in \Theta$ *the sequence of distributions*

$$L(\Delta_{n,\Theta} \mid P_{n,\Theta}), \quad n = 1,2, \dots ,$$

of a random vector $\Delta_{n,\Theta}$ *(appearing in condition (D2)) on page 21 of the Introduction) converges to the distribution* $L_{\Theta}(\Delta)$ *(dependent in general on the value of* Θ*) of a random vector* Δ:

$$L(\Delta_{n,\Theta} \mid P_{n,\Theta}) \to L_{\Theta}(\Delta).$$

2) *For almost all* $\Theta \in \Theta$ *the distribution* $L_{\Theta}(\Delta)$ *coincides with the normal distribution.*
3) *If, in addition, the distribution* $L_{\Theta}(\Delta)$ *is continuous in* Θ*, then for all* $\Theta \in \Theta$ *this distribution coincides with the normal distribution* $N(0,\Gamma_{\Theta})$ *with zero mathematical expectation and covariance matrix* Γ_{Θ} *defined in (D2) on page 21.*

Lemma 5. *Let the family of distributions* $P_{n,\Theta}$, $\Theta \in \Theta$ *be asymptotically differentiable. Then for almost all* $\Theta \in \Theta$,

(13) $$\Delta_{n,\Theta_n} - \Delta_{n,\Theta} + \Gamma_{\Theta}h \to 0, \quad \Theta_n = \Theta + \tau_n^{-1}h_n$$

in $P_{n,\Theta}$ *probability as* $n \to \infty$*, where* h_n*,* $n = 1,2, \dots$*, is a sequence of vectors convergent to* h *in* R_p*. Relation (13) is valid also for all* $\Theta \in \Theta$ *provided* $\Delta_{n,\Theta}$ *is chosen appropriately (below it will always be assumed when necessary that this choice of* $\Delta_{n,\Theta}$ *is being made).*

Lemmas 4 and 5 coincide with the Theorems 1 and 2 on page 62 and Assertion 1 on page 68 of [81] which contains the complete proofs.

Assume now, following LeCam (cf. [81], conditions 5) and 6) on page 79) that along with the conditions (D1)-(D4) of asymptotic differentiability (cf. page 21) the following conditions are also valid.

(D5)   The limiting distribution $L_\Theta(\Delta)$ appearing in the statement of Lemma 4 is continuous in $\Theta \in \Theta$ so that in view of the assertion 3) of Lemma 4 $L_\Theta(\Delta)$ coincides with the normal distribution $N(0,\Gamma_\Theta)$. Moreover, the matrix Γ_Θ is nondegenerate for all $\Theta \in \Theta$.

(D6)   There exists a τ_n-consistent estimator Θ_* of the parameter Θ.

From considerations presented on page 81 of [81] it follows that if the estimator Θ_* is such that for a fixed value of n it may take only a finite number of possible values, then in the relation

$$\Delta_{n,\,\Theta_n} - \Delta_{n,\,\Theta} + \Gamma_\Theta \mathbf{h}_n \to 0 \quad \text{as } n \to \infty$$

where $\Theta_n = \Theta + \tau_n^{-1}\mathbf{h}_n$ (in probability $P_{n,\,\Theta}$) -- which is valid in view of Lemma 5 -- the vector $\mathbf{h}_n$ can be replaced by $\tau_n(\Theta_*-\Theta)$ so that

$$(14) \qquad \Delta_{n,\,\Theta_*} - \Delta_{n,\,\Theta} + \Gamma_\Theta \tau_n(\Theta_*-\Theta) \to 0$$

in $P_{n,\,\Theta}$ probability as $n \to \infty$. If the entries of the random vector $\Delta_{n,\,\Theta}$ are sufficiently smooth functions of Θ, then relation (14) is valid for an arbitrary τ_n-consistent estimator Θ_*. Furthermore, we shall assume that the conditions assuring the validity of relation (14) for some τ_n-consistent estimator Θ_* are fulfilled. Note that under condition (D6) this latter assumption is actually always satisfied since the following lemma is valid (cf. Lemma 4 on page 155 of [82]):

Lemma 6. *There exists a τ_n-consistent estimator $\hat{\Theta}$ of parameter Θ (dependent on Θ_*) such that for every $b \in (0,\infty)$ the number of its possible values contained in a set of diameter b/τ_n does not exceed a certain number k_b.*

A possible method of transition from the estimator Θ_* to its discrete modification $\hat{\Theta}$ is described for example, in [82], pp. 155-156.

It follows directly from relation (14) that in the situation under consideration the random vector $\Delta_{n,\,\Theta} = \Phi_{n,\,\Theta}$ satisfies Assumption 1 in Subsection 1 (cf. page 200) for $\tau_n^* = \tau_n$ and $J_\Theta = \Gamma_\Theta$.

If, furthermore, there exists a τ_n-consistent estimator Γ_* of the matrix Γ_Θ (in the case when the elements of the matrix Γ_Θ are continuous in Θ, the matrix $\Gamma_{\Theta*}$ can clearly be chosen for such an estimator) then in view of Lemma 1

(15) $\Delta_{n,\Theta} - \Gamma_\Theta \tau_n(\check{\Theta}-\Theta) \to 0$ as $n \to \infty$

in $P_{n,\Theta}$ probability, where

(16) $\check{\Theta} = \Theta_* + (\tau_n \Gamma_*^{-1})\Delta_{n,\Theta_*}.$

Thus the following lemma is valid.

Lemma 7. *Under the conditions (D1)-(D6) the estimator $\check{\Theta}$ of the parameter Θ defined by the formula (16) -- where Θ_* and Γ_* are some τ_n-consistent estimators of parameter Θ and matrix Γ_Θ respectively -- is asymptotically normal in the sense that the distribution $L(\tau_n(\check{\Theta}-\Theta) \mid P_{n,\Theta})$ of the random vector $\tau_n(\check{\Theta}-\Theta)$, as n $\to \infty$, tends to the normal distribution $N(0,\Gamma_\Theta^{-1})$ with zero mean and covariance matrix Γ_Θ^{-1}.*

1.5.   The general results presented in the preceding subsection can naturally be utilized in the particular case when $P_{n,\Theta}$, $\Theta \in \Theta$ is a family of Gaussian distributions.

Assume first that X_t, $t = ...,-1,0,1, ...$, is a Gaussian random process with zero expectation and spectral density $f = f_\Theta$, $\Theta \in \Theta$ which satisfies the conditions of Section 2 of the preceding Chapter. In view of the Theorem I.3.5, under these conditions, the family of distributions $P_{n,\Theta}$, $\Theta \in \Theta$, is asymptotically differentiable (in the sense of the definition on page 21, where $\tau_n = \sqrt{n}$, $\Delta_{n,\Theta}$ is a random vector whose k-th entry is given by the formula (II.2.17) and Γ_Θ is a $(p \times p)$-matrix whose $(k \times l)$-th entry is of the form (II.2.14). Also (II.2.18) holds, i.e., the sequence $L(\Delta_{n,\Theta} \mid P_{n,\Theta})$, $n = 1,2, ...$, of the distributions of the random vector $\Delta_{n,\Theta}$ converges to the normal distribution $N(0,\Gamma_\Theta)$. Here the validity of the relation (14) for $\tau_n = \sqrt{n}$ and any $\sqrt{n}$-consistent estimator Θ_* follows from the arguments analogous to those presented in Section 2 of the preceding Chapter in the course of the proof of assertion 1) of theorem 2. Thus the following corollary to Lemma 7 is valid.

Corollary 1. *Under the conditions of Theorem II.2.2 the* *estimator $\tilde{\theta}$ of the parameter θ defined by formula (16) (where $\tau_n = \sqrt{n}$, $\Delta_{n,\theta}$ is a random vector with entries (II.2.17) and Γ_* is a consistent estimator of matrix Γ_θ with entries of the form (II.2.14)) is consistent, asymptotically normal, and asymptotically efficient in the sense that the sequence of distributions*

$$L[\sqrt{n}\,(\tilde{\theta}-\theta) \mid P_{n,\theta}], \quad n = 1, 2, \dots ,$$

converges to the normal distribution $N(0,\Gamma_\theta^{-1})$.

Assume now that the spectral density $f = f_q$ of the process X_t is given by the formula (I.2.11) where the funtion $f_0 = f_{0,\theta}$ satisfies the conditions of Theorem 3.2 of the preceding Chapter. Then, in view of the corollary to I.4.1 the family of distributions $P_{n,\theta}$, $\theta \in \Theta$, is as before asymptotically differentiable, however, now the k-th element of the vector $\Delta_{n,\theta}$ is defined by the formula (II.3.13). Following the argument analogous to the one given in the proof of Lemma II.3.4 it is easy to verify that relation (14) is valid in this case also. Thus the assertions of Lemmas II.3.3 and 7 yield

Corollary 2. *Under the conditions of Theorem II.3.2 the assertion of Corollary 1 is valid if the estimator $\tilde{\theta}$ of the parameter θ is again given by the formula (16) provided the k-th element of the vector $\Delta_{n,\theta}$ is of the form (II.3.13).*

1.6. Assume now that X_t, $t = \dots,-1,0,1, \dots$, is a linear process of the form (II.6.1). Then it follows from the results of Subsection 6.2 in Chapter II that if the spectral density $f_\theta > 0$ of the process X_t satisfies the conditions of Lemma II.6.2 then the distribution $L(\Delta_{n,\theta} \mid P_{n,\theta})$ of a random vector $\Delta_{n,\theta}$ whose k-th entry is given by (II.2.17) approaches, as $n \to \infty$, the normal distribution $N(0,\Gamma_\theta + C_{\kappa_4,\theta})$ with zero expectation and

covariance matrix $\Gamma_\theta + C_{\kappa_4,\theta}$ (cf. Subsection 6.2 of Chapter

II). Thus in the case under consideration, Proposition 3 of Subsection 2 is valid with $\Phi_{n,\theta} = \Delta_{n,\theta}$ and $J = \Gamma_\theta + C_{\kappa_4,\theta}$.

Using the same argument as in the proof of Lemma II.6.6 it is easy to show that under the conditions of Theorem II.6.2, Proposition 1 is also valid with $\Phi_{n,\theta} = \Delta_{n,\theta}$, $J_\theta = \Gamma_\theta$; and $\tau_n =$

$\tau_n^* = \sqrt{n}$.

Since under the above-stated conditions the entries of the matrix Γ_Θ are continuous functions of Θ one can use, for example, the matrix Γ_{Θ^*} as an $\sqrt{n}$-consistent estimator Γ_* of the matrix Γ_Θ.

Thus the following corollary to Lemma 2 is valid.

Corollary 3. *Under the conditions of Theorem II.6.2 the distribution of the random vector*

$$n^{1/2}(\Theta_* - \Theta) + \Gamma_*^{-1}\Delta_{n,\,\Theta_*}$$

(where $\Delta_{n,\,\Theta}$ is defined by the formula (II.2.17), and Θ_* and Γ_* are $\sqrt{n}$-consistent estimators of the parameter Θ and the matrix Γ_Θ respectively) approaches the normal distribution

$$N(0,\Gamma_\Theta^{-1} + \Gamma_\Theta^{-1}C_{\kappa_{4,\,\Theta}}\Gamma_\Theta^{-1})$$

as $n \to \infty$.

From this and Theorem II.6.2 it follows that the estimator

$$\vec{\Theta} = \Theta_* + (n^{1/2}\Gamma_*)^{-1}\Delta_{n,\,\Theta_*}$$

of the parameter Θ is asymptotically equivalent to the least squares estimator $\tilde{\Theta}$ and thus possesses the same asymptotic properties as the latter estimator (cf. Section 6 of Chapter II).

Finally we note that from the results of Section 7 of the preceding Chapter directly follows the asymptotic equivalence of the estimators $\vec{\Theta}$ and $\tilde{\Theta}$ of the parameter Θ provided only the stationary process X_t, $t = ...,-1,0,1, ...,$ satisfies the conditions stipulated in this section.

2. Examples of Preliminary Consistent Estimators

2.1. As it was indicated in the preceding Section, in order to construct simplified estimators $\vec{\Theta}$ using formula (1.16) one should start with a definition of some consistent estimators Θ_*.

The requirement of consistency cannot be considered overly restrictive -- almost all estimators of practical value satisfy this condition. Therefore, the literature contains numerous examples of consistent estimators of various spectral parameters. A substantial amount of these is based on the fact that the values of the empirical covariance function $\beta_n^*(\tau)$ (cf. formula (II.4.5)) under very general conditions converges in probability as $n \to \infty$ to the true values $\beta(\tau)$ (i.e., $\beta_n^*(\tau)$ is a consistent and even $\sqrt{n}$-consistent estimator of $\beta(\tau)$; cf., e.g., [56], Section 5.3). It follows from the $\sqrt{n}$-consistency of $\beta_n^*(\tau)$ that in particular

$$\rho_n^*(\tau) = \beta_n^*(\tau)/\beta_n^*(0)$$

will be a $\sqrt{n}$-consistent estimator of the correlation function $\rho(\tau) = \beta(\tau)/\beta(0)$. Next, under the same condition of consistency of estimator $\beta_n^*(\tau)$ for function $\beta(\tau)$ it is easy to show that the statistic

$$I_n(A) = \int_{-\pi}^{\pi} I_n(\lambda)A(\lambda)d\lambda,$$

where $I_n(\lambda)$ is the periodogram of the process X_t under general conditions on the function $A(\lambda)$ (cf. Appendix 1 to the preceding Chapter), will be a $\sqrt{n}$-consistent estimator of the quantity

$$f(A) = \int_{-\pi}^{\pi} f(\lambda)A(\lambda)d\lambda.$$

Now we can choose, for example, some p values $\beta_n^*(\tau_1)$, ..., $\beta_n^*(\tau_p)$, $\rho_n^*(\tau_1)$, ..., $\rho_n^*(\tau_p)$, or $I_n(A_1)$, ..., $I_n(A_p)$ and equate them to the corresponding values of $\beta(\tau_1)$, ..., $\beta(\tau_p)$, $\rho(\tau_1)$, ..., $\rho(\tau_p)$, or $f(A_1)$, ..., $f(A_p)$ (dependent on the unknown parameters Θ_1, ..., Θ_p). We shall thus obtain a system of p equations with p unknowns Θ_1, ..., Θ_p whose solution (at least in those cases when this solution is uniquely determined) will represent an $\sqrt{n}$-consistent estimator of parameters Θ_1, ..., Θ_p.

As we have seen, the number of different consistent estimators of parameters of a spectrum of stationary processes is very large. Therefore, some criteria are needed to select the most appropriate ones. The most natural criteria are the requirement of simplicity of calculation of corresponding estimators and the requirement of high precision for finite (even if large) values of n.

As far as the precision of the estimator Θ_* is concerned, it is natural to characterize it by the $(p \times p)$-matrix

$$D_n = E(\Theta_* - \Theta)(\Theta_* - \Theta)'.$$

(This matrix which for large values of n differs only slightly from the covariance matrix of random variables Θ_{1*}, ..., Θ_{p*} will be called, as above, the covariance matrix of estimators Θ_*.) Since, as a rule, the estimators Θ_* depend nonlinearly on $X = (X_1, ..., X_n)'$ it is clear that for computing the matrix D_n, information concerning the probability distributions of process X_t is required which is beyond the bounds of its spectral density f (or covariance function $\beta(\tau)$). Next, in order to be able to judge whether a given estimator is "sufficiently nice" it is quite important to be able to estimate the limiting matrix $D_n^{(0)}$ as well, which corresponds to the optimal (i.e., the best possible) estimators of the parameter Θ regardless of how complicated the "best estimators" are. Knowing the limiting matrix $D_n^{(0)}$ and the covariance matrix D_n of the estimator Θ_* one can then decide whether or not it is worthwhile to improve the available estimator Θ_* (since the improvement may not be substantial).

An accurate calculation of the matrices D_n and $D_n^{(0)}$ in most cases is associated with overcoming great difficulties. However, although the investigation of the asymptotic behavior of these matrices as $n \to \infty$ is not easy, nevertheless it is a significantly simpler problem.

2.2. Moving Average Process

Assume that the spectral density f of a Gaussian random process X_t is represented in the form

(1) $$f(\lambda) = \frac{1}{2\pi}|g_r(z)|^2, \qquad g_r(z) = \gamma_0 + \gamma_1 z + \cdots + \gamma_r z^r,$$

$$z = e^{i\lambda},$$

where $\gamma_0, \gamma_1, ..., \gamma_r$ are real numbers such that all the roots of the polynomial $g_r(z)$ are greater than 1 in their absolute value. Assume, furthermore, that the coefficients $\gamma_0, \gamma_1, ..., \gamma_r$ are just the unknown parameters of the spectral density (1) (cf. with the case considered in 4.2 of the preceding Chapter when the quantities $\sigma = \gamma_0$, $\alpha_1 = -\gamma_1/\gamma_0$, ..., $\alpha_r = -\gamma_r/\gamma_0$ were the unknown parameters). Since the covariance matrix $\beta(\tau)$ of the form

$$
(2) \qquad \beta(\tau) = \begin{cases} \sum_{j=0}^{r-|\tau|} y_j y_{j+|\tau|}, & \text{for } |\tau| \leqslant r, \\ \\ 0, & \text{for } |\tau| > r \end{cases}
$$

corresponds to this density (cf. (II.4.14)) it is natural to select the solutions of the following system of nonlinear equations

$$
(3) \qquad \sum_{j=0}^{r-|\tau|} y_{j*} y_{j+\tau*} = \beta_n^*(\tau), \qquad \tau = 0, 1, ..., r,
$$

as the estimators $y_{0*}, y_{1*}, ..., y_{r*}$ of the parameters $y_0, y_1, ..., y_r$.

In principle the system (3) can be very easily solved. Indeed, it is easy to show that if

$$
g_{r*}(z) = y_{0*} + y_{1*} z + \cdots + y_{r*} z^r,
$$

$$
\beta_*(z) = \sum_{\tau=-r}^{r} \beta_n^*(\tau) z^\tau,
$$

then $\beta_*(z) = g_{r*}(z) g_{r*}(z^{-1})$. Therefore, obtaining all the roots of the polynomial $z^r \beta_*(z)$ and selecting those which exceed 1 in their absolute value, we can construct the polynomial $g_{r*}(z)$ to obtain $y_{0*}, y_{1*}, ..., y_{r*}$. The difficulty is that it is not a simple task to obtain all the roots of a polynomial of degree r for $r > 1$. It is therefore desirable to develop a specific method for its solution which would be relatively easily executed on a computer. A simple iteration method for the solution was proposed by Wilson [120] (cf. also [24], Section 6.2).

In certain cases, it may also be convenient to rewrite formula (1) in the form (II.4.13) where $\sigma = y_0$, $\alpha_1 = -y_1/y_0$, ..., $\alpha_r = -y_r/y_0$ and to choose $\alpha_1, ..., \alpha_r$ and σ^2 as the new unknown parameters. In this case, the consistent estimators of these parameters will, for example, be the estimators

$$
(4) \qquad \sigma_*^2 = y_{0*}^2, \qquad \alpha_{1*} = -y_{1*}/y_{0*}, \quad ..., \quad \alpha_{r*} = -y_{r*}/y_{0*},
$$

where $y_{0*}, y_{1*}, ..., y_{r*}$ are arbitrary consistent estimators of parameters $y_0, y_1, ..., y_r$. As it was mentioned in 4.2 of the preceding Chapter, the covariances $\beta(0), \beta(1), ..., \beta(r)$ defined by formula (2) can be chosen as the unknowns. Their simplest consistent estimators will be the empirical covariances $\beta_n^*(0), \beta_n^*(1), ..., \beta_n^*(r)$. In a somewhat more general case when f is defined by the formula (II.4.20), the equations (3) are evidently linear in the unknowns Θ_{1*}, ..., Θ_{p*} and are of the form

(5)
$$\beta_n^*(\tau) = \sum_{k=1}^{p} \Theta_{k*} \tilde{w}_k(\tau), \quad \tau = 0, 1, ..., p-1,$$
$$\tilde{w}_k(\tau) = \int_{-\pi}^{\pi} w_k(\lambda) e^{i\tau\lambda} d\lambda.$$

Example 1. For the case of the spectral density of the form (II.4.23), the relations (4), which define the simple consistent estimators α_{1*} and σ_*^2 of the parameters α_1 and σ^2, are of the form (this case was discussed in the preceding Chapter II, Subsection 4.2)

(6)
$$\rho_n^*(1) = -\frac{\alpha_{1*}}{1-\alpha_{1*}^2}, \quad \alpha_{1*}\sigma_*^2 = -\beta_n^*(1).$$

The quadratic equation (6) possesses a real valued solution only for $|\rho_n^*(1)| \leqslant 1/2$. In this case

(7)
$$\alpha_{1*} = 2\rho_n^*(1)\{1 + [1 - 4\rho_n^*(1)^2]^{1/2}\}^{-1}.$$

It is easy to verify (cf. [121] or [139], Section II.4, or [140] Section VI.4) that

$$D(\alpha_{1*}) = \frac{1+\alpha_1^2+4\alpha_1^4+\alpha_1^6+\alpha_1^8}{n(1-\alpha_1)^2} + o(1/n),$$

so that in view of (II.4.26) the asymptotic efficiency of the estimator α_{1*} equals

$$(1-\alpha_1)^3/(1+\alpha_1^2+4\alpha_1^4+\alpha_1^6+\alpha_1^8).$$

For $\alpha_1 = 1/4$ this value is 0.76 while for $\alpha_1 = 1/2$, it is 0.26.

2.3. Mixed Autoregressive-Moving Averages Process

Here we are dealing with the case already discussed in Subsection 4.3 of the preceding Chapter when the spectral density f is of the form (II.4.28). First we shall assume that the unknown parameters are

(8)
$$\iota_1, ..., \iota_q, \quad \gamma_0 = \sigma, \quad \gamma_1 = -\alpha_1\sigma, ..., \gamma_r = -\alpha_r\sigma.$$

In view of (II.4.29) the estimators ι_{1*}, ..., ι_{q*} of the parameters ι_1, ..., ι_q are the roots of the linear system of equations

$$\beta_n^*(\tau) - \sum_{j=1}^{q} \iota_{j*} \beta_n^*(\tau-j) = 0, \qquad \tau = r+1, ..., r+q,$$

and will be consistent provided only the estimators $\beta_n^*(\tau)$ are consistent estimators of $\beta(\tau)$.

Since the covariance function $\beta_y(\tau)$ of the process (II.4.35) (satisfying (II.4.36)) equals the r.h.s. of the relation (2), after the estimators ι_{1*}, ..., ι_{q*} of the parameters ι_1, ..., ι_q were obtained, the estimators $\gamma_{0*}, \gamma_{1*}, ..., \gamma_{r*}$ of the parameters $\gamma_0, \gamma_1, ..., \gamma_r$ can be found by solving the system of equations of the form (3) in which the r.h.s. $\beta_n^*(\tau)$ are now replaced by

(9) $$\beta_y^*(\tau) = \sum_{k, \ell = 0}^{q} \iota_{k*} \iota_{\ell*} \beta_n^*(\tau + k - \ell), \qquad \iota_{0*} = -1.$$

Here the estimators $\gamma_{0*}, \gamma_{1*}, ..., \gamma_{r*}$ also turn out to be consistent.

If the spectral density is rewritten now in the form (II.4.37) where ι_1, ..., ι_q, $\beta_y(0)$, ..., $\beta_y(r)$ is a new collection of unknown parameters then one can choose the estimators ι_{1*}, ..., ι_{q*}, $\beta_y^*(0)$, ..., $\beta_y^*(r)$ discussed above as their simple consistent estimators.

Finally we note that

$$\alpha_{1*} = -\gamma_{1*}/\gamma_{0*}, \quad ..., \quad \alpha_{r*} = -\gamma_{r*}/\gamma_{0*}, \qquad \sigma_*^2 = \gamma_{0*}^2$$

will be the consistent estimators of the parameters α_1, ..., α_r, σ^2 in view of (8).

Example 2. Let the spectral density f of the process X_t be of the form (II.4.41) where ι_1, α_1 and σ^2 are unknown parameters. It is easy to verify that in this case the "recommendation" of this subsection reduces to the determination of the estimators ι_{1*}, α_{1*}, and σ_*^2 from the solution of the following equations

$$\beta_n^*(2) - \iota_{1*} \beta_n^*(1) = 0, \qquad \sigma_*^2(1+\alpha_{1*}^2) = \beta_n^*(0)(1+\iota_{1*}^2) - 2\iota_{1*}\beta_n^*(1),$$

$$-\alpha_{1*}\sigma_*^2 = \beta_n^*(1)(1+\iota_{1*}^2) - \iota_{1*}[\beta_n^*(0) + \beta_n^*(2)].$$

In view of the relation (II.5.13) the problem of estimating

the parameters ι_1, α_1, and σ^2 discussed here is equivalent to the problem of determining consistent estimators of the parameters ι_1 and σ_S^2 of the signal S_t, which is an autoregressive process of the first order and the intensity σ_N^2 of the white noise which masks the signal (cf. Example II.5.1 and Example 3 below).

2.4. A Signal Observed on a "White Noise" Background

Let X_t represent a sum of two independent processes -- the signal S_t and white noise N_t (cf. Section 5 of Chapter II). Consider the problems of estimating parameters dealing with the Examples 1-5 of Subsection 5.2 in Chapter II which are related to this model. We shall devote special attention to the most important and most frequently considered case (cf., e.g., [39,98,99,128,165]) when an autoregressive process of the q-th order with spectral density

(10) $\qquad f_S(\lambda) = \dfrac{\sigma_S^2}{2\pi}|h_q(z)|^{-2}, \qquad h_q(z) = 1 - \iota_1 z - ... - \iota_q z^q, \quad z = e^{i\lambda},$

depending on unknown parameters ι_1, ..., ι_q, σ_S^2, represents the signal S_t.

Comparing (II.5.3) and (10) in view of (II.5.7) we obtain that in this case (where evidently $r = 0$, $h_0 = 1$) the equations (II.5.10) become

(11) $\qquad \beta_\Theta(k) - \iota_1 \beta_\Theta(k-1) - ... - \iota_q \beta_\Theta(k-q) = c_\Theta(k), \quad k = 0,1,2, ...,$

where $\beta_\Theta(k)$, $\Theta = (\iota_1, ..., \iota_q, \sigma_S^2, \sigma_N^2)'$, is a covariance matrix of the observed process $X_t = S_t + N_t$ (cf. (II.5.51)) and $c_\Theta(k)$ differs from zero only for $k = 0,1, ..., q$, while $c_\Theta(0) = \sigma_S^2 + \sigma_N^2$ and $c_\Theta(k) = -\iota_k$ for $k = 1, ..., q$. Taking into account equation (11) and the consistency of the empirical covariance function $\beta_n^*(\tau)$, Walker [99] suggested to utilize ι_{1*}, ..., ι_{q*}, σ_{S*}^2, and σ_{N*}^2 -- as the estimators of the parameters ι_1, ..., ι_q, σ_S^2, σ_N^2 -- which are the roots of the following equations with respect to ι_1, ..., ι_q, σ_S^2, σ_N^2:

(12)
$$\beta_n^*(k) - \iota_1 \beta_n^*(k-1) - \cdots - \iota_q \beta_n^*(k-q) = c_\Theta(k),$$
$$k = 0,1, ..., q+1.$$

It is easy to show that these roots under very general conditions turn out to be consistent estimators of the parameters $\iota_1, ..., \iota_q, \sigma_S^2, \sigma_N^2$. It is evident that in the case of large values of q the system of equations (12) is rather complicated. However in the case when $q = 1$ or $q = 2$ the roots of equation (12) can be easily obtained (cf. below).

Example 3. For $q = 1$ the consistent estimators $\iota_{1*}, \sigma_{S*}^2$ and σ_{N*}^2 for the unknown parameters ι_1 and σ_S^2 of the spectrum of "the signal" S_t and the variance σ_N^2 of "the noise" N_t are given by the following simple formulas

(13)
$$\iota_{1*} = \beta_n^*(2)/\beta_n^*(1), \quad \sigma_{S*}^2 = ([\beta_n^*(1)]^2-[\beta_n^*(1)]^2)/\beta_n^*(2),$$

$$\sigma_{N*}^2 = \beta_n^*(0) - [\beta_n^*(1)]^2/\beta_n^*(2).$$

Since the covariance matrix of the estimators $\iota_{1*}, \sigma_{S*}^2$ and σ_{N*}^2 is

$$B[\text{cov}\{\beta_n^*(k), \beta_n^*(j)\}]_{k, j=0, 1, 2} B',$$

with

$$B = \begin{bmatrix} 0 & -(1-\iota_1^2)/\sigma_S^2 & (1-\iota_1^2)/\sigma_S^2 \, \iota_1 \\ 0 & 2/\iota_1 & -(1+\iota_1^2)/\iota_1^2 \\ 1 & -2/\iota_1 & 1/\iota_1^2 \end{bmatrix},$$

utilizing Bartlett's formulas [10] for the $\text{cov}\{\beta_n^*(k),\beta_n^*(j)\}$ (cf. Appendix 1 to Chapter II) it is quite simple to write out the formulas for its entries which are valid up to a summand of the form $o(1/n)$.

Example 4. In the case when an autoregressive process of the second order represents the signal S_t, whose spectrum depends on the unknown parameters $\iota_1, \iota_2, \sigma_S^2$, the roots $\iota_{1*}, \iota_{2*}, \sigma_{S*}^2$ and σ_{N*}^2 of equation (12) for $q = 2$ are given by the following formulas

$$\iota_{1*} = \frac{x\beta_n^*(3) - \beta_n^*(1)\beta_n^*(2)}{x\beta_n^*(2) - [\beta_n^*(1)]^2}, \quad \iota_{2*} = \frac{[\beta_n^*(2)]^2 - \beta_n^*(1)\beta_n^*(3)}{x\beta_n^*(2) - [\beta_n(1)]^2},$$

(14)

$$\sigma_{S*}^2 = x - \iota_{1*}\beta_n^*(1) - \iota_{2*}\beta_n^*(2), \quad \sigma_{N*}^2 = \beta_n^*(0) - x,$$

where x is the approprite root of the quadratic equation

$$\beta_n^*(3)x^2 - 2\beta_n^*(1)\beta_n^*(2)x + [\beta_n^*(1)]^3 - \beta_n^*(3)[\beta_n^*(1)]^2$$

$$+ \beta_n^*(1)[\beta_n^*(2)]^2 = 0.$$

Utilizing the formula (II.5.10), one can generalize Walker's method [128] for determining simple consistent estimators of unknown parameters of the autoregression $\iota_1, ..., \iota_q, \sigma_S^2$ and the variance of the noise σ_N^2 to the case when the spectral density f, of the signal S_t is of the form (II.5.3), where $0 \leqslant r < q$ and the unknown parameters are $\iota_1, ..., \iota_q, \alpha_1, ..., \alpha_r, \sigma_S^2$ (together with the noise variance σ_N^2). However, we shall confine ourselves here to the consideration of a single, rather simple, particular case (cf. Example II.5.3) when $r = 1$, $q = 2$.

Example 5. Let f_S be of the form (II.5.18). It follows from the consistency of $\beta_n^*(\tau)$ and the form of the five equations (II.5.10) for $r = 1$ and $q = 2$ that the roots $\iota_{1*}, \iota_{2*}, \alpha_{1*}, \sigma_{S*}^2$, and σ_{N*}^2 of the equations

$$\beta_n^*(0) - \iota_{1*}\beta_n^*(1) - \iota_{1*}\beta_n^*(2) = \sigma_{S*}^2[1 + \alpha_{1*}(\alpha_{1*} - \iota_{1*})] + \sigma_{N*}^2,$$

$$\beta_n^*(1) - \iota_{1*}\beta_n^*(0) - \iota_{2*}\beta_n^*(1) = -\sigma_{S*}^2\alpha_{1*} - \sigma_{N*}^2\iota_{1*},$$

$$\beta_n^*(2) - \iota_{1*}\beta_n^*(1) - \iota_{2*}\beta_n^*(0) = -\sigma_{N*}^2\iota_{2*},$$

$$\beta_n^*(3) - \iota_{1*}\beta_n^*(2) - \iota_{2*}\beta_n^*(1) = 0,$$

$$\beta_n^*(4) - \iota_{1*}\beta_n^*(3) - \iota_{2*}\beta_n^*(2) = 0$$

are consistent estimators of the parameters $\iota_1, \iota_2, \alpha_1, \sigma_S^2$, and σ_N^2. The last two equations easily yield the values of the estimators ι_{1*} and ι_{2*}, while the third gives the value of the estimator σ_{N*}^2:

$$\iota_{1*} = \frac{\beta_n^*(1)\beta_n^*(4) - \beta_n^*(2)\beta_n^*(3)}{\beta_n^*(1)\beta_n^*(3) - [\beta_n^*(2)]^2},$$

$$\iota_{2*} = \frac{[\beta_n^*(3)]^2 - \beta_n^*(2)\beta_n^*(4)}{\beta_n^*(1)\beta_n^*(3) - [\beta_n^*(2)]^2},$$

$$\sigma_{N*}^2 = \beta_n^*(0) + \frac{\iota_{1*}}{\iota_{2*}}\beta_n^*(1) - \frac{1}{\iota_{2*}}\beta_n^*(2).$$

The estimator α_{1*} is a root of the quadratic equation

$$\frac{1 - \alpha_{1*}(\iota_{1*} - \alpha_{1*})}{\alpha_{1*}}$$

$$= \frac{\beta_n^*(1)\iota_{1*}(1 + \iota_{2*}) - (1 - \iota_{2*})\beta_n^*(2)}{(\iota_{1*}^2 + \iota_{2*} - \iota_{2*}^2)\beta_n^*(1) - \iota_{1*}\beta_n^*(2)},$$

and

$$\sigma_{S*}^2 = [(\iota_{2*}^2 - \iota_{2*} - \iota_{1*}^2)\beta_n^*(1) + \iota_{1*}\beta_n^*(2)]/\alpha_{1*}\iota_{2*}.$$

Example 6. Now let the spectral density f_S be of the form (II.5.20). Then if in the first three equations (II.5.18) the values of the covariance function $\beta_\Theta(k)$ of the observed process $X_t = S_t + N_t$ are replaced by their consistent estimators $\beta_n^*(k)$ then the corresponding roots ι_{q*}, σ_{S*}^2, and σ_{N*}^2 will be consistent estimators of the parameters ι_q, σ_S^2, and σ_N^2. It is easy to verify that in this case

$$\iota_{q*} = \beta_n^*(2q)/\beta_n^*(q), \qquad \sigma_{S*}^2 = \frac{[\beta_n^*(q)]^2 - |\beta_n^*(2q)|^2}{\beta_n^*(2q)}$$

(15)

$$\sigma_{N*}^2 = \beta_n^*(0) - \frac{[\beta_n^*(q)]^2}{\beta_n^*(2q)}.$$

Example 7. In the case considered in the Example II.5.5 the roots (with respect to ι_1, σ_S^2, σ_N^2) of the first three equations (II.5.21) will be consistent estimators ι_{1*}, σ_{S*}^2 and σ_{N*}^2 of the parameters ι_1, σ_S^2, and σ_N^2 provided one replaces the values of the covariance function $\beta_\Theta(k)$, $k = 0,1,2$, by the values of the empirical covariance function $\beta_n^*(k)$, $k = 0,1,2$, in these equations. Consequently, (cf. (II.5.23))

$$\iota_{1*} = \beta_n^*(2)/\beta_n^*(1), \qquad \sigma_{S*}^2 = -[\beta_n^*(1)]^2/\beta_n^*(2),$$

(16)

$$\sigma_{N*}^2 = \beta_n^*(0) - 2\beta_n^*(2) - \frac{[\beta_n^*(1)]^2}{\beta_n^*(2)}.$$

2.5. Processes with an Exponential Spectral Density

Consider a model of the random process X_t which was discussed in Subsection 4.4 of Chapter II assuming again that the values of the parameters y_1, ..., y_r and σ^2 are unknown and must be estimated based on n consecutive observations over the random variables X_1, ..., X_n.

In Section 3 of paper [23] simple consistent estimators of the parameters y_j, $j = 1$, ..., r, are proposed for this case

(17) $y_{j*} = \dfrac{2}{n} \sum\limits_{k=1}^{m} \log I_n(\lambda_k) \cos(j\lambda_k), \quad j = 1, ..., r,$

here $\lambda_k = 2\pi k/n$ and $m = [(n-1)/2]$. The distribution of the vector $\sqrt{n}(y_* - y)$ with the j-th element $\sqrt{n}(y_{j*} - y_j)$ as $n \to \infty$

approaches the normal distribution $N(0, \pi^2 I_r/6)$ with zero expectation and covariance matrix $\pi^2 I_r/6$ (cf. [23]). Since the limit of the corresponding Fisher's information matrix as $n \to \infty$ equals I_r (cf. Subsection 4.4 of Chapter II), the asymptotic efficiency of the estimators y_{j*}, $j = 1$, ..., r, is approximately equal to $6/\pi^2$ 0.6. As a consistent estimator σ_*^2 of the parameter σ^2 the estimator

(18) $\sigma_*^2 = 2\pi \exp\left\{ y + \dfrac{2}{n} \sum\limits_{j=1}^{m} \log I_n(\lambda_j) \right\}$

is utilized in [23], where $y = 0.57722...$ is the Euler's constant (cf. also [62]).

2.6. Processes with Spectral Densities Possessing Fixed Zeroes

We shall consider two simple examples when the spectral density $f = f_q$ satisfies the conditions of Section 3 in Chapter II.

Example 8. In the case when the process X_t satisfies the stochastic difference equation (II.4.51) the asymptotic m.l. estimators $\tilde{\iota}_1$ and $\tilde{\sigma}^2$ of the parameters ι_1 and σ^2 are defined by relatively simple formulas (II.4.52).

We shall consider here the even simpler consistent estimators ι_{1*} and σ_*^2 of these parameters determined by the conditions

$$\beta_n^*(0) = \frac{\sigma_*^2}{2\pi} \int_{-\pi}^{\pi} |1-z|^2 |1-\iota_{1*}z|^{-2} d\lambda,$$

$$\beta_n^*(1) = \frac{\sigma_*^2}{2\pi} \int_{-\pi}^{\pi} |1-z|^2 |1-\iota_{1*}z|^{-2} z\, d\lambda, \quad z = e^{i\lambda}.$$

Clearly,

$$\iota_{1*} = 1 + 2\, \frac{\beta_n^*(1)}{\beta_n^*(0)}$$

and

$$\sigma_*^2 = \beta_n^*(0) + \beta_n^*(1).$$

Utilizing the well-known Bartlett's formulas [10] for $\operatorname{cov}[\beta_n^*(k),\beta_n^*(j)]$ we obtain

$$\lim_{n\to\infty} nD(\iota_{1*}) = 2(1+\iota_1), \quad \lim_{n\to\infty} nD(\sigma_*^2) = 3\sigma^4,$$

$$\lim_{n\to\infty} n\, \operatorname{cov}(\iota_{1*},\sigma_*^2) = \sigma^2(1 + \iota_1).$$

Defining the asymptotic efficiency of the estimators ι_{1*} and σ_*^2 as the limit of the relations $D(\tilde{\iota}_1)/D(\iota_{1*})$ and $D(\tilde{\sigma}^2)/D(\sigma_*^2)$ with $n \to \infty$, we obtain that their values are $(1-\iota_1)/2$ and $2/3$ respectively.

Example 9. As it was indicated in Example (II.4.6) the asymptotic m.l. estimator $\tilde{\alpha}_1$ of the parameters α_1 in (II.4.53) is the root of quite a complicated equation (cf. (II.4.25), where $\beta_n^*(j)$ is replaced by $r_{j,y}$). It seems reasonable here to consider a simple consistent estimator α_{1*} which is the root of the equation

$$2\beta_n^*(1)(1 + \alpha_{1*}-\alpha_{1*}^2) - \beta_n^*(0)(1+\alpha_{1*})^2 = 0.$$

It is easy to verify that

$$\lim_{n\to\infty} nD(\alpha_{1*}) = 2 \sum_{j=0}^{8} d_j \alpha_1^j /(1-\alpha_1^2)^2,$$

where $d_0 = d_8 = 1$, $d_1 = d_7 = 2$, $d_2 = ... = d_6 = 3$. Since

$$\lim_{n\to\infty} nD(\tilde{\alpha}_1) = 1 - \alpha_1^2,$$

the asymptotic (as $n \to \infty$) efficiency of the estimator α_{1*}

equals $(1-\alpha_1^2)^3/2\sum_{j=0}^8 d_j \alpha_1^j$; it approaches 0.5 or 0 as α_1 approaches 0 or 1 respectively.

3. Examples of Constructing Simplified Estimators

3.1. Based on the results of Section 1 and the examples presented in Sections 4 and 5 of Chapter II and Section 2 of this Chapter, we are going to attempt to propose several specific "recommendations" for constructing estimators of parameters which will be asymptotically equivalent to the Whittle estimators $\tilde{\theta}$ in all cases of the examples considered.

3.2. Moving Average Process

We start with the case when the spectral density $f = f_\theta$ depends linearly on the unknown parameters $\theta = (\theta_1, ..., \theta_p)$ (cf. (II.4.20)). As we have seen in Subsection 4.2 of Chapter II even in this simple case the asymptotic m.l. equations (II.4.21) are rather complicated. On the other hand, the results of Corollary 1 presented in Section 1 (pp. 208-209) permit us by utilizing formula (1.16) to construct simple estimators $\vec{\theta} = (\vec{\theta}_1, ..., \vec{\theta}_p)$ asymptotically equivalent to the roots of these equations. Indeed, the elements of the vector $\Delta_{n,\theta}$ and of the matrix Γ_θ -- as we know -- are given in this case by formulas (II.4.21) (more precisely, up to a multiplier, $\Delta_{n,\theta}$ coincide with the l.h.s. of this formula) and (II.4.22) provided f_θ is of the form (II.4.20). Therefore, assuming in (1.16) that $\tau_n = n^{1/2}$, $\Gamma_* = \Gamma_{\theta *}$, where θ_* is a p-dimensional random vector whose components are roots of a linear system of equations (2.5), after some simple manipulations we obtain that $\vec{\theta}_j$, $j = 1, ..., p$, are the roots of the following system of linear equations

(1)
$$\int_{-\pi}^{\pi} w_j(\lambda) \left[\sum_{k=1}^p w_k(\lambda)\vec{\theta}_k - I_n(\lambda) \right] \left[\sum_{k=1}^p w_k(\lambda)\theta_{k*} \right]^{-2} d\lambda = 0,$$
$$j = 1, ..., p.$$

Equations (1) actually coincide[1] with the basic equations of the paper [74] reproduced also in [100]. These equations permit us to obtain the asymptotically efficient estimators $\vec{\theta}_1$, ..., $\vec{\theta}_p$, based on given consistent estimators Θ_{1*}, ..., Θ_{p*}, by solving a simple system of p linear equations. In the particular case when X_t is a moving average process of the r-th order with spectral density of the form (II.4.19), where covariances $\beta(0), \beta(1)$, ..., $\beta(r)$ are unknown parameters, the system of linear equations for determining the asymptotically efficient estimators

$$(\vec{\beta}_n(0), \vec{\beta}_n(1), ..., \vec{\beta}_n(r))' = \vec{\beta}_n$$

of these parameters is of the form

(2) $\Gamma_* \vec{\beta}_n = \mathbf{b},$

where $\mathbf{b}$ and Γ_* are $(r+1)$-dimensional column-vector and the $(r+1) \times (r+1)$-dimensional matrix with the entries

$$\int_{-\pi}^{\pi} I_n(\lambda) w_j(\lambda) f_*^{-2}(\lambda) d\lambda, \quad \int_{-\pi}^{\pi} w_k(\lambda) w_j(\lambda) f_*^{-2}(\lambda) d\lambda,$$

$$k, j = 0, 1, ..., r,$$

respectively.

Here the functions $w_0, w_1, ..., w_r$ are the same as in the formula (II.4.19), and f_* is a consistent estimator of f obtained from the formula (II.4.19), where the values of the covariances $\beta(k)$ are replaced by their consistent estimators $\beta_n^*(k)$.

In relation to the estimators of the parameters $\gamma_0, \gamma_1, ..., \gamma_r$ or $\alpha_1, ..., \alpha_r, \sigma^2$, given by formulas (2.1) and (II.4.12)-(II.4.13), it was suggested in the paper [74] to obtain first -- by solving equations (2) -- the consistent estimators $\vec{\beta}_n(0), \vec{\beta}_n(1), ..., \vec{\beta}_n(r)$, and then utilize some method of solving the corresponding equations (2.3) (for instance the method described in the

---

[1]Actually in the papers [5, 39, 74, 100, 141] and the book [140] in place of integrals with respect to λ in the limits $-\pi$ to π, the Riemann sums are used which correspond to the subdivision of the interval $[-\pi, \pi]$ into n equal parts. Replacement of the integral by finite sums is obviously an advantage for implementing corresponding estimation methods on computers. A rather substantial part of the paper [74] is devoted to the justification of such a replacement. We shall, however, not dwell on this problem and confine ourselves to a reference to [74] (see also [26]).

paper [120]). This suggestion seems to us not to be very reasonable since equations analogous to (2) can actually be obtained directly also for estimating parameters $\gamma_0, \gamma_1, ..., \gamma_r$ or $\alpha_1, ..., \alpha_r$.

Indeed, consider the problem of estimating the parameters $\alpha_1, ..., \alpha_r, \sigma^2$ of a moving average process (II.4.11) with the spectral density (II.4.13). If we utilize equations of the form (1.16) for $\vec{\theta} = \vec{\alpha} = (\vec{\alpha}_1, ..., \vec{\alpha}_r)'$ where $\tau_n = n^{1/2}$, $\theta_* = \alpha_*$ is a consistent estimator of α, $\Delta_{n,\theta}$ is a random r-dimensional column-vector, the k-th component of which coincides with the l.h.s. of the equation (II.4.15) multiplied by $-1/\sigma^2$, and $\Gamma_* = \Gamma_{r,\alpha^*}$ (cf. (II.4.17)), we then obtain equations of the form

$$\frac{\sigma_*^2}{2\pi} \int_{-\pi}^{\pi} \sum_{k=1}^{r} \frac{\cos(k-j)d\lambda}{|g_r^*(z)|^2} (\vec{\alpha}_k - \alpha_{k*})$$

(3)
$$+ \int_{-\pi}^{\pi} I_n(\lambda)|g_r^*(z)|^{-4} \left[\cos j\lambda - \sum_{k=1}^{r} \alpha_{k*}\cos(j-k)\lambda \right] d\lambda = 0$$

$$j = 1, ..., r,$$

where

and
$$g_r^*(z) = 1 - \alpha_{1*}z - ... - \alpha_{r*}z^r, \qquad z = e^{i\lambda},$$
$$\sigma_*^2 = \int_{-\pi}^{\pi} I_n(\lambda)|g_r^*(z)|^{-2}d\lambda, \qquad z = e^{i\lambda}.$$

One can, however proceed differently, i.e., to utilize the quantity

$$\sigma_*^{-2} \int_{-\pi}^{\pi} \cos(k-\ell)\lambda I_n(\lambda)|g_r^*(z)|^{-4}d\lambda$$

as a consistent estimator of the (k,ℓ)-th entry of the matrix $\Gamma_{r,\alpha}$. In this case, as it is easy to see, instead of equation (3) we shall arrive at

$$\int_{-\pi}^{\pi} I_n(\lambda)\cos j\lambda|g_r^*(z)|^{-4}d\lambda$$

$$- \sum_{k=0}^{r} \int_{-\pi}^{\pi} I_n(\lambda)|g_r^*(z)|^{-4}\cos(j-k)d\lambda(\alpha_k - 2\alpha_{k*}) = 0,$$

$$j = 1, ..., r.$$

where $\overset{\vee}{\alpha}_1, ..., \overset{\vee}{\alpha}_r$ are the required estimators asymptotically equivalent to Whittle estimators. These equations, in fact, coincide with the equations suggested in the paper [141] and the book [140], Chapter VI, Section 5. After the estimators of the parameters $\alpha_1, ..., \alpha_r$ possessing good asymptotic properties were obtained, for estimating the parameter σ^2 one can simply utilize the formula (II.4.16).

Example 1. In the particular case when $r = 1$, in view of (II.4.25), (2.6), and (3) we have

(5)
$$\vec{\alpha}_1 = \alpha_{1*} \left[1 - \frac{1-\alpha_{1*}^2}{\beta_n^*(1)} \frac{\partial}{\partial \alpha_{1*}} \left\{ \frac{1}{1-\alpha_{1*}^2} \left[\frac{1}{2} \beta_n^*(0) \right. \right. \right.$$
$$\left. \left. \left. + \sum_{j=1}^{n-1} \alpha_{1*}^j \beta_n^*(j) \right] \right\} \right],$$

where α_{1*} is given by the formula (2.7). It follows from (4) that the estimator $\overset{\vee}{\alpha}_1$ using Hannan's method is defined by the simple formula

$$\overset{\vee}{\alpha}_1 = 2\alpha_{1*} + \frac{A_1}{A_0}, \quad A_k = \int_{-\pi}^{\pi} I_n(\lambda)|1-\alpha_{1*}z|^{-4}\cos k\lambda \, d\lambda,$$
$$k = 0,1.$$

3.3. Mixed Autoregressive-Moving Average Process

Assume that f is a general rational function in $z = e^{i\lambda}$ (II.4.28) and the unknown parameters Θ are the parameters $\iota_1, ..., \iota_q, \alpha_1, ..., \alpha_r, \sigma^2$. The problem of estimating the parameter σ^2 will be postponed for a while and we shall first deal with estimation of the parameters $(\iota, \alpha)' = (\iota_1, ..., \iota_q, \alpha_1, ..., \alpha_r)'$. Utilizing the general formula (1.16) we can reduce the determination of the estimators of these parameters to a solution of $(q+r)$ linear equations. Utilizing, however, the fact that the first q components of the $(q+r)$-th dimensional random vector $\Delta_{n,\Theta}$, coinciding with the l.h.s. of equations (II.4.31) (up to the multiplier $\sqrt{n}/\sigma^2$) are linear in $\iota_1, ..., \iota_q$, we can further simplify the determination of the estimators possessing "nice" asymptotic properties and confine ourselves to a solution of several systems of q and r linear equations. For this purpose it is necessary, starting with some consistent estimators $\iota_{1*}, ..., \iota_{q*}, \alpha_{1*}, ..., \alpha_{r*}$ (for example, those discussed in Subsection 2.3), to consider at the second step the

estimators $\iota_{1**}, ..., \iota_{q**}$ obtained in the course of the solution
of q equations (II.4.31) in which $\alpha_1, ..., \alpha_r$ are replaced by $\alpha_{1*},$
$..., \alpha_{r*}$. Now if we use as consistent estimators of Θ_* the
estimators $\iota_{1**}, ..., \iota_{q**}, \alpha_{1*}, ..., \alpha_{r*}, \sigma_*^2$ then the
$(q+r)$-dimensional vector $\Delta_{n,\Theta}$ will be of the form $(0, ...,0,$
$\Delta_{n,\Theta*}^{(q+1)}, ..., \Delta_{n,\Theta*}^{(q+r)})'$, i.e., it will contain only f nonzero

components $\Delta_{n,\Theta*}^{(q+k)}$, $k = 1, ..., r$, where $\Delta_{n,\Theta*}^{(q+k)}$ coincide with the
l.h.s. of the relations (II.4.30) (multiplied by $\sqrt{n}/\sigma_*^2$ and
evaluated at $\Theta = \Theta_*$). The equations (1.16) which determine
the estimators $(\vec{\delta},\vec{\alpha})$ of the parameters (ι,α) became here, in
view of (II.4.33) and $\tau_n = \sqrt{n}$, of the form

$$
(6) \qquad
\begin{bmatrix} \vec{\iota} \\ \vec{\alpha} \end{bmatrix}
=
\begin{bmatrix} \iota_{**} \\ \alpha_* \end{bmatrix}
+
\begin{bmatrix} \Gamma_{\iota}^{(q)} & -(\Omega_*)^* \\ -\Omega_* & \Gamma_{\iota}^{(r)} \end{bmatrix}^{-1}
\begin{bmatrix} 0 \\ n^{-1/2}\Delta_{n,\Theta_*} \end{bmatrix},
$$

where

$$\iota_{**} = (\iota_{1**}, ..., \iota_{q**})', \qquad \alpha_* = (\alpha_{1*}, ..., \alpha_{r*})',$$

$$\Delta_{n,\Theta} = (\Delta_{n,\Theta}^{(q+1)}, ..., \Delta_{n,\Theta}^{(q+r)})',$$

and $\Gamma_{\iota}^{(q)}, \Gamma_{\iota}^{(r)},$ and Ω_* are certain consistent estimators of the
matrices $\Gamma_{\iota}^{(q)}, \Gamma_{\alpha}^{(r)},$ and Ω respectively (for example, those

obtained from $\Gamma_{\iota}^{(q)}, \Gamma_{\alpha}^{(r)},$ and Ω by replacing the values of the
parameters ι and α by their consistent estimators). Applying
the formula for inverting block matrices (see the footnote on
page 41) we obtain from (6):

$$(7) \qquad \vec{\alpha} = \alpha_* + [\Gamma_{\iota}^{(r)} - \Omega_*[\Gamma_{\iota}^{(q)}]^{-1}(\Omega_*)^*]^{-1}n^{-1/2}\Delta_{n,\Theta_*}.$$

    After the estimator $\vec{\alpha}$ of the parameter α has been
obtained, it is then easy to determine the estimator $\vec{\iota}$ of the
parameter ι asymptotically equivalent to the estimator $\tilde{\iota}$; for
example, let $\vec{\iota}$ be a root with respect to ι of a system of q
linear equations (II.4.31) where $g_r(z)$ is replaced by $\vec{g}_r(z) =$
$1-\vec{\alpha}_1 z - ... - \vec{\alpha}_r z^r$. Finally the estimator $\vec{\sigma}^2$ of the parameter σ^2
can be obtained from (II.4.32) where $\tilde{\iota}$ and $\tilde{\alpha}$ are replaced by
$\vec{\iota}$ and $\vec{\alpha}$ respectively.

Now following the papers [53] and [54] (cf. also [2]) we shall show that under the particular choice of the consistent estimators $\Gamma_*^{(r)}$, $\Gamma_*^{(q)}$, and Ω_* of the matrices $\Gamma_\alpha^{(r)}$, $\Gamma_l^{(q)}$, and Ω respectively, the estimator $\vec{\alpha}$ of the parameter α coincides with the estimator suggested in Hannan's book [140] (cf. also [141]). Let

$$\Gamma_*^{(r)} = \left[\sigma_*^{-2} \int_{-\pi}^{\pi} I_n(\lambda) |h_q^{**}(z)|^2 |g_r^*(z)|^{-4} \cdot \right.$$
$$\left. \cdot \cos(j-k)\lambda d\lambda \right]_{j,\,k=1,\ldots,r}$$

where

$$h_q^{**}(z) = 1 - l_{1**}z - \cdots - l_{q**}z^q$$

and

$$g_r^*(z) = 1 - \alpha_{1*}z - \cdots - \alpha_{r*}z^r$$

and $\alpha^{(1)}$ be a new consistent estimator of α in which a root with respect to α of a linear system of equations obtained from (II.4.30) with $g_r(z)$ and $h_q(z)$ replaced by $g_r^*(z)$ and $h_q^{**}(z)$ respectively. It is easy to verify that $\Gamma^{(r)} n^{1/2}(\alpha^{(1)} - \alpha_*) = \Delta_{n,\,\Theta^*}$ so that in view of (6)

$$\vec{\alpha} = [I_r - [\Gamma_*^{(r)}]^{-1}\Omega_*[\Gamma_*^{(q)}]^{-1}(\Omega_*)^*]^{-1}(\alpha^{(1)} - \alpha_*) + \alpha_*.$$

Now if we follow Hannan [140] and utilize $A = (2\pi/\sigma^2)\Gamma_\alpha^{(r)}$ and $B = (\sigma^2/2\pi)\Gamma_l^{(q)}$ in place of $\Gamma_\alpha^{(r)}$ and $\Gamma_l^{(q)}$ respectively we shall arrive at Hannan's equation[2]

(8) $$\vec{\alpha} = [I_r - A_*^{-1}\Omega_*B_*^{-1}(\Omega_*)^*]^{-1}(\alpha^{(1)} - \alpha_*) + \alpha_*,$$

where

$$A_* = \left[\int_{-\pi}^{\pi} \frac{I_n(\lambda)|h_q^{**}(z)|^2 \cos(k-j)\lambda}{2\pi f_y^*(\lambda)} d\lambda \right]_{k,\,j=1,\ldots,r}$$

and

---

[2]The estimator of the form (8) coincides with the estimator proposed in [140] only if all the integers with respect to λ in the limits $-\pi$ to π are replaced by the corresponding Riemann sums (cf. footnote on p. 223).

$$B_* = \left[\int_{-\pi}^{\pi} \frac{I_n(\lambda)\cos(k-j)\lambda}{2\pi|g_r^*(z)|^2} d\lambda\right]_{k,\,j=1,\,\ldots,\,r}$$

are chosen as the consistent estimators A_* and B_* of A and B respectively (here

$$f_y^*(\lambda) = \left[\beta_y^*(0) + 2\sum_{k=1}^{r} \beta_y^*(k)\cos k\lambda\right]/2\pi;$$

see (2.9)), and Ω_* coincides with Ω for $\iota = \iota_*$ and $\alpha = \alpha_*$.

An analogous procedure can be applied also for determining the estimators of the unknown parameters ι_1, ..., ι_q, $\beta_y(0), \beta_y(1)$, ..., $\beta_y(r)$ in the spectral density of the form (II.4.37). Here one can start with some consistent estimators $\beta_y^* = (\beta_y^*(0), \beta_y^*(1), ..., \beta_y^*(r))'$ and then compute the consistent estimators ι_{1*}, ..., ι_{q*} of parameters ι_1, ..., ι_q from the system of linear equations in the first row of (II.4.38) in which $f_y(\lambda)$ is replaced by $f_y^*(\lambda)$. Since the components of the $(q+r+1)$-dimensional vector $\Delta_{n,\Theta}$, $\Theta = (\iota_1, ..., \iota_q, \beta_y(0), ..., \beta_y(r))'$, up to a constant factor, coincide with the l.h.s. of the equations (II.4.38) then clearly

$$\Delta_{n,\Theta_*} = (0, ..., 0, \Delta_{n,\Theta_*}^{(q+1)}, ..., \Delta_{n,\Theta_*}^{(q+r+1)}),$$

where $\Theta_* = (\iota_{1*}, ..., \iota_{q*}, \beta_y^*(0), \beta_y^*(1), ..., \beta_y^*(r))$ (and $\Delta_{n,\Theta}^{(q+k)}$ coincides with the r.h.s. of the $(q+k)$-th equation (II.4.38) multiplied by $1/4\pi$). Consequently, in view of the arguments analogous to those leading to (7) we obtain that

(9) $\tilde{\beta}_y = \beta_y^* + (\Gamma_*^{(r+1)} - \Psi_*[\Gamma_*^{(q)}]^{-1}(\Psi_*)^*)^{-1}n^{-1/2}\Delta_{n,\Theta}$

(where $\Gamma_*^{(q)}\,\Gamma_*^{(r+1)}$ and Ψ_* are certain consistent estimators of the matrices $\Gamma_\iota^{(q)}$, $\Gamma_{\beta_y}^{(r+1)}$, and Ψ respectively, and $\Delta_{n,\Theta} = (\Delta_{n,\Theta}^{(q+1)}, ..., \Delta_{n,\Theta}^{(q+r+1)})'$; cf. (II.4.40)) is an estimator of β_y asymptotically equivalent to the Whittle estimator β_y.

Consider now a new consistent estimator $\beta_y^{(1)} = (\beta_y^{(1)}(0), ..., \beta_y^{(1)}(r))'$ of the parameter β_y satisfying the linear system of equations in the second row of (II.4.38) where $f_y(\lambda)$ and $h_q(z)$ are replaced by $f_y^*(\lambda)$ and $h_q^*(z) = 1 - \iota_{1*}z - \cdots - \iota_{q*}z^q$ respectively. If we choose

$$\Gamma_*^{(r+1)} = \left[\int_{-\pi}^{\pi} \frac{w_k(\lambda)w_\ell(\lambda)}{4\pi[f_y^*(\lambda)]^2} d\lambda\right]_{k,\ell=0,\ldots,r},$$

then it is easy to verify that

$$\Gamma_*^{(r+1)} n^{1/2}(\beta_y^{(1)} - \beta_y^*) = \underline{\Delta}_{n,\Theta_*} .$$

Consequently, equation (9) can be written in the form

$$\tilde{\beta}_y = (I_{r+1} - \Gamma_*^{(r+1)} \Psi_* [\Gamma_*^{(q)}]^{-1}(\Psi_*))^{-1}(\beta_y^{(1)} - \beta_y^*) + \beta_y^*,$$

analogous to Hannan's equation (8). The last result represents a simplification and generalization of Parzen's suggestion [100]. Parzen proposed in fact the very same equations but used for β_y^* a very special estimator of β_y obtained from a complicated three-step procedure.

After the estimator $\tilde{\beta}_y$ of parameters β_y is obtained, one can determine an estimator of ι asymptotically equivalent to the estimator $\tilde{\iota}$ by means of linear equations in the first row of (II.4.38). It is also clear that all the arguments presented above can be applied without any modification in a somewhat more general case of the spectral density of the form

$$f(\lambda) = \sum_{k=0}^{r} w_k(\lambda)\Theta_k \; |1 - \sum_{j=1}^{q} \iota_j z^j|^{-2},$$

where w_k, $k = 0,1, ..., r$, are arbitrarily given functions of λ, and $\Theta_0, \Theta_1, ..., \Theta_r$ and $\iota_1, ..., \iota_q$ are unknown parameters.

Example 2. Let the spectral density f of the process X_t be of the form (II.4.41) where ι_1, α_1, and σ^2 are unknown parameters. To determine the estimators $\tilde{\Theta} = (\tilde{\Theta}_1, \tilde{\Theta}_2)' = (\tilde{\iota}_1, \tilde{\alpha}_1)'$ of the parameters $\Theta = (\Theta_1, \Theta_2)' = (\iota_1, \alpha_1)'$ we shall use the formula (1.16), where $\tau_n = \sqrt{n}$ and the components of a 2-dimensional random vector $\Delta_{n,\Theta}$ are equal to the l.h.s. of the equations (II.4.42) and (II.4.43) multiplied by $n^{1/2}/\sigma^2$ and $-n^{-1/2}/\sigma^2$ respectively,

$$\Gamma_{\Theta_*}^{-1} = \frac{1-\iota_{1*}\alpha_{1*}}{(\iota_{1*}-\alpha_{1*})^2} \begin{bmatrix} (1-\iota_{1*}^2)(1-\iota_{1*}\alpha_{1*}) & (1-\iota_{1*}^2)(1-\alpha_{1*}^2) \\ (1-\iota_{1*}^2)(1-\alpha_{1*}^2) & (1-\alpha_{1*}^2)(1-\iota_{1*}\alpha_{1*}) \end{bmatrix},$$

and $\Theta_* = (\Theta_{1*}, \Theta_{2*})' = (\iota_{1*}, \alpha_{1*})'$ and σ_*^2 are the estimators of the parameters $\Theta = (\Theta_1, \Theta_2)' = (\iota_1, \alpha_1)'$ and σ^2 determined in Example 2.2. After some simple manipulations, we obtain

$$\vec{\theta}_k = \theta_{k*} + (1-\theta_{k*}^2)(1-\theta_{1*}\theta_{2*})(\theta_{1*}-\theta_{2*})^{-2}\sigma_*^{-2}$$

$$\cdot \int_{-\pi}^{\pi} I_n(\lambda)\Big\{|1-\theta_{2*}z|^{-2}(\cos\lambda-\theta_{1*})\Big[1-\frac{\theta_{1*}^2\theta_{2*}}{\theta_{k*}}\Big]$$

$$- (\cos\lambda-\theta_{2*})\frac{|1-\theta_{1*}z|^2}{|1-\theta_{2*}z|^4}\Big[1-\frac{\theta_{1*}\theta_{2*}^2}{\theta_{k*}}\Big]\Big\}d\lambda, \quad k = 1,2.$$

As far as the estimator $\vec{\sigma}^2$ of the parameter σ^2 is concerned, it can be determined by the formula (II.4.44) provided $\tilde{\iota}_1$ and $\tilde{\alpha}_1$ are replaced by $\vec{\iota}_1$ and $\vec{\alpha}_1$ respectively.

Note also that Hannan's equation (8) becomes

$$\vec{\alpha}_1 = \alpha_{1*} + \frac{A_*B_*}{A_*B_*-|\Omega_*|^2}(\alpha_1^{(1)}-\alpha_{1*}),$$

where

$$\alpha_1^{(1)} = A_1^{(1)}/A_0^{(1)},$$

$$A_k^{(1)} = \int_{-\pi}^{\pi} I_n(\lambda)|1-\iota_{1*}z|^2|1-\alpha_{1*}z|^{-4}\cos k\lambda\,d\lambda,$$

$$\iota_{1**} = B_1/B_0, \qquad B_k = \frac{1}{2\pi}\int_{-\pi}^{\pi} I_n(\lambda)|1-\alpha_{1*}z|^{-1}\cos k\lambda\,d\lambda$$

$$A_* = \frac{1}{2\pi}\int_{-\pi}^{\pi} I_n(\lambda)|1-\iota_{1**}z|^2(f_y^*(\lambda))^{-1}\,d\lambda,$$

$$B_* = B_0, \qquad \Omega_* = (1-\alpha_{1*}\iota_{1*})^{-1}.$$

3.4. "Signal" Observed on a "White Noise" Background

Based on the results of Section 5 in Chapter II and the Examples 3-7 of the preceding Section we shall present specific suggestions here for constructing simplified estimators of parameters possessing "nice" asymptotic properties in all the cases of the examples considered.

Example 3. We shall first return to the case already discussed in Example 1 of Section 5 of Chapter II and Example 3 in the preceding section.

In accordance with the results of Section 1 for determining the simplified estimators $\vec{\iota}_1$, $\vec{\sigma}_S^2$, and $\vec{\sigma}_N^2$ of the parameters ι_1, σ_S^2, and σ_N^2 by means of the formula (1.16), it is first required to determine some consistent estimators of the parameters ι_1,

σ_S^2, and σ_N^2. As we have seen above the estimators ι_{1*}, σ_{S*}^2, and σ_{N*}^2 defined by the formulas (2.13) (cf. page 217) can be chosen to be such estimators. Next the matrix $D(\iota_{1*}, \sigma_{S*}^2, \sigma_{N*}^2)$ (cf. Example 1 in Section 5 of Chapter II) can be used for Γ_*^{-1}. However, as it was observed in Example II.5.1 in the case under consideration it is more convenient in place of σ_S^2 and σ_N^2 to introduce the new parameters σ^2 and α_1 connected with ι_1, σ_S^2, and σ_N^2 by the equations (II.5.12). After that it is convenient to obtain the estimators $\vec{\iota}_1$, $\vec{\alpha}_1$, and $\vec{\sigma}^2$ of the parameters ι_1, α_1, and σ^2 using one of the procedures described in the preceding example, next one must replace in equation (II.5.12) the values ι_1, α_1, and σ^2 by their estimators $\vec{\iota}_1$, $\vec{\alpha}_1$, and $\vec{\sigma}^2$ and finally determint the roots $\vec{\sigma}_S^2$ and $\vec{\sigma}_N^2$ of the obtained equations σ_S^2 and σ_N^2.

Example 4. In the case considered in Examples 2 in Section 5 of Chapter II and Example 4 of the preceding section, the preliminary estimators $\Theta_* = (\iota_{1*}, \iota_{2*}, \sigma_{S*}^2, \sigma_{N*}^2)$ can be determined by the formulas (2.14) with $\Gamma_*^{-1} = \Gamma_{\Theta*}^{-1}$ (cf. (5.17)). The l.h.s. of equations (II.5.15) and (II.5.16) are the components of the 4-dimensional vector $\Delta_{n,\Theta}$ multiplied by $n^{1/2}\sigma_S^2$ and $n^{1/2}/2$ respectively. Unfortunately, unlike the case in the preceding example, one is not able to simplify the expressions (1.16) by choosing beforehand the estimators ι_{1*}, ι_{2*}, σ_{S*}^2 and σ_{N*}^2 such that at least one component of the vector $\Delta_{n,\Theta*}$ is equal to zero. One should, however, note that to determine the estimators which are asymptotically equivalent to the estimators $\vec{\Theta} = (\vec{\iota}_1, \vec{\iota}_2, \vec{\sigma}_S^2, \vec{\sigma}_N^2)$ the general procedure can be applied here (as well as in the case considered in the next example). This procedure is essentially a generalization of the iteration method of Davidon, Fletcher and Powell of solving nonlinear systems of equations to the general case considered in Section 1 (cf. the Appendix to this Chapter).

Example 5. In the case considered in Example 3 in Section 5 of Chapter II and Example 5 of the preceding section we have five unknown parameters $\Theta = (\iota_1, \iota_2, \alpha_1, \sigma_S^2, \sigma_N^2)$ whose preliminary estimators $(\iota_{1*}, \iota_{2*}, \alpha_{1*}, \sigma_{S*}^2, \sigma_{N*}^2) = \Theta_*$ are described in Example 2.5. If, now, one chooses the matrix $\Gamma_{\Theta*}^{-1}$ for Γ_*^{-1} (cf. Example II.5.3) then in view of (1.16) (where $\tau_n = \sqrt{n}$ and the components of the 5-dimensional vectors $\Delta_{n,\Theta}$ are given by the l.h.s. of the equations (II.5.19) multiplied by $-1/2$ the simplified estimators $(\vec{\iota}_1, \vec{\iota}_2, \vec{\alpha}_1, \vec{\sigma}_S^2, \vec{\sigma}_N^2) = \vec{\theta}$ will be the roots of a system of linear equations of the fifth order.

Example 6. Return now to the case considered in Example 4 in Section 5 of Chapter II and Example 6 of the preceding section. As the unknown parameters one can choose here ι_q, σ_S^2, σ_N^2 or the parameters ι_q, α_q, and σ^2 determined by the formula (II.5.12), where $\iota_1 = \iota_q$ and $\alpha_1 = \alpha_q$. The preliminary estimators ι_{q*}, σ_{S*}^2, σ_{N*}^2 are given in Example 2.6 (cf. formula (2.15)). The preliminary estimators α_{q*} and σ_*^2 can be determined as the roots with respect to α_1 and σ^2 respectively of the equations obtained by replacing in (II.5.12) the values ι_1, σ_S^2, and σ_N^2 by ι_{q*}, σ_{S*}^2, and σ_{N*}^2. Now to determine the

estimators $\vec{\iota}_q$, $\vec{\alpha}_q$, and $\vec{\sigma}^2$ of the parameters ι_q, α_q, and σ^2 using one of the procedures described in Example 2, one should substitute in all the formulas of this example $\cos q\lambda$ and $z^q = e^{iq\lambda}$ in place of $\cos\lambda$ and $z = e^{i\lambda}$ respectively. In view of (II.5.12) with $\iota_1 = \iota_q$ and $\alpha_1 = \alpha_q$, the estimators $\vec{\sigma}_S^2$ and $\vec{\sigma}_N^2$ can be easily obtained.

Example 7. In the case considered in Example 5 of Section 5, Chapter II and Example 7 of the preceding section, the preliminary estimators ι_{1*}, σ_{S*}^2, and σ_{N*}^2 of the parameters ι_1, σ_S^2, and σ_N^2 can be determined in the manner indicated in Example 2.7 (cf. formula (2.16)). Taking the relations (II.5.24) into account, based on the values ι_{1*}, σ_{S*}^2, and σ_{N*}^2 it is easy to determine also the consistent estimators α_{1*} and σ_*^2 of the parameters α_1 and σ^2. Following now the suggestion of Example 3 it is easy to obtain the simplified estimators $\vec{\iota}_1$, $\vec{\alpha}_1$, and $\vec{\sigma}^2$ of the parameters ι_1, α_1, and σ^2 and in view of (II.5.24) the simplified estimators $\vec{\sigma}_S^2$, $\vec{\sigma}_N^2$ of the parameters σ_S^2 and σ_N^2.

3.5. Process with Exponential Spectral Density

In the case when the spectral density of the process X_t is of the form (II.4.47), where y_1, ..., y_r, σ^2 are unknown parameters, the formula (1.46) applied to the determination of the simplified estimators $(\vec{y}_1, ..., \vec{y}_r) = \vec{y}$ of the parameters $(y_1, ..., y_r) = y$ becomes of a specially simple form. Indeed, in this particular case, we may set (in accordance with the formula (II.4.50)) $\Gamma_* = I_r$. Since the components of the vector $\Delta_{n,\theta}$ coincide with the l.h.s. of the equations (II.4.48) (for $y = \tilde{y}$) multiplied by $\sqrt{n}/\sigma^2$ and $\tau_n = \sqrt{n}$ then in view of (1.16)

$$\vec{y}_k = y_{k*} + \sigma_*^{-2} \int_{-\pi}^{\pi} I_n(\lambda)\cos k\lambda \, \exp\left[-2 \sum_{j=1}^{r} y_{j*}\cos j\lambda\right] d\lambda$$

$$k = 1, ..., r,$$

where $y_{1*}, ..., y_{r*}$ and σ_*^2 are consistent estimators of parameters $y_1, ..., y_r$ and σ^2, for example, those which are defined by the formulas (2.17) and (2.18). As far as the parameter σ^2 is concerned its estimators $\vec{\sigma}^2$ coincide with the r.h.s. of the equation (II.4.49) for $\tilde{y} = \vec{y}$.

3.6. Processes with Spectral Densities Possessing Fixed Zeros

In all the examples discussed in II.4.5 of Chapter II except for the last one (cf. Example 6 in Subsection 4.4), the asymptotic m.l. estimators $\tilde{\theta}$ of the parameters θ, appearing in the expression for spectral densities f possessing fixed zeros, are of a rather simple form. Therefore we shall now concentrate our attention on the case considered in Example 6 of Section 4 of Chapter II when the spectral density is of the form (II.4.53) where α_1 and σ^2 are unknown parameters.

Example 8. As we have seen above (cf. Example 6 of Section 4 in Chapter II and Example 9 of the preceding section) the estimator $\tilde{\alpha}_1$ of the parameter α_1 of the spectral density of the form (II.4.53), which is a root of equation (II.4.25) where $\beta_n^*(j)$ is replaced by $r_{j,y}$, is very complex as compared with the very simple consistent estimator α_{1*} which is a root of a quadratic equation (2.19). Unfortunately, however, the asymptotic efficiency of the latter estimator is rather low (cf. Example 9 on page 221). Therefore it makes sense here to apply the formula (1.16) for the determination of the estimator $\vec{\alpha}_1$ asymptotically equivalent to the estimator $\tilde{\alpha}_1$. It is easy to verify that the estimator $\vec{\alpha}_1$ coincides with the r.h.s. of (5) here, provided $\beta_n^*(j)$ is replaced by $r_{j,y}$.

Appendix 1. Remarks and Bibliography

Section 1

1. The problems discussed in Subsection 1.3 are dealt with further in the note [197]. The purpose of this note is an informal substantiation of the solution of the system of equations of the form (6) by other iteration methods (in addition to the Newton-Raphson method or its stochastic modification, the scoring method) possessing the property of quadratic termination such as quasi-Newton methods or their alternatives, the conjugate gradient methods ([94,136]). The discussion is based on the material of the paper [196] as well as on the Ph.D. thesis by G. Beinicke (1979): Application of iterative methods to estimation of the parameters of stochastic models, Dept. of Probability Theory, Tbilisi State University, Tbilisi, USSR.

2. Subsection 1.4 which is relatively descriptive in nature, consists of the ascertainment of asymptotic efficiency in Fisher's sense of the estimator $\tilde{\theta}$ of the form (16) (Lemma 7). Obviously, LeCam's investigations reveal much more refined properties of the estimator $\tilde{\theta}$. Readers interested in studying further general properties of $\tilde{\theta}$ may find useful material in the paper [198], Section VII.

3. When checking the conditions (D5) in the particular case of a family of Gaussian distributions $P_{n,\theta} = P_n(f_\theta)$, $\theta \in \Theta$, discussed in Subsection 1.5, it is simpler to demonstrate the continuity of the limiting distribution of the vector $\Delta_{n,\theta}$ as it is done in [61], Theorem 4.6, rather than using the results of Subsection 1 in Appendix 1 to the preceding Chapter.

Sections 2-3

1. Application of Newton-Raphson's and Powell's iterative processes to a determination of the roots of the system of equations (II.4.39) and (II.4.31) was recommended as early as 1966 and 1967 by Aström and Bohlin [9] and by Tretter and Steiglitz [115] respectively. However, the fact that Newton-Raphson's method (or more precisely the generalized "scoring" method) after the first iteration already results in an

estimator asymptotically equivalent to Whittle's estimator was first stated in the papers [42,46,53,54], where it was directly applied to a more general problem of estimating the parameter θ of a spectral density f_θ. Moreover, in the above-mentioned papers, cases of the processes X_t with discrete as well as with continuous time t were considered. The results of these papers are easily generalized to the cases of the multidimensional process X_t and random field X_t (cf. Technical reports [6] and [32] respectively; the results of these works are briefly presented in the papers [7] and [33] respectively). In the above-mentioned papers of the author of this book, it was also stated for the first time that in the particular case of a mixed autoregressive-moving average process, the general class of estimators $\vec{\theta}$, contains also estimators proposed earlier by Clevenson, Hannan, and Parzen [74,100,140,141] (cf. also [2,6,7]).

Chapter IV
TESTING HYPOTHESES ON SPECTRUM PARAMETERS OF A GAUSSIAN TIME SERIES

1. Testing Simple Hypotheses

1.1. Following the general ideas of LeCam [80-82] (cf. also [110]) we shall consider a sequence of experiments

$$E_n = \{X_n, \mathfrak{U}_n, P_{n,\Theta}, \Theta \in \Theta\}, \quad n = 1, 2, ...,$$

where the family of distributions $P_{n,\Theta}, \Theta \in \Theta$ for some choice of random vector $\Lambda_{n,\Theta} = \Lambda_{n,\Theta}(x)$, $x \in X_n$, and the nonrandom matrices Γ_Θ satisfy the conditions (D1)-(D4) for $\tau_n = \sqrt{n}$ of asymptotic differentiability as well as the condition (D5) which assures the asymptotic normality of the vector $\Lambda_{n,\Theta}$ (cf. the Introduction, page 21 and Section 1 of Chapter III). Assume for definiteness that the set $\Theta \in R_p$ of possible values of the vector-valued parameter Θ contains the origin and consider the problem of testing the hypothesis H_0 that the parameter Θ takes on the value 0. A test for this hypothesis is given by a sequence of test functions $\Phi_n = \Phi_n(x)$ defined on the sample space $x \in X_n$. Any measurable function taking on values $0 \leqslant \Phi_n \leqslant 1$ may serve as a test function which determines the probability $\Phi_n(x)$ that hypothesis H_0 will be rejected when x is observed.

As in [80-82] we shall assume that the alternative hypothesis H_1 is that the parameter Θ takes on the value $n^{-1/2}h$ where h is a nonzero p-dimensional vector such that $n^{-1/2}h \in \Theta$. In such a case the power $M_n(h,\Phi_n)$ of the test (the probability of rejecting H_0 when it is true) is determined by the relation

(1) $M_n(h, \phi_n) = E_{n, h} \phi_n(X),$

where $E_{n, h}$ denotes the mathematical expectation corresponding to the value $\theta = n^{-1/2}h$. We fix the significance level α, $0 < \alpha < 1$, and consider only those tests for which the probability of rejecting H_0, when it is true (which is evidently equal to $M_n(0, \phi_n)$), converges to α as $n \to \infty$.

The test defined by a sequence of test (critical) functions

(2) $\hat{\phi}_n(x) = \begin{cases} 1, & x \in \hat{\Phi}_n, \\ 0, & x \notin \hat{\Phi}_n, \end{cases}$

where

(3) $\hat{\Phi}_n = \{x: \Delta'_{n, 0}(x) \Gamma_0^{-1} \Delta_{n, 0}(x) > d_\alpha\}$

is the critical region and d_α determined by the relation

(4) $\int_{d_\alpha}^{\infty} \ell_p(x)dx = \alpha$

may serve as an example of such a test; (here ℓ_p is the density of a χ^2-distribution with p degrees of freedom). Indeed it follows immediately from condition (D5) that the distribution $L\{\Delta'_{n, 0}(X)\Gamma_0^{-1}\Delta_{n, 0}(X) \mid P_{n, 0}\}$ of a random variable $\Delta'_{n, 0}(X)\Gamma_0^{-1}\Delta_{n, 0}(X)$ (under the condition that the random vector X possesses distribution $P_{n, 0}$) tends as $n \to \infty$ to a χ^2-distribution with p degrees of freedom. Therefore, in view of (1)-(4) we have

$$M_n(0, \hat{\phi}_n) = E_{n, 0} \hat{\phi}_n(X)$$

$$= P_{n, 0}\{\Delta'_{n, 0}(X)\Gamma_0^{-1}\Delta_{n, 0}(X) > d_\alpha\}$$

$$\to \int_{d_\alpha}^{\infty} \ell_p(x)dx = \alpha.$$

Now observe that in view of the contiguity of sequences of distributions $P_{n, 0}$, $n = 1, 2, \ldots,$ and $P_{n, n^{-1/2}h}$, $n = 1, 2, \ldots,$ and condition (D5) the sequence of distributions

$$L(\Delta_{n,0} \mid P_{n,n^{-1/2}\mathbf{h}}), \qquad n = 1, 2, \ldots,$$

converges as $n \to \infty$ to the normal distribution $N(\Gamma_0\mathbf{h}, \Gamma_0)$. (Cf. Appendix 2 to Chapter II; the proof of this fact is actually the same as the proof of Theorem 4.6 on page 66 of the book [110] dealing with a particular case.) This implies that the sequence

$$\{\Delta'_{n,0}\Gamma_0^{-1}\Delta_{n,0} \mid P_{n,n^{-1/2}\mathbf{h}}\}, \qquad n = 1, 2, \ldots,$$

converges to a noncentral χ^2-distribution with p degrees of freedom and noncentrality parameter $\mathbf{h}'\Gamma_0\mathbf{h}$. Consequently, if one denotes the density of this limiting distribution by $\ell_p(x; \mathbf{h}'\Gamma_0\mathbf{h})$ then the power of the test $\hat{\phi}_n$ will satisfy the relation

(5)
$$\begin{aligned} M_n(\mathbf{h}, \hat{\phi}_n) &= P_{n,n^{-1/2}\mathbf{h}}\{\Delta'_{n,0}\Gamma_0^{-1}\Delta_{n,0} > d_\alpha\} \\ &\to \int_{d_\alpha}^{\infty} \ell_p(x; \mathbf{h}'\Gamma_0\mathbf{h})dx. \end{aligned}$$

1.2.   As it is known (see, e.g., the book [110] -- results presented therein are straightforwardly carried out to the more general case considered herein) a test defined by the critical function $\hat{\phi}_n$ possesses several properties of asymptotic optimality (as $n \to \infty$). Theorems describing these properties are presented in Chapter 6 of [110] and are in fact a generalization to a wide class of experiments (and not only the case of a stationary Markov sequence considered in [110]) of results presented in the basic paper by A. Wald [29] which dealt with the particular case where the variables $X_1, \ldots, X_n$ are mutually independent and identically distributed. (It was LeCam who, as early as 1960 in the paper [80], indicated the possibility of such a generalization; see also [81] and [82]). As an example, we shall present only one of the "optimality properties" here (cf. [110], Chapter VI, Theorem 2.1) which, in the general case under consideration (and using our notation) can be stated as follows:

Theorem 1. *Let* E_n, $n = 1, 2, \ldots$, *be a sequence of experiments possessing the properties stipulated above. Next, let*

$$\overline{M}_{n,\,c}(\Phi_n) = \int M_n(\mathbf{h},\Phi_n)\pi_c(d\mathbf{h})$$

be the average power of the test defined by the critical (test) function Φ_n, where π_c, $c > 0$, is a family of measures situated on the surfaces

$$S_c = \{x \in R_p, \; x'\Gamma_0 x = c\}$$

and defined on S_c by the densities

$$\xi_c(x) = \xi(x)/\overline{\xi}_c, \qquad c > 0, \qquad x \neq 0$$

where

$$\xi(x) = [\det(\Gamma_0)x'\Gamma_0 x]^{1/2}/|\Gamma_0 x|$$

and

$$\overline{\xi}_c = \int_{S_c} \xi(x)\, dA$$

is the integral of $\xi(x)$ over the surface S_c.
    Then the test defined by the critical function $\hat{\Phi}_n$ possesses asymptotically the best average power in the sense that

$$\lim_{n\to\infty} \inf\{\inf[\,\overline{M}_{n,\,c}(\hat{\Phi}_n) - \overline{M}_{n,\,c}(\Phi_n)];\;\; c \in K\} \geqslant 0,$$

where Φ_n, $n = 1,2, ...,$ is a sequence of arbitrary critical functions defining a test with an asymptotic level of significance α and K is an arbitrary compact subset of the interval $(0,\infty)$.

Since in the next subsection the proof of Theorem 1 will be extended to the case of testing the composite hypothesis it makes sense to briefly indicate the basic ideas of this proof. The proof of Theorem 1 is based on the following two propositions due to LeCam (cf. [81] Chapter III).

Proposition 1. *For any n and $\theta \in \Theta$ there exists a family of distributions $Q_{n,\,\theta,\,\mathbf{h}}$ on $\mathfrak{U}_n$ satisfying the following conditions:*

(a) *for any $b > 0$*

$$\sup_{|h|<b} \; \sup_{A\in \mathfrak{U}_n} |Q_{n,\Theta,h}(A) - P_{n,\,\Theta+n^{-1/2}h}(A)| \to 0$$

as $n \to \infty$;

(b)   *the random vector* $\mathbf{\Delta}_{n,\Theta}$ *is sufficient for the family* $Q_{n,\Theta,h}$ *(cf. Appendix 2 to Chapter II, and Appendix 1 to this Chapter).*

Proposition 2. *Let the conditions (D1)-(D6) be fulfilled for* $\tau_n = \sqrt{n}$ *(cf. the Introduction, page 21 and Section 1 in Chapter III). Then for any* $\Theta \in \Theta$ *there exists a function* $\iota_n \colon R_p \to R_p$ *such that for* $n \to \infty$

$$n^{1/2} \sup_{v} |\iota_n(v) - v| \to 0$$

and for any $b > 0$

$$\sup_{|h|<b} \; \sup_{A\in \mathfrak{U}_n} |N_{n,\Theta,h}(\iota_n^{-1}(A)) - F_{n,\Theta+n^{-1/2}h}| \to 0$$

where $N_{n,\Theta,h}$ *is a Gaussian measure on* R_p *with mathematical expectation* $\Theta + n^{1/2}h$ *and covariance matrix* $(n\,\Gamma_\Theta)^{-1}$, *and* $F_{n,\Theta}$

is a probability measure corresponding to the random vector $\tilde{\Theta}$ *defined by the formula (III.1.16) for* $\tau_n = \sqrt{n}$ *(under the assumption that* $\mathbf{X}$ *possesses the distribution* $P_{n,\Theta}$*).*

From Propositions 1 and 2 the validity of Theorem 1 is easily deduced if one utilizes yet another proposition proved by Wald [29] (cf. also [110], Theorem 4.4A).

Proposition 3. *Let* $\mathbf{Z}$ *be a p-dimensional random vector, possessing the normal distribution* $N(\Theta,\Sigma)$ *with unknown expectation* Θ *and known nondegenerate covariance matrix* Σ. *Consider the problem of testing the hypothesis* H_0: $\Theta = 0$ *based on the results of one observation on each of the components of vector* $\mathbf{Z}$. *In this case, the test-statistic* Φ *defined by the critical region*

$$\Phi = \{z \colon z' \, \Sigma^{-1}z > d_\alpha\}$$

(d_α *is defined here by the relation (4) so that the level of significance of the test equals* α*) possesses the best average*

power

$$\bar{M}_c(\phi) = \int P_\Theta(\phi) \pi_c(d\Theta)$$

(where the family of measures π_c, $c > 0$, is determined as in Theorem 1 but with only one difference, that the matrix Γ_0 is now replaced by Σ^{-1}) in the sense that if ψ is an arbitrary test with level of significance α, then $\bar{M}_c(\phi) \geqslant \bar{M}_c(\psi)$.

In view of the Propositions 2 and 3 for testing the hypothesis H_0: $\Theta = 0$ the test with critical region $\{x: \vec{\Theta}'(x)\Gamma_0\vec{\Theta}(x) > d_\alpha\}$ possesses asymptotically the best average power (cf. Remark 1 below). From here, the formulas (3), (III.1.15) for $\tau_n = \sqrt{n}$, $\Theta = 0$, and the contiguity of sequences of distributions $P_{n,0}$, $n = 1,2, ...$, and $P_{n,n^{-1/2}h}$, $n = 1,2, ...$ the validity of Theorem 1 follows.

1.3. Below, for convenience of exposition, we shall refer to the test-statistic $\hat{\phi}_n$ as Rao's test since it can be viewed as a generalization of the test statistics proposed by Rao in 1948 (cf. [105] or [106] page 417) in application to a particular case of independent observations.

Definition 1. Two test-statistics $\{\phi_n$, $n = 1,2, ...\}$ and $\{\psi_n$, $n = 1,2, ...\}$ are called asymptotically equivalent if for any $\Theta \in \Theta$

$$\lim_{n \to \infty} |M_n(\Theta, \phi_n) - M_n(\Theta, \psi_n)| = 0.$$

Remark 1. Let $\tilde{\Theta} = \tilde{\Theta}_n(X)$ be an asymptotically efficient estimator of Θ satisfying the relation

(6) $\Gamma_0 n^{1/2}(\tilde{\Theta}-\Theta) - \Delta_{n,0} \to 0$

in $P_{n,\Theta}$ probability as $n \to \infty$ (cf. (II.2.20) or (III.1.15)). Then it is clear that

$$\Delta'_{n,\Theta}\Gamma_0^{-1}\Delta_{n,\Theta} - n(\tilde{\Theta}-\Theta)'\Gamma_0(\tilde{\Theta}-\Theta) \to 0$$

in $P_{n,\Theta}$ probability as $n \to \infty$ and hence, in view of the contiguity of sequences of distributions $P_{n,\Theta}$, $n = 1,2, ...$, and $P_{n,\Theta+n^{-1/2}h}$, $n = 1,2, ...$, Rao's statistic $\hat{\phi}_n$ for testing the hypothesis H_0: $\Theta = 0$ is asymptotically equivalent to the statistic $\phi_n^{(1)}$ determined by the critical region

(7) $\Phi_n^{(1)} = \{x: n\tilde{\theta}_n'(x)\Gamma_0\tilde{\theta}_n(x) > d_\alpha\}.$

In the case of independent observations when $\tilde{\theta}$ is a maximum likelihood estimator, the asymptotic properties of statistic $\Phi_n^{(1)}$ were studied in the above-mentioned paper by Wald [29] (so that there is a reason to refer to the statistic $\Phi_n^{(1)}$ in the general case under consideration as Wald's statistic).

Remark 2. If in condition (D2) on page 21 of the Introduction $\tau_n = \sqrt{n}$ and vector **h** can be replaced by the random vector $n^{1/2}(\tilde{\theta}-\theta)$ where $\tilde{\theta}$ is an asymptotically efficient estimator satisfying relation (6) (the general conditions under which this replacement is possible are discussed in [81], Appendix 1), then in view of the obvious relation

$$\Lambda_{0,\tilde{\theta}} - \frac{1}{2}\Delta_{n,\theta}'\Gamma_0^{-1}\Delta_{n,\theta} \to 0,$$

$$\Lambda_{\theta_1,\theta_2} = \log(dP_{n,\theta_2}/dP_{n,\theta_1})$$

(here the approach to the limit as $n \to \infty$ is in the sense of convergence in $P_{n,\theta}$ probability), Rao's and Wald's statistics are asymptotically equivalent to the likelihood ratio test $\Phi_n^{(2)}$ for testing the hypothesis H_0, $\theta = 0$ determined by the critical region

$$\Phi_n^{(2)} = \{x: 2\Lambda_{0,\tilde{\theta}}(x) > d_\alpha\}.$$

Remark 3. Above we have considered the case of the vector-valued unknown parameter $\theta \in \Theta$. The case when $p = 1$ and Θ is a subset of the real line (containing the origin) should be dealt with separately. However, we are not going to dwell on this case (since the results in this case for a Markovian stationary sequence are discussed in detail in Chapter 4 of [110] and the Markovian assumption has no effect on the argument in this case). We shall observe only that if the hypothesis H_0: $\theta = 0$ versus the alternative H_1: $\theta > 0$ is tested then, as it is shown in Section 3 of Chapter 4 of [110] the asymptotically uniformly most powerful test (in the sense of Definition 3.1 presented in [110]) is determined by the critical region

$$\{x:\ \Gamma_0^{-1/2}\Delta_{n,\,0}(x) > d_\alpha\},$$

where d_α satisfies

(8) $$\alpha = \frac{1}{(2\pi)^{1/2}} \int_{d_\alpha}^{\infty} e^{-t^2/2}\, dt.$$

1.4. The general results presented in the preceding subsections can be applied to the case which is of special interest to us (but which apparently was not considered previously from this aspect) when $P_{n,\Theta}$, $\Theta \in \Theta$ is a family of Gaussian distributions corresponding to a stationary random process.

If spectral density f_Θ, $\Theta > 0$ corresponding to Gaussian distribution $P_{n,\Theta}$ satisfies the condition of Theorem II.2.2, then the family of distributions $P_{n,\Theta}$, $\Theta \in \Theta$ will be asymptotically differentiable in the sense of the Definition presented on page 21 of the Introduction, while the distribution $L(\Delta_{n,\Theta} \mid P_{n,\Theta})$ of the random vector $\Delta_{n,\Theta}$ given by formula (II.2.17) converges as $n \to \infty$ to the normal distribution $N(0,\Gamma_\Theta)$ with zero expectation and covariance matrix Γ_Θ of the form (II.2.14). If, however, f_Θ satisfies the conditions of Theorem II.3.2 then the preceding assertion is also valid provided the vector $\Delta_{n,\Theta}$ is defined by the formula (II.3.8) and Γ_Θ by the formula (II.3.9).

Thus, under these conditions the arguments of the preceding subsections could be applied; this permits us to construct several different asymptotically equivalent tests for testing the hypothesis H_0: $\Theta = 0$ which are "optimal" in the sense defined in Theorem 1.

Example 1.   Let X_t, $t = \dots,-1,0,1,\ \dots$, be a Gaussian autoregressive process of the q-th order satisfying the difference equation (II.4.1). Assume that under the hypothesis H_0 the coefficients ι_j, $j = 1, \dots, q$ in (II.4.1) take on the values $\iota_j = \iota_j^{(0)}$, while under the alternative H_1: $\iota_j = \iota_j^{(0)} + n^{-1/2}\iota_j^{(1)}$ (for the time being we shall asume that the value σ^2 of the variance of random variables ε_t in (II.4.1) is known).

It follows from the results of Section 4 in Chapter II that in the particular case under consideration, the k-th component of the random matrix $\Delta_{n,\Theta}$ is of the form

$$\sqrt{n}[\ \beta_n^*(k) - \iota_1^{(0)}\beta_n^*(k-1) - \cdots - \iota_q^{(0)}\beta_n^*(k-q)\ ]\ /\sigma^2$$

and $\Gamma_0 = [\beta_0(k-\ell)/\sigma^2; \; k,\ell = 1, ..., q]$ where $\beta_0(k)$ is the value of the covariance function at $\iota_j = \iota_j^{(0)}$, while the asymptotic m.l. estimators $\tilde{\iota} = (\tilde{\iota}_1, ..., \tilde{\iota}_q)'$ of the parameters $\iota = (\iota_1, ..., \iota_q)'$ are the roots with respect to $\iota_1, ..., \iota_q$ of a linear system of equations (II.4.6). Utilizing the formulas (3) and (7) we may construct two asymptotically equivalent tests -- Rao's and Wald's tests for testing the hypothesis H_0: $\iota_j^{(1)} = 0$ versus the alternative H_1: $\iota_j^{(1)} \neq 0$. If the value of σ^2 is unknown it can be replaced by the estimator $\tilde{\sigma}^2$ defined by the formula (II.4.7). The power of these tests satisfies relation (5) where $p = q$ and $\mathbf{h} = (\iota_1^{(1)}, ..., \iota_q^{(1)})$.

Example 2. Assume now that X_t, $t = ...,-1,0,1, ...$, is a mixed autoregressive-moving average process considered in II.4.3 of Chapter II. Assume also that under H_0 the coefficients ι_j, $j = 1, ..., q$, α_k, $k = 1, ..., r$, take on the value $\iota_j = \iota_j^{(0)}$ and $\alpha_k = \alpha_k^{(0)}$ and under

and
$$H_1: \; \iota_j = \iota_j^{(0)} + n^{-1/2}\iota_j^{(1)}$$
$$\alpha_k = \alpha_k^{(0)} + n^{-1/2}\alpha_k^{(1)}.$$

In this case the Rao test for testing the hypothesis H_0 versus H_1 is defined by the critical region of the form (3), where the first q components of the $(q+r)$-dimensional random vector $\Delta_{n,\theta}$ coincided with the l.h.s. of the equations (II.4.31) multiplied by $-\sqrt{n}/\sigma^2$ and the last r components with the l.h.s. of the equations (II.4.30) multiplied by $\sqrt{n}/\sigma^2$ provided we set $\iota_j = \iota_j^{(0)}$, $\alpha_k = \alpha_k^{(0)}$ in these equations and the $(q+r) \times (q+r)$-matrix Γ_0 coincides with the matrix

(9)
$$\begin{bmatrix} \Gamma_\iota^{(q)} & -\Omega^* \\ -\Omega & \Gamma_\alpha^{(r)} \end{bmatrix}$$

for $\iota_j = \iota_j^{(0)}$ and $\alpha_k = \alpha_k^{(0)}$ (cf. formula (II.4.33); if the value of the parameter σ^2 is unknown then it can be replaced by its consistent estimator σ_*^2).

If the asymptotically efficient estimators $\tilde{\iota}_1, ..., \tilde{\iota}_q, \tilde{\alpha}_1, ..., \tilde{\alpha}_r$ of the parameters $\iota_1, ..., \iota_q, \alpha_1, ..., \alpha_r$ are known (cf. Subsection 4.3 of Chapter II and Subsection 3.3 of Chapter III concerning the methods for determining such an estimator), utilizing formula (7) the hypothesis H_0: $\iota_j^{(1)} = 0$, $\alpha_k^{(1)} = 0$ versus the alternative H_1: $\iota_j^{(1)} \neq 0$, $\alpha_k^{(1)} \neq 0$ can be tested also by means of Wald's test. The power of these tests satisfies relation (5) where $p = q+r$ and $\mathbf{h} = (\iota_1^{(1)}, ..., \iota_q^{(1)}, \alpha_1^{(1)}, ..., \alpha_r^{(1)})'$.

Remark 4. Since under the conditions of this subsection the relations

$$n^{1/2} \sum_{k=1}^{p} (\Theta_{k*}-\Theta_k) \frac{\partial}{\partial\Theta_k} \log f_\Theta \approx n^{1/2} \log(f_{\Theta_*}/f_\Theta)$$

$$\approx n^{1/2} \frac{f_{\Theta_*}-f_\Theta}{f_\Theta} ,$$

hold (up to terms tending to zero in $P_{n,\Theta}$ probability) provided only $\Theta_* = (\Theta_{1*}, ..., \Theta_{p*})'$ is a $\sqrt{n}$-consistent estimator of Θ then Wald's test $\Phi_n^{(1)}$ determined in this case by the critical region

$$\Phi_n^{(1)} = \left\{x: n\tilde{\Theta}_n'(x)\Gamma_0\tilde{\Theta}_n(x) \right.$$

$$\left. = \frac{n}{4\pi}\int_{-\pi}^{\pi} \left\{\sum_{k=1}^{p} \tilde{\Theta}_{nk}(x) \left[\frac{\partial}{\partial\Theta_k}\log f_\Theta(\lambda)\right]_{\Theta=0}\right\}^2 d\lambda > d_\alpha\right\}$$

is asymptotically equivalent to two different tests with critical regions

$$\left\{x: \frac{n}{4\pi} \int_{-\pi}^{\pi} [\log(f_{\tilde{\Theta}_n(x)}(\lambda)/f_0(\lambda))]^2 d\lambda > d_\alpha\right\} ,$$

and

$$\left\{x: \frac{n}{4\pi} \int_{-\pi}^{\pi} \left[\frac{f_{\tilde{\Theta}_n(x)}(\lambda)-f_0(\lambda)}{f_0(\lambda)}\right]^2 d\lambda > d_\alpha\right\}.$$

Remark 5. In view of the formulas (I.2.7) and (I.2.31) the likelihood ratio test (and thus all the tests considered above) are asymptotically equivalent to the test defined by the critical region

$$\left\{x: 2n\left[U_{n,0}(x) - U_{n,\tilde{\Theta}_n(x)}(x)\right] > d_\alpha\right\}$$

(the case of a strictly positive spectral density, $U_{n,\Theta}$ is given by the formula (II.2.1), while in the case of the spectral density possessing fixed zeros by the formula (II.3.1)).

Remark 6. Let the spectral density f_Θ of the Gaussian random process X_t be such that

(10) $\left[\dfrac{\partial}{\partial \theta_j} \log f_{\Theta}(\lambda) \right]_{\Theta=0} = \Phi_j(\lambda), \qquad j = 1,2, ..., p,$

where $\Phi_j(\lambda)$ are orthogonal functions satisfying the conditions

(11)
$$\Phi_j(\lambda) = \Phi_j(-\lambda), \qquad \int_{-\pi}^{\pi} \Phi_j(\lambda)d\lambda = 0,$$
$$\frac{1}{4\pi} \int_{-\pi}^{\pi} \Phi_j(\lambda)\Phi_k(\lambda)d\lambda = \delta_{jk}.$$

Then in view of the formula (II.2.14) and (II.2.17) the Rao test for testing the hypothesis H_0: $\Theta = (\Theta_1, ..., \Theta_p)' = 0$ is of a particularly simple form since in this case $\Gamma_0 = I_p$ and the k-th entry of the vector $\mathbf{A}_{n,\Theta}$ is of the form

(12) $\dfrac{n^{1/2}}{4\pi} \int_{-\pi}^{\pi} \Phi_k(\lambda)\dfrac{I_n(\lambda)}{f(\lambda)} d\lambda.$

Consequently, $\hat{\phi}_n$ is determined here by the critical region

(13)
$$\hat{\phi}_n = \{x: \Delta'_{n,0}\Delta_{n,0} > d_\alpha\},$$
$$\Delta'_{n,0}\Delta_{n,0} = n\sum_{k=1}^{n} \left\{\frac{1}{4\pi} \int_{-\pi}^{\pi} \Phi_k(\lambda)\frac{I_n(\lambda)}{f(\lambda)}d\lambda\right\}^2$$

(cf. (3)).

Example 3. Assume that the spectral density f of a Gaussian random process X_t is of the form (II.4.47) where the parameters y_j, $j = 1, ..., r$ under H_0 take on the values $y_j = y_j^{(0)}$ and under H_1 the values $y_j = y_j^{(0)} + n^{-1/2}y_j^{(1)}$. Then for $\Theta = (y_1^{(1)}, ..., y_r^{(1)})'$, $r = p$, the relation (10) is valid, where $\Phi_j(\lambda) = 2 \cos i\lambda$ and thus, in view of the formulas (II.4.47), (12), and (13) the test statistic $\hat{\phi}_n$ is determined here by the critical region of the form

$$\phi_n = \left\{x: \sum_{k=1}^{r} \left[\frac{n^{1/2}}{2\sigma^2} \int_{-\pi}^{\pi} I_n(\lambda)\cos k\lambda \right.\right.$$
$$\left.\left. \cdot \exp\left[-2 \sum_{j=1}^{r} y_j^{(0)}\cos j\lambda\right] d\lambda\right]^2 > d_\alpha\right\}$$

(as is the case in Examples 1 and 2 if the value of parameter σ^2 is unknown here it can be replaced by a consistent estimator σ_*^2 of this parameter). The power of this test also satisfies relation (5), where, however, $p = r$ and the noncentrality parameter is equal to $|y_1^{(1)}|^2 + \cdots + |y_r^{(1)}|^2$.

2. Testing Composite Hypotheses (The Case of a Sequence of General "Asymptotically Differentiable Experiments")

2.1. We now again return to the general case of the sequence of experiments

$$E_n = \{X_n, \mathfrak{U}_n, P_{n,\Theta}, \Theta \in \Theta\}, \qquad n = 1,2, ...,$$

considered in Subsection 1 of the preceding Section. For convenience of exposition we denote $y = (\Theta_1, ..., \Theta_s)'$ and $\delta = (\Theta_{s+1}, ..., \Theta_{s+k})'$ where $k = p\text{-}s > 0$ so that $\Theta = (y,\delta)'$ and $P_{n,\Theta} = P_{n,(y,\delta)}$. Let Γ and D be sets of vectors y and δ respectively, corresponding to all $\Theta \in \Theta$; assume that D contains the value $\delta = 0$.

Consider the problem of testing the composite hypothesis H_0 that the distribution on the space X_n belongs to the family of distributions $P_{n,(y,0)}$, $y \in \Gamma$ versus the alternative H_1 that it belongs to the family of distributions $P_{n,(y,\delta)}$, $y \in \Gamma$, $\delta \in D$, $\delta \neq 0$ assuming as usual that δ is related to n by $\delta = \delta_n = n^{-1/2}d$.

Let the family of distributions $P_{n,\Theta}$, $\Theta \in \Theta$ be asymptotically differentiable for $\tau_n = \sqrt{n}$ in the sense of the definition on page 21 of the Introduction. Observe that the first two conditions of this definition in the particular case where $\Theta = (y,0)$, $y \in \Gamma$, can be stated as follows:

(D1') The sequences of distributions $P_{n,(y,0)}$, $n = 1,2, ...,$ and $P_{n,(y_n,\delta_n)}$, $n = 1,2, ...,$ are contiguous for any g and d

such that

$$y_n = y + n^{-1/2}g \in \Gamma \quad \text{and} \quad \delta_n = n^{-1/2}d \in D.$$

(D2') For any $h = (g,d)'$ where g and d are as in the preceding condition

(1)
$$\Lambda_{(y,0),(y_n,\delta_n)} - h'\Delta_{n,(y,0)} + \frac{1}{2}h'\Gamma_{(y,0)}h \to 0$$

in $P_{n,(y,0)}$ probability as $n \to \infty$.

Let the condition (D5) stated in Section 1 of the preceding chapter be fulfilled also; this condition for $\Theta = (y,0)$ is of the form:

(D5') A sequence of distributions

$$L(\Delta_{n,(y,0)}|P_{n,(y,0)}), \quad n = 1, 2, ...,$$

of a random vector $\Delta_{n,(y,0)}$ converges to the normal distribution $N(0, \Gamma_{(y,0)})$ with zero expectation and nondegenerate covariance matrix $\Gamma_{(y,0)}$.

It follows from the condition (D1') that for any $\mathfrak{U}_n$-measurable random variables X_n, the covergence $X_n \to 0$ in $P_{n,(y,0)}$ probability holds if and only if $X_n \to 0$ in $P_{n,(y_n,\delta_n)}$ probability.

In particular, it follows from the conditions (D1') and (D5') that

$$(2) \qquad L\left[\Delta_{n,(y,0)}|P_{n,(y_n,\delta_n)}\right] \to N(\Gamma_{(y,0)}\mathbf{h}, \Gamma_{(y,0)})$$

as $n \to \infty$ (see the related Theorem 4.6 in Chapter II of [110]).

In view of Lemma III.1.5 we have

$$(3) \qquad \Delta_{n,(y_n,\delta_n)} - \Delta_{n,(y,0)} + \Gamma_{(y,0)}\mathbf{h} \to 0$$

in $P_{n,(y,0)}$ probability as $n \to \infty$ (and in view of (D1') also in $P_{n,(y_n,\delta_n)}$ probability). Denote

$$\Delta_{n,(y,0)}(\mathbf{X}) = (\mathbf{L}_{n,y}(\mathbf{X}), \mathbf{Y}_{n,y}(\mathbf{X}))',$$

where $\mathbf{L}_{n,y}$ and $\mathbf{Y}_{n,y}$ are random column-vectors of dimension s and k respectively, and represent the matrix $\Gamma_{(y,0)}$ in the form

$$\Gamma_{(y,0)} = \begin{bmatrix} J_y & H_y \\ H'_y & G_y \end{bmatrix},$$

where J, H, and G are matrices of orders $(s{\times}s)$, $(s{\times}k)$, and $(k{\times}k)$ respectively. We obtain from (3) for $\mathbf{d} = 0$

$$(4) \qquad \mathbf{L}_{n,y+n^{-1/2}g} - \mathbf{L}_{n,y} + J_y g \to 0$$

and

$$(5) \qquad \mathbf{Y}_{n,y+n^{-1/2}g} - \mathbf{Y}_{n,y} + H'_y g \to 0.$$

Below we shall assume that the conditions are valid under which the vector g can be replaced in relations (4) and (5) by bounded in $P_{n,(y,0)}$ probability random vector $n^{1/2}(y_*-y)$ where y_* is an arbitrary $\sqrt{n}$-consistent estimator of y (these conditions are quite general in nature and are discussed, for example, in [80], Appendix 1). Carrying out this substitution we obtain

(6) $$L_{n,y_*} - L_{n,y} + J_y n^{1/2}(y_*-y) \to 0$$

and

(7) $$Y_{n,y_*} - Y_{n,y} + H'_y n^{1/2}(y_*-y) \to 0$$

in $P_{n,(y,0)}$ probability as $n \to \infty$.

2.2. Below, when necessary, it will also always be assumed that there exists a consistent estimator Γ_* of the matrix $\Gamma_{(y,0)}$ (in case of continuous dependence of the entries of the matrix $\Gamma_{(y,0)}$ on y the matrix $\Gamma_{(y^*,0)}$ can obviously be chosen for such an estimator).

Under the above-stated conditions the following lemma holds:

Lemma 1. *Let matrix J_y be nondegenerate and*

(8)
$$Z_{n,y} = Y_{n,y} - H'_y J_y^{-1} L_{n,y},$$
$$Z_{n*} = Y_{n,y_*} - H'_* J_*^{-1} L_{n,y_*},$$

where H_* and J_* are consistent estimators of the matrices H and J respectively. Then

1) *For any $y \in \Gamma$*

(9) $$|Z_{n,y} - Z_{n*}| \to 0$$

in $P_{n,(y,0)}$ probability as $n \to \infty$.

2) *The sequence of distributions*

$$L\{Z_{n,y} \mid P_{n,(y,n^{-1/2}d)}\}, \qquad n = 1,2, ...,$$

of a random vector $Z_{n,y}$ *converges to the normal distribution* $N(C_y d, C_y)$ *where*

(10) $C_y = G_y - H_y' J_y^{-1} H_y$.

Proof. It follows from the consistency of H_* and J_* and the boundedness in $P_{n,(y,0)}$ probability of the random vector $L_{n,(y,0)}$ that

(11) $|(Z_{n,y} - Z_{n*}) - (Y_{n,y} - Y_{n,y*}) - H_*' J_*^{-1}(L_{n,y*} - L_{n,y})| \to 0$

in $P_{n,(y,0)}$ probability as $n \to \infty$. Utilizing the relations (6) and (7) we easily obtain relation (9) from (11).

Assertion 2) follows immediately from relation (2) with $g = 0$. Lemma 2 is thus proved. $\square$

Corollary 1. *Let the matrix* C_y *defined by the formula* (10) *be nondegenerate and the random matrix* C_* *be its consistent estimator. Then the sequence of distributions*

$$L\{Z_{n*}' C_*^{-1} Z_{n*} \mid P_{n,(y,\,n^{-1/2}d)}\}, \quad n = 1, 2, \ldots,$$

converges to the noncentral χ^2-*distribution* $\chi_k^2(d'C_y d)$ *with* k *degrees of freedom and noncentrality parameter* $d'C_y d$.

Proof. Since C_* is consistent and the sequence of distributions $P_{n,(y,0)}$, $n = 1, 2, \ldots$ and $P_{n,(y,\,n^{-1/2}d)}$, $n = 1, 2, \ldots$, are contiguous we have

(12) $|Z_{n*}' C_*^{-1} Z_{n*} - Z_{n*}' C_y^{-1} Z_{n*}| \to 0$

as $n \to \infty$ in $P_{n,(y,\,n^{-1/2}d)}$-probability. From here and the assertions 1) and 2) of Lemma 1 the assertion to Corollary 1 easily follows. $\square$

The results of Corollary 1 allow us to construct the test Φ_n for testing the hypothesis H_0 of the preceding subsection determined by the critical region

(13) $\Phi_n = \{x : Z_{n*}' C_*^{-1} Z_{n*} > d_\alpha\}$.

Since d_α is defined here by the relation (1.4) also (where, however $p = k$), the level of significance of the test Φ_n tends to α as $n \to \infty$ and the power

$$M_{n,y}(d,\Phi_n) = P_{n,(y,n^{-1/2}d)}(\Phi_n)$$

satisfies relation

(14) $\qquad M_{n,y}(d,\Phi_n) \to \int_{d_\alpha}^{\infty} \textbf{l}_k(x;\ d'C_y d)dx$

($\textbf{l}_k(x;a)$ is here, as above, the density of a noncentral χ^2-distribution with k degrees of freedom and noncentrality parameter a).

2.3.   Based on the results of Propositions 1 and 2 of the preceding Section, it is easy to verify that for testing the composite hypothesis H_0: $\mathfrak{b} = 0$ the test Φ_n possesses the same "optimal" properties as does the Wald test in the case of independent observations (cf. [29], Sections 8-11). We shall, however, present only one of these properties here by stating the following theorem here which shows that for testing the hypothesis H_0: $\mathfrak{b} = 0$ the test-statistic Φ_n possesses the uniformly best average power

(15) $\qquad \overline{M}_{n,y,c}(\Phi_n) = \int M_{n,y}(z,\Phi_n)\pi_c(dz)$

as $n \to \infty$; here π_c, $c > 0$ is a family of measures situated on the surface $S_c = \{z \in R_k, z'C_y z = 0\}$ and possessing on S_c the densities

$$\zeta_c(z) = \xi(z)/\overline{\xi}_c, \quad z \neq 0, \quad c > 0,$$

where

$$\xi(z) = |\det(C_y)z'C_y z|^{1/2}/|C_y z|,$$

and

$$\overline{\xi}_c = \int_{S_c} \xi(z)dA.$$

Theorem 1.   *Let ψ_n be a test-statistic[1] such that for each $y \in \Gamma$*

---
[1] A test with such a property is called differentially asymptotically similar of size α, on a set Γ (Appendix 1, Remark 2 to this section).   Cf. [80], Definition 2 on p. 84 and Proposition 7.3 on p. 86 that specifies Theorem 1 for the particular case of $k = 1$ (Subsection 2.7 below).

and $b > 0$

$$\sup_{|g|<b} |M_{n,y}(0,\psi_n) - \alpha| \to 0.$$

Then

$$\lim_{n\to\infty} \inf\{\inf[\overline{M}_{n,y,c}(\Phi_n) - \overline{M}_{n,y,c}(\psi_n)];$$

$$c \in K, \quad y \in \Gamma\} \geqslant 0,$$

where K *is an arbitrary compact subset of the interval* $(0,\infty)$.

As in the particular case of independent observations, the proof of Theorem 1 is based on the validity of the following propositions (cf. [29], Appendix V, p. 456).

Proposition 1. *Let* **Z** *be a p-dimensional random vector possessing normal* $N(\Theta,\Sigma)$ *distribution with unknown expectation* Θ *and a known nondegenerate covariance matrix*

$$\Sigma = \begin{bmatrix} \Sigma_{11} & \Sigma_{12} \\ \Sigma_{21} & \Sigma_{22} \end{bmatrix}$$

(where Σ_{11}, Σ_{12} *and* Σ_{22} *are matrices of orders* $s \times s$, $k \times s$, *and* $k \times k$ *respectively).   To test the hypothesis* $\delta = (\Theta_{s+1}, ..., \Theta_{s+k})'$ $= 0$, $s = p-k > 0$ *based on a single observation from each of the components of vector* **Z** *the test-statistic* Φ *determined by the critical region*

$$\Phi = \{z \in R_k; \; z' \; \Sigma_{22}^{-1} \; z > d_\alpha\}$$

(the matrix Σ_{22} *is nondegenerate), where* d_α *as in (13) (so that the level of significance of* Φ *equals* α*) possesses the uniformly best average power*

$$\overline{M}_{y,c}(\Phi) = \int P_{(y,\delta)}(\Phi)\pi_c(d\delta),$$

where $y = (\Theta_1, ..., \Theta_s)'$ *(and* π_c, $c > 0$, *is a family of measures defined as in (15) with the only difference that here* $C_y = \Sigma_{22}^{-1}$*) in the sense that if* ψ *is an arbitrary test-statistic with the level of significance* α, *then* $\overline{M}_{y,c}(\Phi) \geqslant \overline{M}_{y,c}(\psi)$ *for any* $y \in \Gamma$ *and* $c > 0$.

Let $\overset{\vee}{\theta} = (\overset{\vee}{\gamma}, \overset{\vee}{\delta})'$ be an estimator of the parameter $\theta \in \Theta$, $\Theta = (\gamma, \delta)$ satisfying the asymptotic relation

(16) $\qquad \Gamma_0 n^{1/2}(\overset{\vee}{\theta}-\theta) - \Delta_{n,\theta} \to 0$

in $P_{n,\theta}$ probability as $n \to \infty$. An example of such an estimator $\overset{\vee}{\theta}$ under quite general conditions may serve the m.l. estimator determined by the condition

$$\max_{\theta} \log p_{n,\theta} = \log p_{n,\overset{\vee}{\theta}},$$

where $p_{n,\theta}$ is the density of distribution $P_{n,\theta}$; in place of an m.l. estimator one can utilize some asymptotically equivalent estimators, for example, the root of equation $\Delta_{n,\theta} = 0$ or the estimator $\overset{\to}{\delta}$ defined by the formula (III.1.16) for $\tau_n = \sqrt{n}$. Then as $n \to \infty$ the distribution of the random vector $n^{1/2}(\theta-\theta)$ under the condition that $\mathbf{X}$ possesses the distribution $P_{n,\theta}$ tends to the normal distribution $N(0,\Gamma_\theta^{-1})$.

It follows from Proposition 1 and the self-evident formula

(17) $\qquad \Gamma_{(\overset{\vee}{\gamma}, 0)}^{-1} = \begin{bmatrix} J_\gamma^{-1}+J_\gamma^{-1}H_\gamma C_\gamma^{-1}H_\gamma' J_\gamma^{-1} & -J_\gamma^{-1}H_\gamma C_\gamma^{-1} \\ -C_\gamma^{-1}H_\gamma' J_\gamma^{-1} & C_\gamma^{-1} \end{bmatrix},$

that should the random vector $n^{1/2}(\overset{\vee}{\theta}-\theta)$ possess the distribution $N(0,\Gamma_\theta^{-1})$ not in the limit (as $n \to \infty$) but for finite values of n, then when applied to the problem of testing the hypothesis H_0: $\delta = 0$, the test-statistic which differs slightly from the test-statistic $\phi_n^{(1)}$ with the critical region

(18) $\qquad \phi_n^{(1)} = \{x: n\overset{\vee}{\delta}'C_*\overset{\vee}{\delta} > d_\alpha\},$

would possess the uniformly best average power -- namely one defined by the critical region of form (18) but with C_γ in place of its consistent estimator C_*. Actually, however, in the case under consideration this test-statistic as well as the test-statistic $\phi_n^{(1)}$, in view of Proposition 2 of the preceding section, possesses this property only asymptotically (as $n \to \infty$).

For $\theta = (\gamma,0)'$ it follows from (8), (10), (16), and (17) that

(19) $\qquad n^{1/2}(\overset{\vee}{\gamma}-\gamma) - J_\gamma^{-1}L_{n,\gamma} + J_\gamma^{-1}H_\gamma C_\gamma^{-1}Z_{n,\gamma} \to 0$

and

(20) $n^{1/2}\overset{\vee}{\delta} - C_{y}^{-1}Z_{n,\,y} \to 0$

in $P_{n,\,(y,\,0)}$ probability as $n \to \infty$ and it follows -- from (9), (20), the consistency of C_*, and the contiguity of the sequence of distributions $P_{n,\,(y,\,0)}$, $n = 1,2, ...,$ and $P_{n,\,(y,\,n^{1/2}d)}$, $n = 1,2,$ ..., -- that

(21) $n^{1/2}\overset{\vee}{\delta} - C_{*}^{-1}Z_{n*} \to 0$

in $P_{n,\,(y,\,d)}$ probability as $n \to \infty$. The assertion of Theorem 1 now becomes a corollary of the asymptotic equivalence of the tests Φ_n and $\Phi_n^{(1)}$ (which follows from (13), (18), and the relation (21)).

Remark 1. Simultaneously we have proved the asymptotic equivalence of the tests Φ_n and $\Phi_n^{(1)}$. In the particular case of independent observations (and under the assumption that $\overset{\vee}{d}$ is an m.l. estimator of the parameter d) the test $\Phi_n^{(1)}$ coincides with Wald's criterion [29].

2.4. Remark 2. Let

$$D_V = E(\mathbf{Y}_{n,\,y} - VL_{n,\,y})(\mathbf{Y}_{n,\,y} - VL_{n,\,y})',$$

where V is an arbitrary matrix of order $(k \times s)$.
It is easy to verify that

$$D_{H_y'J_y^{-1}} = G_y - H_y'J_y^{-1}H_y = C_y$$

and that the matrix $D_V = D_{H_y'J_y^{-1}}$ is nonnegatively

definite. Indeed, in view of the equality

$$D_V = G_y - VH_y - H_y'V' + VJ_yV'$$

and the positive definiteness of J_y we have

$$D_V - D_{H_y'J_y^{-1}} = D_V - C_y = (H_y' - VJ_y)J_y^{-1}(H_y - J_yV') \geqslant 0.$$

It follows from Remark 2 that for $k = 1$ the entries of the row-vector $H_y'J_y^{-1}$ are the coefficients of the regression of a random variable $\mathbf{Y}_{n,\,y}$ on the random vector $\mathbf{L}_{n,\,y}$ and that the quantity $C_y > 0$ is the minimal variance. Taking into account this fact and the explicit expressions for the entries of the vector $\mathbf{A}_{n,(y,0)}$ and matrix $\Gamma_{(y,0)}$ in the case of independent observatoins (cf., e.g., [44], Section 3) we conclude that in the latter case the test-statistic Φ_n coincides[2] with Neyman's $C(\alpha)$-test [90] for testing the hypothesis $\delta = 0$.

2.5.   Assume that matrix $J_{y_\wedge}$ is nondegenerate and that there exists the estimator $\hat{y} = \hat{y}_n(X)$ of the parameter $y \in \Gamma$ (evidently this estimator is in general different from the estimator $\check{y}$ introduced above) such that

(22) $J_y n^{-1/2}(\hat{y}-y) - \mathbf{L}_{n,y} \to 0$

in $P_{n,(y,0)}$ as $n \to \infty$ probability (and in view of condition (D1) also in $P_{n,(y,n^{-1/2}d)}$ probability). The m.l. estimator defined by the condition

$$\max_y \log p_{n,(y,0)} = \log p_{n,(\hat{y},0)}$$

may serve as an example of such an estimator $\hat{y}$ under general conditions where $p_{n,(y,0)}$ is the density of the distribution $P_{n,(y,0)}$. Instead of the maximum likelihood estimator one may also utilize a simpler but asymptotically equivalent estimator for example, the root of equation $\mathbf{L}_{n,y} = 0$ relative to the unknown y or the estimate

$$\vec{y} = y_* + n^{-1/2}J_*^{-1}\mathbf{L}_{n*}$$

(cf. Chapter III, Section 1).
The following lemma is valid.

Lemma 2. Let $\hat{Y}_n = Y_{n,\hat{y}}$. Then

(24) $|\hat{Y}_n - Z_{n*}| \to 0$

in $P_{n,(y,0)}$ probability as $n \to \infty$.

Proof. In view of (8) and (9) it is sufficient to show that

---

[2] In paper [90] it is assumed that $k = 1$; however, the $C(\alpha)$ test is easy to generalize to the case $k > 1$ (cf., e.g., [15, 28]).

$$|\hat{\mathbf{Y}}_n - \mathbf{Y}_{n,\,y} + H'_y J_y^{-1} \mathbf{L}_{n,\,y}| \to 0$$

in $P_{n,\,(y,\,0)}$ probability as $n \to \infty$, and this follows from (7) with $y_* = y$ and (22).

Consider the test-statistic $\hat{\phi}_n$ determined by the critical region

(25) $\hat{\phi}_n = \{x: \hat{\mathbf{Y}}_n' C_*^{-1} \hat{\mathbf{Y}}_n > d_\alpha\}.$

Comparing (13) with (25) and taking (24) and condition (D1') into account we conclude that the test-statistic $\hat{\phi}_n$ and $\hat{\phi}_n$ are asymptotically equivalent in the sense of Definition 1.

Remark 3. In view of (17), if $\hat{y}$ coincides with the root of equation $\mathbf{L}_{n,\,y} = 0$ we obtain

(26) $\hat{\mathbf{Y}}_n' C_*^{-1} \hat{\mathbf{Y}}_n = \Delta'_{n,\,(y,\,0)} \Gamma_{(y,\,0)}^{-1} \Delta_{n,\,(y,\,0)} |_{y=\hat{y}}.$

If, however, $\hat{y}$ is an arbitrary estimator of the parameter y satisfying the relation (22), then in view of (6) and (22), the equation $\mathbf{L}_{n,\,y} = 0$ and consequently, also equation (26) in general are valid only up to a term approaching zero in $P_{n,\,(y,\,0)}$ probability.

Comparing the formula (26) with the formula (6e.3.6) on page 350 of [106] and taking into account the form of the components of the vector $\Delta_{n,\,(y,\,0)}$ and matrix $\Gamma_{(y,\,0)}$ in the case of independent observations, we obtain that in the latter case the test-statistic $\hat{\phi}_n$ coincides with Rao's test for testing the hypothesis $\delta = 0$.

2.6. In this Subsection we shall show that the tests for testing the composite hypothesis H_0: $\delta = 0$ are all asymptotically equivalent to the likelihood ratio test $\phi_n^{(2)}$ with the critical region

(27) $\phi_n^{(2)} = \{x: 2\Lambda_{(\hat{y},\,0),\,(\check{y},\,n^{-1/2}\check{d})}(x) > d_\alpha\},$

where $\hat{y}$ and $\check{\theta} = (\check{y}, n^{-1/2}\check{d})$ are estimators of the parameters y and θ determined by the formulas (22) and (16) respectively. Indeed, this fact follows easily from (25), condition (D2') and the assertion of the following lemma.

Lemma 3. *Let the conditions be fulfilled under which the vector* **g** *can be replaced by a random vector* $n^{1/2}(y_*-y)$ *in relation (1), where* y_* *is a* $\sqrt{n}$*-consistent estimator of parameter* y *and vector* **d** *can be replaced by its* $\sqrt{n}$*-consistent estimator* d_* *so that*

(28)
$$\Lambda_{(y,0),(y_*,n^{-1/2}d_*)} - h_*'\Delta_{n,(y,0)}$$
$$+ \frac{1}{2}h_*'\Gamma_{(y,0)}h_* \to 0, \quad h_* = (n^{1/2}(y_*-y),d_*)$$

in $P_{n,(y,0)}$ *probability as* $n \to \infty$. *Then*

(29)
$$\hat{Y}_n C_*^{-1}\hat{Y}_n + 2\Lambda_{(\check{y},n^{-1/2}\check{d}),(\hat{y},0)} \to 0 \quad as \; n \to \infty$$

in $P_{n,(y,0)}$ *probability.*

Proof. Taking into account that

$$\Lambda_{(\check{y},n^{-1/2}\check{d}),(\hat{y},0)} = \Lambda_{(y,0),(\hat{y},0)} - \Lambda_{(y,0),(\check{y},n^{-1/2}\check{d})},$$

and that in view of (28)

$$\Lambda_{(y,0),(\hat{y},0)} - n^{1/2}(\hat{y}-y)'L_{n,y} + \frac{1}{2}(\hat{y}-y)'J_y(\hat{y}-y) \to 0$$

and

$$\Lambda_{(y,0),(\check{y},n^{-1/2}\check{d})} - \check{h}'\Delta_{n,(y,0)}$$
$$+ \frac{1}{2}\check{h}'\Gamma_{(y,0)}\check{h} \to 0, \quad \check{h} = (n^{1/2}(\check{y}-y),\check{d})'$$

in $P_{n,\theta}$ probability as $n \to \infty$, it follows from (22) and (16) that for $\theta = (y,0)'$ we have

(30)
$$2\Lambda_{(\check{y},n^{-1/2}\check{d}),(\hat{y},0)} - L_{n,y}'J_y^{-1}L_{n,y}$$
$$+ \Delta_{n,(y,0)}'\Gamma_{(y,0)}^{-1}\Delta_{n,(y,0)} \to 0$$

in $P_{n,(y,0)}$ probability as $n \to \infty$.
Since, in view of (8) and (17)

$$\Delta_{n,(y,0)}'\Gamma_{(y,0)}^{-1}\Delta_{n,(y,0)} - L_{n,y}'J_y^{-1}L_{n,y} =$$

$$(31) \qquad = \Delta'_{n,\,(y,\,0)} \begin{bmatrix} J_y^{-1}H_y C_y^{-1}H'_y J_y^{-1} & -J_y^{-1}H_y C_y^{-1} \\[2mm] -C_y^{-1}H'_y J_y^{-1} & C_y^{-1} \end{bmatrix} \Delta_{n,\,(y,\,0)}$$

$$= (Y_{n,\,y} - H'_y J_y^{-1} L_{n,\,y})' C_y^{-1} (Y_{n,\,y} - H'_y J_y^{-1} L_{n,\,y})$$

$$= Z'_{n,\,y} C_y^{-1} Z_{n,\,y},$$

(9), (12), (24), (30), and (31) imply (29). Thus Lemma 3 is proved.

2.7. The case when $k = 1$ and $\mathbf{D}$ is a subset of the real line requires separate consideration. Here we shall only observe that for testing the hypothesis H_0: $\delta = 0$ against the one-sided alternative H_1: $\delta > 0$ the test-statistic Φ_n determined by the critical region

$$(32) \qquad \Phi_n = \{x\colon C_{y^*}^{-1/2} Z_{n,\,y^*} > d_\alpha\},$$

where d_α is defined by the relation (1.8), is an asymptotically uniformly most powerful test since its power

$$M_{n,\,y}(d,\Phi_n) = P_{n,\,(y,\,n^{-1/2}d)}(\Phi_n)$$

satisfies the following conditions:

1) $\lim\limits_{n\to\infty} \sup\limits_{d\in\mathbf{D}'} |M_{n,y}(d,\Phi_n) - \dfrac{1}{(2\pi)^{1/2}} \int_{d_\alpha - C_y^{1/2}d}^{\infty} e^{-t^2/2} dt| = 0,$

2) if ψ_n is a test-statistic, such as in Theorem 1, then

$$\lim\limits_{n\to\infty} \sup\limits_{d\in\mathbf{D}'} \{M_{n,\,y}(d,\psi_n) - M_{n,\,y}(d,\Phi_n)\} \leqslant 0$$

where $\mathbf{D}'$ is a bounded subset of $\mathbf{D}$ (cf. [80], Proposition 7.3).

3. Testing of Composite Hypothesis about a Parameter of a Spectrum of a Gaussian Time Series

3.1. We now apply the results of the preceding section to the important particular case when $P_{n,\,\theta}$, $\theta \in \Theta$ is a family of Gaussian distributions with zero expectation and spectral density $f_\Theta = f_{(y,\delta)}$, $\Theta = (y,\delta) \in \Theta$ satisfying the conditions of Section 2 of Chapter II. In view of the formulas (17) and (14) of that Section, the components of the vectors $\mathbf{L}_{n,y}$ and $\mathbf{Y}_{n,y}$ are of the form

(1) $$\frac{n^{1/2}}{4\pi} \int_{-\pi}^{\pi} \frac{I_n(\lambda)-f_{(y,0)}(\lambda)}{f_{(y,0)}(\lambda)} \, \Phi_{j,y}(\lambda)d\lambda, \quad j = 1,2,...,s,$$

and respectively,

(2) $$\frac{n^{1/2}}{4\pi} \int_{-\pi}^{\pi} \frac{I_n(\lambda)-f_{(\cdot,0)}(\lambda)}{f_{(y,0)}(\lambda)} \, \Psi_{j,y}(\lambda)d\lambda, \quad j = 1,2, ..., k,$$

and the (k,ℓ)-th entries of the matrices J_y, H_y, and G_y are respectively of the form

(3) $$(\Phi_{k,y}\cdot\Phi_{\ell,y}), \; (\Phi_{k,y}\cdot\Psi_{\ell,y}), \; (\Psi_{k,y}\cdot\Psi_{\ell,y})$$

where

(4) $$\Phi_{j,y} = \frac{\partial}{\partial y_j} \log f_{(y,0)}, \quad \Psi_{j,y} = \left[\frac{\partial}{\partial\delta_j} \log f_{(y,\delta)}\right]_{\delta=0},$$

and

(5) $$(f,g) = \frac{1}{4\pi} \int_{-\pi}^{\pi} f(\lambda)g(\lambda)d\lambda.$$

Assume that a $\sqrt{n}$-consistent estimator y_* of the parameter y is available as well as a consistent estimator C_* of a nondegenerate matrix $C_y = G_y - H_y'J_y^{-1}H_y$ (under the conditions of Section 2 of Chapter II the matrix C_{y_*} can be chosen for such an estimator). Then utilizing the formulas (1)-(5) we obtain that for testing the hypothesis H_0: $\delta = 0$ one can construct a generalized Neyman's $C(\alpha)$-test Φ_n for the case considered herein, with the critical region Φ_n defined by the formula (2.13).

Example 1. Assume that X_t, $t = ...,-1,0,1, ...$ is a Gaussian autoregressive process of the q-th order satisfying the difference equation (II.4.1) and it is required to test the hypothesis H_0 that $q = s$ and $y = (\iota_1, ..., \iota_s)'$ belongs to an open subset $\Gamma \in R_s$ versus the alternative H_1 that $q = s+k$ and

$$y = (\iota_1, ..., \iota_s)' \in \Gamma, \quad \delta = (\iota_{s+1}, ..., \iota_{s+k})' \in D,$$

where $\mathbf{D}$ is an open subset in R_k, and $\iota_{s+j} = n^{-1/2}\iota_j^{(1)}$. We shall assume for the time being that the value of the parameter σ^2 is known.

Taking into consideration that in this particular case the j-th component of the vector $\mathbf{\Delta}_{0,(\gamma,0)} = (\mathbf{L}_{n,\gamma}, \mathbf{Y}_{n,\gamma})'$, in view of (1), (2), and (4), is of the form

$$n^{1/2}[\beta_n^*(j) - \iota_1\beta_n^*(j-1) - \cdots - \iota_s\beta_n^*(j-s)]/\sigma^2$$

and

(6)
$$\Gamma_{(\gamma,0)} = \begin{bmatrix} J_\gamma & H_\gamma \\ H_\gamma' & G_\gamma \end{bmatrix}$$

$$= \left[\frac{1}{4\pi}\int_{-\pi}^{\pi} e^{i\lambda(j-\ell)}|h_q(z)|^{-2}d\lambda\right]_{j,\ell=1,\ldots,s+k},$$

for testing the hypothesis H_0: $\iota_j^{(1)} = 0$ versus the alternative

$$H_1: \iota_j^{(1)} \neq 0$$

one can utilize the test ι_n determined by the critical region (2.13) where the estimators $\iota_{1*}, \ldots, \iota_{s*}$ of the parameters $\iota_1, \ldots, \iota_s$ are roots of the equations (II.4.6) with $q = s$ and unknown $\iota_1, \ldots, \iota_s$ while J_*, H_*, and G_* are either $J_{\gamma*}$, $H_{\gamma*}$, and $G_{\gamma*}$, $\gamma_* = (\iota_{1*}, \ldots, \iota_{s*})'$ or the matrices obtained from J_γ, H_γ and G_γ if $\beta(k)$ is replaced by $\beta_n^*(k)$. In the case when the value of the parameter σ^2 is unknown it can be replaced by the corresponding estimator (II.4.7).

The power of this test-statistic ϕ_n satisfies relation (2.14) where $\mathbf{d} = (\iota_1^{(1)}, \ldots, \iota_k^{(1)})'$.

Example 2. Consider now a more general case where X_t, $t = \ldots, -1, 0, 1, \ldots$, is a mixed autoregressive-moving average process (cf. II.4.3, Chapter II).

Let the hypothesis H_0, that $q+r = s$ and $\gamma = (\iota_1, \ldots, \iota_q, \alpha_1, \ldots, \alpha_r)$ belong to an open subset $\Gamma \in R_s$ versus the alternative H_1 that

$$q = q', \quad r = r', \quad q' + r' = s+k$$

and

$$\gamma = (\iota_1, ..., \iota_q, \alpha_1, ..., \alpha_r)' \in \Gamma,$$

$$\delta = (\iota_{q+1}, ..., \iota_{q'}, \alpha_{r+1}, ..., \alpha_{r'})' \in D$$

is to be tested (here D is an open subset R_k, $k = (q'-q) + (r'-r)$ and $\iota_{q+1} = n^{-1/2}\iota_{q+1}^{(1)}, ..., \iota_{q'} = n^{-1/2}\iota_{q'}^{(1)}, \alpha_{r+1} = n^{-1/2}\alpha_{r+1}^{(1)}, ...,$ $\alpha_{r'} = n^{-1/2}\alpha_{r'}^{(1)}$.

The first q components of the $(q+r)$-dimensional random vector $\mathbf{L}_{n, \gamma}$ and the first $q' = q$ components of the $(q'-q) +$ $(r'-r)$-dimensional vector $\mathbf{Y}_{n, \gamma}$ correspond here to the left-hand-sides of the equations (II.4.31) with $k = 1, ..., q$ and respectively with $k = q+1, ..., q'$ (multiplied by $-\sqrt{n}/\sigma^2$) while the last $(r'-r)$ components of the vector $\mathbf{Y}_{n, \gamma}$ coincides with the l.h. sides (multiplied by $\sqrt{n}/\sigma^2$) of the equations (II.4.30) for $k = 1, ..., r$ and $k = r+1, ..., r'$ respectively.

Clearly, the $(q'+r') \times (q'+r')$-matrix which coincides with (1.9) for $q = q'$ and $r = r'$ can be represented in the form

$$
\begin{bmatrix}
A_{q, q} & A_{q, q'-q} & A_{q, r} & A_{q, r'-r} \\
A_{q'-q, q} & A_{q'-q, q'-q} & A_{q'-q, r} & A_{q'-q, r'-r} \\
A_{r, q} & A_{r, q'-q} & A_{r, r} & A_{r, r'-r} \\
A_{r'-r, q} & A_{r'-r, q'-q} & A_{r'-r, r} & A_{r'-r, r'-r}
\end{bmatrix},
$$

where $A_{r, s}$ is the matrix of dimensionality $r \times s$. Then

$$
J_\gamma = \begin{bmatrix}
A_{q, q} & A_{q, r} \\
A_{r, q} & A_{r, r}
\end{bmatrix},
$$

$$
G_\gamma = \begin{bmatrix}
A_{q'-q, q'-q} & A_{q'-q, r'-r} \\
A_{r'-r, q'-q} & A_{r'-r, r'-r}
\end{bmatrix},
$$

$$
H_\gamma = \begin{bmatrix}
A_{q, q'-q} & A_{q, r'-r} \\
A_{r, q'-q} & A_{r, r'-r}
\end{bmatrix}.
$$

To test the hypothesis H_0: $\mathbf{d} = 0$ versus the alternative H_1: $\mathbf{d} \neq 0$ where

$$\mathbf{d} = (\iota_{q+1}^{(1)}, \ldots, \iota_q^{(1)}, \alpha_{r+1}^{(1)}, \ldots, \alpha_r^{(1)})'$$

one can utilize the test-statistic ϕ_n determined by the critical region (2.13) where

$$\mathbf{y}_* = (\iota_{1*}, \ldots, \iota_{q*}, \alpha_{1*}, \ldots, \alpha_{r*})',$$

and ι_{j*} and α_{k*} are consistent estimators of parameters ι_j and α_k (cf. Section 2.3 of Chapter III concerning the methods for determining such an estimator). In the case when the parameter σ^2 is unknown its value evidently can be replaced by an arbitrary consistent estimator.

The power of the test-statistic ϕ_n evidently satisfies the relation (2.14) in this case where $k = (q'-q) + (r'-r)$.

Example 3. Consider the case when the spectral density f of a Gaussian random process X_t, $t = \ldots-1,0,1, \ldots$, is of the form (II.4.47). Assume that under the null hypothesis H_0: $r = s$ and $\mathbf{y} = (y_1, \ldots, y_s)' \in \Gamma$ and under the alternative H_1:

$$r = s+k, \quad \mathbf{y} = (y_1, \ldots, y_s)' \in \Gamma,$$

and
$$\delta = (y_{s+1}, \ldots, y_{s+k})' \in \mathbf{D},$$

where Γ and $\mathbf{D}$ are subsets of R_s and R_k respectively and

$$y_{s+j} = n^{-1/2} y_j^{(1)}, \quad j = 1, \ldots, k.$$

Here in view of (4)

$$\phi_{j,\mathbf{y}} = 2 \cos j\lambda, \quad j = 1, \ldots, s,$$

$$\psi_{j,\mathbf{y}} = 2 \cos(s+j)\lambda, \quad j = 1, \ldots, k,$$

the j-th entry of the $(k+s)$-dimensional random vector

$$\Delta_{n,(\mathbf{y},0)} = (L_{n,\mathbf{y}}, Y_{n,\mathbf{y}})'$$

is of the form

$$\frac{n^{1/2}}{\sigma^2} \int_{-\pi}^{\pi} I_n(\lambda)\cos j\lambda \, \exp\left[-2 \sum_{i=1}^{s} y_i\cos i\lambda\right]d\lambda,$$

and $\Gamma_{(y,0)} = I_{k+s}$. Therefore the critical region (2.13) here is of the form

(7)
$$\left\{ x: \sum_{j=1}^{k} \left[\frac{n^{1/2}}{\sigma^2}\int_{-\pi}^{\pi} I_n(\lambda)\cos(s+j)\lambda \right. \right.$$
$$\left. \left. \cdot \exp\left[-2 \sum_{i=1}^{s} y_{i*}\cos i\lambda\right]d\lambda\right]^2 \right\},$$

where $y_{1*}, ..., y_{s*}$ are consistent estimators of the parameters $y_1, ..., y_s$ (cf. Subsection 2.4 of Chapter III). When the value of the parameter σ^2 is unknown it can be replaced in (7) by the consistent estimator (III.2.18).

The power of the test-statistic ϕ_n determined by the critical region (7) satisfies the relation (2.14) where the noncentrality parameter equals $[y_1^{(1)}]^2 + \cdots + [y_k^{(1)}]^2$.

In the case when we know the value of the asymptotically efficient $\hat{y}$ of the parameter y for $\delta = 0$ (satisfying the condition (2.22)), for testing the hypothesis H_0: $\delta = 0$ one can utilize the test-statistic $\hat{\phi}_n$ determined by the critical region (2.25). As it was shown in the preceding Section this test-statistic (which is actually a generalization of Rao's statistic to the case under consideration) is asymptotically equivalent in the sense of Definition 1.1 to the generalized $C(\alpha)$-test ϕ_n.

Yet two other tests asymptotically equivalent to the above ones can be constructed provided we have available the asymptotically efficient estimator $\hat{\theta} = (\hat{y},\hat{\delta})'$ of the parameter $\theta = (y,\delta)$ satisfying condition (2.16). These two new test-statistics are generalizations of Wald's statistic $\phi_n^{(1)}$ with the critical region (2.18) and the likelihood ratio test $\phi_n^{(2)}$ determined by the critical region $\phi_n^{(2)}$ of the form (2.27), where the log of the likelihood ratio $\Lambda_{\theta_1,\theta_2} = \Lambda(f_{\theta_1}, f_{\theta_2})$ is

given by the formula presented at the beginning of Section 3 of Chapter I.

Remark 1. It follows from (2.10), (2.19), (2.20), (2.22), (3)-(5), and the mean value theorem that up to a term converging to zero in $P_{n,(y,0)}$ probability (and in view of the contiguity of sequences

$$P_{n,(y,0)}, \qquad n = 1, 2, ..., \text{ and}$$

$$P_{n,\,(y,\,n^{-1/2}d)}, \quad n = 1,2, \dots,$$

in $P_{n,(y,n^{-1/2}d)}$ probability as well) the relations

$$\frac{n}{4\pi} \int_{-\pi}^{\pi} \left[\log \frac{f_{(\hat{y},\,0)}(\lambda)}{f_{(y,\,n^{-1/2}d)}(\lambda)}\right]^2 d\lambda$$

$$\approx \frac{n}{4\pi} \int_{-\pi}^{\pi} \left[\frac{f_{(y,\,0)}(\lambda) - f_{(y,\,n^{-1/2}d)}(\lambda)}{f_{(\hat{y},\,0)}(\lambda)}\right]^2 d\lambda$$

$$\approx Z'_{n,\,y} C_y^{-1} Z_{n,\,y}$$

are valid.

Taking (2.9), (2.12), and (2.13) into account we conclude that the two test statistics with the critical regions

$$\left\{x: \frac{n}{4\pi} \int_{-\pi}^{\pi} [\log(f_{(\hat{y},\,0)}(\lambda)/f_{(y,\,n^{-1/2}d)}(\lambda))]^2 d\lambda > d_\alpha\right\}$$

and

$$\left\{x: \frac{n}{4\pi} \int_{-\pi}^{\pi} \left[\frac{f_{(\hat{y},\,0)}(\lambda) - f_{(y,\,n^{-1/2}d)}(\lambda)}{f_{(\hat{y},\,0)}(\lambda)}\right]^2 d\lambda > d_\alpha\right\}$$

are asymptotically equivalent to the statistics introduced above.

Remark 2. In view of the relation (I.2.7), the likelihood ratio test (and thus all the tests considered above) are asymptotically equivalent to the test with the critical region

$$\{x: 2n[U_{n(\hat{y},\,0)}(x) - U_{n(y,\,n^{-1/2}d)}] > d_\alpha\},$$

where $U_{n,\,\theta}$ is defined by the formula (II.2.1).

3.2.   Assume now that the spectral density $f_0 = f_{q,\,0}$ satisfies the conditions of Section 3 of Chapter II. Then the random variable $U_{n,\,\theta}$ is now given by the formula (II.3.1) so that in view of (II.3.8) the components of the vectors $\mathbf{L}_{n,\,y}$ and $\mathbf{Y}_{n,\,y}$ are of the form

$$\frac{n^{1/2}}{4\pi} \int_{-\pi}^{\pi} \frac{I_n(\lambda,\widetilde{Y})-f_{0(\cdot,0)}(\lambda)}{f_{0(y,0)}(\lambda)} \; \hat{\pi}_{i,y}(\lambda)d\lambda,$$

$$i = 1,2, ..., s$$

and respectively

$$\frac{n^{1/2}}{4\pi} \int_{-\pi}^{\pi} \frac{I_n(\lambda,\widetilde{Y})-f_{0(\cdot,0)}(\lambda)}{f_{0(y,0)}(\lambda)} \; \hat{\psi}_{i,y}(\lambda)d\lambda,$$

$$i = 1,2, ..., k,$$

where $\phi_{i,y}$ and $\psi_{i,y}$ are defined by the formula (4) for $f = f_0$. Taking this fact into account and carrying out the appropriate modifications in the arguments of the preceding subsection, we can also construct in the case under consideration several asymptotically equivalent criteria for testing the hypothesis H_0: $\delta = 0$ which possess "asymptotically optimal" properties.

Appendix 1. Remarks and Bibliography

Section 1

1. We shall comment on the assertion (b) of Proposition 1. Since, in view of (3) of Appendix 2 to Chapter II

$$\frac{dQ_{n,\theta_0,h}}{dP_{n,\theta_0}} = \exp(h'\hat{\Delta}_{n,\theta_0})/E_{n,\theta_0}\{\exp(h'\hat{\Delta}_{n,\theta_0})\},$$

$$h = n^{1/2}(\theta-\theta_0)$$

for each n the statistic $\hat{\Delta}_{n,\theta}$, which is a truncation of vector $\Delta_{n,\theta}$, is sufficient in θ_0 for the family $\{Q_{n,\theta}, \theta \in \Theta\}$ where $Q_{n,\theta} = Q_{n,\theta,n^{1/2}(\theta-\theta_0)}$.

In case the relation (4) in Appendix 2 to Chapter II is verified it is usually stated that the sequence of families $\{Q_{n,\theta}, \theta \in \Theta\}$, $n = 1,2, ...$ and $\{P_{n,\theta}, \theta \in \Theta\}$, $n = 1,2, ...$ are differentially asymptotically equivalent at point θ_0 and the sequence $\hat{\Delta}_{n,\theta}$, $n = 1,2, ...$ is differentially asymptotically sufficient at point θ_0 for the family $\{P_{n,\theta}, \theta \in \Theta\}$ ([110], Chapter III, Section 4).

2. In application to the general problem of testing the simple hypothesis $H_0\colon \Theta = 0$ considered in Subsections 1.1-1.3 the sufficiency of $\hat{\Delta}_{n,\Theta}$ means that from the aspect of comparing the asymptotic values of powers of tests as presented in Theorem 1 it is sufficient to search for the best test among all possible tests dependent on $\hat{\Delta}_{n,\Theta}$ only.

More precisely, the following two assertions are valid:

1) With an arbitrary sequence of test functions ϕ_n, $n = 1,2, \ldots$ we associate the sequence $\bar{\phi}_n = E_{n,0}(\phi_n \mid \Delta_{n,0})$, $n = 1,2, \ldots$ which depends only on $\Delta_{n,0}$. Then for any $b > 0$

$$\sup_{|h|<b} |M_n(h,\phi_n) - M_n(h,\bar{\phi}_n)| \to 0$$

(cf. (1)) where the sup for each n is also taken over all possible test functions ϕ_n.

2) Let $\phi_n(\Delta_{n,0})$, $n = 1,2, \ldots$ be a sequence of test functions dependent on $\Delta_{n,0}$ only. Then for any $b > 0$

$$\sup_{|h|<b} |M_n(h,\phi_n(\Delta_{n,0})) - M_n(h,\phi_n(\hat{\Delta}_{n,0}))| \to 0$$

([110], Chapter III, Theorems 5.1 and 5.2; cf. [146], Theorem 2.2).

Both these assertions are clearly valid also for average powers in Theorem 1. Thus in view of the key assertion 1) it is not necessary here to consider tests which are not based on $\Delta_{n,0}$. Assertion 2) allows us to proceed from a test expressed in terms of $\Delta_{n,0}$ to a test which is expressed in terms of $\hat{\Delta}_{n,0}$ -- such passage is evidently performed for purely theoretical purposes, namely in the proof of Theorem 1, one must utilize an approximation of the initial family of distributions by an exponential family determined in terms of $\hat{\Delta}_{n,0}$ (Remark 1). For the very same purposes it is shown that one can indeed consider only the tests defined by critical regions, i.e., by test functions which are indicators of certain closed and convex sets (as it was shown by Chibisov [146] these tests are admissible only in the sense that for such a test there is no test dominating it[3]; moreover, such a class is

---

[3] We say that a test ϕ_1 dominates ϕ_2 if $M_n(h,\phi_1) \geqslant M_n(h, \phi_2)$ for all h with the equality for $h = 0$ and a strict inequality for some values of h.

essentially complete in the sense of Definition 4.1A, [110], p. 234, cf. also Theorem 4.2A, p. 236). This reduces the problem to checking that there is no critical region which is better (in the sense of Theorem 1) than (3).

Section 2

1. There is an extensive literature devoted to the problem of testing composite hypotheses (which however deals mostly with the case of independent observations X_1, ..., X_n). We shall consider first this case and assume that all observations are sample values of the very same random variable **X** with the probability density $p(x) = p(x,\Theta)$, $x \in R_1$, dependent on an unknown $(s+k)$-dimensional parameter $\Theta = (\Theta_1, ..., \Theta_{s+k})'$. We shall assume that we are dealing with testing a composite hypothesis H_0: $\delta = (\Theta_{s+1}, ..., \Theta_{s+k})' = 0$ versus the alternative H_1: $\delta \neq 0$. This problem was studied as early as 1938 by Wilks [119] who utilized the naturally defined "likelihood ratio test" and investigated the asymptotic distribution of this test under the null hypothesis H_0. However the likelihood ratio test is usually of a very complex form which renders it inapplicable to a majority of applied problems[4]; therefore several authors, in particular Wald [29] and Rao [105], suggested other simple tests. As it is well-known at present (cf. also [106], p. 370-372) the likelihood ratio, Wald's and Rao's tests are asymptotically (as $n \to \infty$ and $\delta = O(n^{-1/2})$) equivalent to each other. Moreover, these three tests possess as $n \to \infty$ a maximal average power (in the sense defined in [29]; cf. also [83, 199], Chapter 7, Problem 5, where the maximal average power is actually defined in application to a somewhat different problem related to Gaussian random variables, however, in view of the results of Section 13 of Chapter 7 of the book [83] this definition of the maximal average power can, in principle, be extended also to the case considered herein). Later Neyman [90] proposed for the solution of this problem yet another test, simpler than all the previous tests, called the $C(\alpha)$-test.[5] As it was noted by

---

[4]This circumstance is valid even to a greater extent for the case of dependent observations under consideration.

[5](See next page.)

LeCam [79] the $C(\alpha)$-test is also asymptotically equivalent to the likelihood ratio test. We note that Bartlett [12] proposed earlier a test which is actually equivalent to the Neyman's $C(\alpha)$-test and indicated the connection between his test and the likelihood ratio test. Moreover, Bartlett also pointed out the applicability of his method in the case of dependent observations but he considered only the simplest class of such observations -- the case of a stationary autoregressive sequence of the first order.

We shall now deal with the general case of observations which will generally be dependent. In this case, an investigation of the asymptotic properties of various tests encounters very substantial difficulties. However, the general asymptotic theory developed in recent years by LeCam [80-82] which is based on the utilization of broad conditions of "asymptotic differentiability" (imposed on the n-dimensional probability distributions $P_{n,0}$) allows us to simplify substantially the investigation of the asymptotic behavior of a wide class of statistical estimators and tests. Developing one of LeCam's remarks (cf. [80], pp. 86-87) we generalize in Section 2 of this work[6] the results of the publications mentioned above [29,90,105] to the case of a general sequence of "asymptotically differentiable families of experiments" E_n, $n = 1,2, \dots$. The basic results of this section were published in the author's paper [44]. In that paper it is also noted that in the particular case of independent observations, for a certain class of alternative hypotheses (cf. [44], Section 3.2, case a) the test determined by the critical region (2.13) coincides with the generalization for the case of composite hypotheses of Neyman's ψ_k^2-test [89] (this generalization was suggested by Chibisov [145]); for another class of alternatives (cf. [44], Section 3.2, cases b and c) a test with the critical region (2.25) coincides with the test suggested by Bol'shev and Nikulin at the International Conference on Probability Theory and Mathematical Statistics (June 1973; cf. also [91]).

---

[5]In the paper [90] it is assumed that $k = 1$, however the $C(\alpha)$-test can easily be generalized to the case $k > 1$; cf., for example, [15,18].

[6]Here we do not mention Bayesian tests, and recommend publication [201] to interested readers.

2. We fix the level of significance α, $0 < \alpha < 1$, and consider only those tests which are differentially asymptotically similar of size α in the sense of the following definition (LeCam [80], Definition 2, p. 84):

Definition. A sequence of tests defined by a sequence of test functions ϕ_n, $n = 1, 2, \ldots$, is called differentially asymptotically similar of size α on the set Γ if for any $y \in \Gamma$ and any $b > 0$

(*) $$\sup_{|g| < b} |M_{n, y_n}(0, \phi_n) - \alpha| \to 0 \quad \text{as } n \to \infty,$$

where $y_n = y + n^{-1/2} g$, and

$$M_{n, y_n}(d, \phi_n) = E(\phi_n \mid P_{n,(y_n, \delta_n)})$$

with $\delta_n = n^{-1/2} d$.

The basic difficulty that arises when one passes from the problem of testing a simple hypothesis to the problem of testing a composite hypothesis is that a direct extension of assertion 1) presented on p. 266 to the composite case leads to a sequence

$$\bar{\phi}_n(y) = E_{n,(y,0)}(\phi_n \mid \Delta_{n,(y,0)}), \qquad n = 1, 2, \ldots,$$

which depends on a nuisance parameter y and is, therefore, not suitable for a definition of a sequence of corresponding tests.[7]

Nevertheless, it will be shown below that under our conditions $\bar{\phi}_n(y_*)$, $n = 1, 2, \ldots$, where y_* is the usual $\sqrt{n}$-consistent estimator of y, can serve as a sequence of critical functions which determine a sequence of asymptotically differentially similar tests of size α. Moreover, it will be shown that for any sequence of test functions ϕ_n, $n = 1, 2, \ldots$

(**) $$\sup_{|d| < b} |M_{n, y}(d, \phi_n) - M_{n, y}(d, \bar{\phi}_n(y_*))| \to 0$$

for any $b > 0$ and $y \in \Gamma$ where the sup for any n is also taken over all possible test functions ϕ_n. Since this assertion is

---

[7] Cf. the note [200] dealing with the case of i. i. d. observations.

valid also for average powers (cf. (15)) it follows from here that in Theorem 1 it is sufficient to consider only those tests which are based solely on y_* and $\Delta_{n,(0,y^*)}$. Evidently a test defined by the critical region (13) is such a test and Theorem 1 states that it is, in a certain sense, the best among them.

The following Proposition serves as the basis for the results presented in this Remark.

Proposition. *Introduce along with* $\delta_n = n^{-1/2}d$ *and* $y_n = y + n^{-1/2}g$ *the quantity* $y_{1n} = y + n^{-1/2}g_1$. *Then under the conditions of Section 2 for any* $y \in \Gamma$

(***) $\sup|M_{n,y_n}(d,\phi_n) - M_{n,y_n}(d,\bar{\phi}_n(y_{1n}))| \to 0,$

where for each n *the* sup *is taken over all possible test functions* ϕ_n *and over all* $|d| < b$, $|g| < b$, *and* $|g_1| < b$ *for some* b > 0.

The proof of this proposition is postponed until the final part of this remark. Meanwhile we shall discuss the necessary corollaries following from it.

Let ϕ_n define a differentially asymptotically similar test of size α. Then it follows from (*) and (***) with $d = 0$ that

$$\sup_{|g| < b, |g_1| < b} |M_{n,y_n}(0,\bar{\phi}_n(y_{1n})) - \alpha| \to 0.$$

Now we shall assume, without stating further details, that the conditions, under which the vector g_1 can be replaced in the last relation by a random vector $n^{1/2}(y_*-y)$ bounded in $P_{n,(y,0)}$ probability, where y_* is an arbitrary $\sqrt{n}$-consistent estimator of y, are also fulfilled. Carrying out such a replacement we obtain

$$\sup_{|g| < b} |M_{n,y_n}(0,\bar{\phi}_n(y_*)) - \alpha| \to 0$$

which shows that the test defined by the test function $\bar{\phi}_n(y_*)$ is differentially asymptotically similar of size α. Analogously from (***) with $g = 0$ one derives the required relation (**).

Proof of the Proposition. The proof is based on approximating $P_{n,\Theta+n^{-1/2}h}$ by an exponential measure $Q_{n,\Theta,h}$ which constitutes the content of Proposition 1 presented in Section 1 of this Chapter. Namely, arguing in the same manner as in the proof of Theorem 5.1 on pp. 81-82 in the book [110] (Assertion 1) presented on page 266) one can justify the transition in (***) from integration with respect to $dP_{n, \Theta+n^{-1/2}h}$ to integration with respect to $dQ_{n, \Theta, h}$ which reduces the problem to the proof of the following relation

(****) $\sup |\int \{\phi_n - \bar{\phi}_n(y_{1n})\} dQ_{n, \Theta, h}| \to 0, \quad \Theta = (y, 0), \quad h = (g, d).$

Clearly in view of (3) and (4') of Appendix 2 to Chapter II we have for $h_1 = (g_1, 0)$:

$$\frac{dQ_{n, \Theta, h}}{dQ_{n, \Theta, h_1}} = C_n(1-\eta_n) \exp\{(h-h_1)'\hat{\Delta}_{n, (y_{1n}, 0)}\},$$

where

$$C_n \to \exp\left\{-\frac{1}{2}(h-h_1)'\Gamma_\Theta(h-h_1)\right\},$$

and (from (3) and the definition of truncation $\hat{\Delta}_{n, \Theta}$ of $\Delta_{n, \Theta}$)

$$\eta_n = 1 - \exp\left\{(h_1-h)'[\hat{\Delta}_{n, (y_{1n}, 0)} \right.$$
$$\left. - \hat{\Delta}_{n, (y, 0)} - \Gamma_{(y, 0)}h_1]\right\} \to 0$$

in $P_{n, (y, 0)}$ probability. Therefore if in the r.h.s. of the equation

$$\int \bar{\phi}_n(y_{1n}) dQ_{n, \Theta, h} = \int \bar{\phi}_n(y_{1n}) \frac{dQ_{n, \Theta, h}}{dQ_{n, \Theta, h_1}} dQ_{n, \Theta, h_1}$$

one replaces the integration with respect to dQ_{n, Θ, h_1} by

integration with respect to the differentially asymptotically equivalent measure $dP_{n, (y_{1n}, 0)}$ then we obtain the following

relations up to a summand which converges to zero in $P_{n, (y, 0)}$ probability

$$\int \bar{\phi}_n(y_{1n}) dQ_{n, \Theta, h}$$
$$\approx C_n \int \phi_n \exp\{(h-h_1)'\hat{\Delta}_{n, (y_{1n}, 0)}\} dP_{n, (y_{1n}, 0)}$$

$$= \int \Phi_n \frac{dQ_{n, \Theta, h}}{dQ_{n, \Theta, h_1}} dP_{n, (Y_{1n}, 0)} = \int \Phi_n \, dQ_{n, \Theta, h}$$

(in the last relatin we replace the integration with respect to $dP_{n, (Y_{1n}, 0)}$ by the integration with respect to dQ_{n, Θ, h_1}).

Informal arguments leading to the last asymptotic relation can be rigorously substantiated so that it will yield the required relation (****).   □

Section 3

1. Wald's test generalized to the case considered herein (as well as the test appearing in Remark 2) was considered already in Whittle's papers [122,125] (as it is stated in the author's Ph.D. dissertation, the results of these papers are easily carried over to the case of the process X_t with continuous time t; in the last work, the tests appearing in Remark 1 are also considered). Rao's and Neyman's tests generalized to the case of Gaussian time series discussed herein (as well as to the case of a time-continuous Gaussian process) were treated in the author's paper [44]. In the same paper asymptotic values of powers of tests studied herein under general "contiguous alternatives" were obtained.

Chapter V
GOODNESS-OF-FIT TESTS FOR TESTING THE HYPOTHESIS ABOUT THE SPECTRUM OF LINEAR PROCESSES

1. A Class of Goodness-of-Fit Tests for Testing a Simple Hypothesis about the Spectrum of Linear Processes

1.1. In this chapter the problem of testing the hypothesis H_0 concerning the form of spectral density f of a linear process X_t of the form (II.6.1) is considered. Unlike in the preceding chapter more general assumptions on the nature of the process X_t are imposed. Namely, it is assumed that the coefficients $g_1, g_2, \ldots$ and the sequence of identically distributed random variables ε_t, $t = \ldots, -1, 0, 1, \ldots$, satisfy the following conditions which are more stringent than those in Chapter II, Subsection 6.1: for some $\delta > 0$, $\sum_{j=1}^{\infty} j^{1/2+\delta} |g_j| < \infty$ and for some $r > 4$, $E(|\varepsilon_t|^r) < \infty$.

Then, in accordance with the assertion about a linear process of Corollary A1.3 at the end of Appendix 1 to Chapter II we have the following proposition.[1]

---

[1] From this corollary actually follows only the asymptotic normality of finite-dimensional distributions of the sequence of random functions $\xi_n(x)$, $n = 1, 2, \ldots$, so it would seem necessary to provide for its tightness. However, for the derivation of Proposition 1 we simply follow the route suggested by Grenander and Rosenblatt [36] which requires the introduction of the above stated more stringent conditions (cf. Appendix 1, Remark 1 to this section).

Proposition 1. *Let* $V[c] = V[c(\tau), 0 \leqslant \tau \leqslant 1]$ *be a continuous functional defined on the space* $C[0,1]$ *of continuous functions* $c = c(\tau), 0 \leqslant \tau \leqslant 1$. *Then under the conditions stated above, the sequence of distributions* $L[V(\zeta_n)], n = 1,2, ...,$ *where*

$$\zeta_n = \zeta_n(\tau) = \sqrt{2}\,[\xi_n(\pi\tau) - \tau\xi_n(\pi)],$$

(1)

$$\xi_n(x) = \frac{\sqrt{n}}{2\pi} \int_0^x \frac{I_n(\lambda)}{f(\lambda)}\,d\lambda, \qquad 0 \leqslant \tau \leqslant 1,$$

is weakly convergent as $n \to \infty$ *to the distribution* $[V(\zeta)]$ *of the functional* V *of the Brownian bridge* $\zeta(\tau), 0 \leqslant \tau \leqslant 1$, *which is a Gaussian random process with zero mean and covariance matrix*

(2) $E\zeta(t)\zeta(s) = \min(t,s) - ts, \qquad 0 \leqslant t,s \leqslant 1.$

Fix an asymptotic level of significance α and consider the class of goodness-of-fit tests for testing the simple hypothesis H_0 about the form of the spectral density f with asymptotic level of significance α determined by critical regions of the form

(3) $\{x: V(\zeta_n) < d_\alpha\};$

here d_α is the quantile of distribution $[V(\zeta)]$ determined from relation (IV.1.4) where $\mathbf{1}_p$ is the density of this distribution.

Observe that tests included in this class are related to the tests for testing the hypothesis about the form of the distribution of the random variable X from an indepenent sample based on statistics of the Kolmogorov-Smirnov type (cf., for example, [25], [31], Chapter V, Section 3), i.e., test statistics such as the Kolmogorov-Smirnov test, Renyi's test, and the ω^2 test. Indeed, for $V(c)$ one can choose in particular the functionals continuous on $C[0,1]$ such as

$$\max_{0 \leqslant \tau \leqslant 1} |c(\tau)|, \quad \max_{a \leqslant \tau \leqslant 1} |c(\tau)|/\tau, \quad 0 < a < 1,$$

and

$$\int_0^1 c^2(\tau)d\tau.$$

Consequently, under the conditions of Proposition 1 the

following relations are valid:

$$\lim_{n\to\infty} P\{ \max_{0\leqslant\tau\leqslant1} |\zeta_n(\tau)| < \varepsilon\} = \sum_{k=-\infty}^{\infty} (-1)^k e^{-2k^2 \varepsilon^2},$$

$$\lim_{n\to\infty} P\{ \max_{a\leqslant\tau\leqslant1} |\zeta_n(\tau)| < \varepsilon\} = 4\Phi(\varepsilon\, a') - 3$$

$$+ \sum_{k=1}^{\infty} (\Phi[(2k+1)\,\varepsilon a'] - \Phi[2k\,\varepsilon a']), \quad a' = \left[\frac{a}{1-a}\right]^{1/2},$$

$$\lim_{n\to\infty} P\left\{\int_0^1 \zeta_n^2(\tau)d\tau < \varepsilon\right\} = G(\varepsilon),$$

where $\Phi(x)$ and $G(x)$ are distribution functions with

characteristic functions equal to $e^{-t^2/2}$ and

$$\left[\prod_{k=1}^{\infty} 1 - \frac{2it}{k^2\pi^2}\right]^{-1}$$

respectively.

1.2. One can judge the advantages and drawbacks of various goodness-of-fit tests based on statistics of the form $V(\zeta_n)$ by comparing the asymptotic (as $n \to \infty$) values of powers of these tests corresponding to a sufficiently wide class of alternative hypotheses.

Following the considerations utilized in the preceding chapter, we shall confine ourselves here only to a discussion of those alternatives H_1 which "approach" the null hypothesis H_0 with the rate $n^{-1/2}$. More precisely, we shall asume that under the hypothesis H_1 the linear process X_t is the same as in the preceding subsection, however, here the coefficients of this representation g_{sn} (satisfying the conditions $g_{0n} = 1$ and

$$\sum_{j=1}^{\infty} j^{1/2+\delta}|g_{jn}| < C, \quad n = 1,2, ...,$$

for $C < \infty$ independent of n) depend on n so that the sequence of the outer functions

$$g_n(z) = \sum_{s=0}^{\infty} g_{sn}z^s, \quad n = 1,2, ...,$$

converges to the outer function

$$g(z) = \sum_{s=0}^{\infty} g_s z^s$$

uniformly in $z = e^{i\lambda}$ and, moreover, the finite limit

(4) $$\lim_{n\to\infty} n^{1/2} \left[\frac{\tilde{g}_n(z) - \tilde{g}(z)}{\tilde{g}(z)} \right] = h(z)$$

exists. If the alternative hypothesis H_1 is valid the spectral density is of the form

(5) $$f_n(\lambda) = \frac{\sigma^2}{2\pi} |\tilde{g}_n(z)|^2,$$

while from (4) and (5) it follows that

(6) $$n^{1/2} \frac{f_n - f}{f} - a \to 0$$

as $n \to \infty$ uniformly in λ where $a = a(\lambda) = 2\mathrm{Re}\, h(z)$.

Under the above stated conditions the following proposition is valid.[2]

Proposition 2. *Under the alternative hypothesis* H_1 *the sequence* $L[V(\zeta_n)|\ H_1]$, $n = 1,2,\ \dots$ *of distributions* $V(\zeta_n)$ *converges to the distribution* $L[V(\zeta+A)]$, *where as above* $\zeta(\tau)$, $0 \leqslant \tau \leqslant 1$, *is a Brownian bridge and*

(7) $$A(\tau) = \frac{\sqrt{2}}{2\pi} \left[\int_0^{\pi\tau} a(\lambda)d\lambda - \tau\int_0^{\pi} a(\lambda)d\lambda \right], \quad 0 \leqslant \tau \leqslant 1.$$

In view of this proposition, the asymptotic (as $n \to \infty$) power of the test with the critical region (3) equal to

$$\int_{d_\alpha}^{\infty} \boldsymbol{l}_p(x,A)dx,$$

where $\boldsymbol{l}_p(x,A)$ is the density of the distribution $L[V(\zeta+A)]$.

1.3. In the next section we shall discuss in more detail the important particular case most often used in practice when

(8) $$V[c] = 2 \sum_{j=1}^{m} \left\{ \int_0^1 \tilde{\Phi}_j(\tau)dc(\tau) \right\}^2,$$

where $\tilde{\Phi}_j$, $j = 1, \dots, m$, are continuous and bounded functions satisfying the conditions

---

[2] See Appendix 1, Remark 2 to this section.

$$\tilde{\Phi}_j(\tau) = \tilde{\Phi}_j(-\tau), \qquad \int_{-1}^{1} \tilde{\Phi}_j(\tau)d\tau = 0,$$

(9)

$$\int_{-1}^{1} \tilde{\Phi}_j(\tau)\tilde{\Phi}_k(\tau)d\tau = \delta_{kj}.$$

It follows from (2) and (9) that

(10) $$E\left[\int_0^1 \tilde{\Phi}_j(\tau)d\zeta(\tau)\right] = 0$$

and

(11) $$2E\left[\int_0^1 \tilde{\Phi}_j(\tau)d\zeta(\tau) \int_0^1 \tilde{\Phi}_k(\tau)d\zeta(\tau)\right] = \delta_{jk} .$$

Thus $L[V(\zeta)]$ is a χ^2-distribution with m degrees of freedom. It also follows from (6), (8), and (9) that

(12) $$V[\zeta+A] = \sum_{j=1}^{m} \left\{\sqrt{2} \int_0^1 \tilde{\Phi}_j(\tau)d\zeta(\tau) + \int_0^1 \tilde{\Phi}_j(\tau)a(\pi\tau)d\tau\right\}^2.$$

Consequently, in view of (10)-(12) the distribution $L[V(\zeta+A)]$ is a noncentral χ^2-distribution with m degrees of freedom and noncentrality parameter $\mu'\mu$, where μ is a column-vector, the k-th entry of which equals

(13) $$\mu_k = \frac{1}{4\pi}\int_{-\pi}^{\pi} \Phi_k(\lambda)a(\lambda)d\lambda, \qquad k = 1, ..., m.$$

and $\Phi_k(\lambda) = 2\tilde{\Phi}_k(\lambda/\pi)$, $k = 1, ..., m$, in view of (9) satisfies the conditions (IV.1.11).

2. χ^2-Test for Testing a Simple Hypothesis about the Spectrum of a Linear Process

2.1. Under the assumptions of the preceding section for testing the simple hypothesis H_0 about the form of spectral density $f > 0$ of a linear process consider the class of goodness-of-fit tests Φ_n determined by the critical region of the form

(1) $$\{x: \Phi_n'(x)\Phi_n(x) > d_\alpha\},$$

where $\Phi_n = \Phi_n(x)$ is an m-dimensional random column-vector whose k-th entry is of the form (cf. Remark 6 in Section 1 of Chapter IV):

$$\frac{n^{1/2}}{4\pi} \int_{-\pi}^{\pi} \Phi_k(\lambda) \frac{I_n(\lambda)}{f(\lambda)} d\lambda.$$

Moreover, we shall assume that the functions Φ_k, $k = 1,2, ...,$ m, are as in the last subsection of the preceding section, and d_α is defined by the relation (IV.1.4) with $p = m$ (since under H_ρ the distribution of the quantity $\Phi_n' \Phi_n$ tends to the χ^2-distribution with m degrees of freedom as $n \to \infty$).

If we confine ourselves here to the situation in which Proposition 2 is valid (Subsection 1.2), then the asymptotic value (as $n \to \infty$) of the power of the test Φ_n which, in view of the discussion presented in Subsection 1.3, equals

$$(2) \qquad \int_{d_\alpha}^{\infty} \mathbf{1}_m(x,\mu'\mu)dx,$$

(where $\mathbf{1}_m(x,a)$ is the density of a χ^2-distribution with m degrees of freedom and the noncentrality parameter a), will substantially depend on m and the form of the functions $\Phi_1,$ $..., \Phi_m$. The orthogonal functions $\Phi_1, ..., \Phi_m$ should be chosen in such a manner that the quantity (2) will attain the largest possible value.

In connection to this, we note that for a fixed m the asymptotic value of the power (2) evidently increases with the increase in the value of the noncentrality parameter $\mu'\mu$ (however, one should keep in mind that

$$(3) \qquad \sum_{k=1}^{m} \mu_k^2 \leqslant \frac{1}{4\pi} \int_{-\pi}^{\pi} a^2(\lambda)d\lambda,$$

in view of (IV.1.11), (1.13), and the Bessel inequality). On the other hand, it is easy to verify that the value of the quantity (2) decreases with an increase in the number of degrees of freedom $\dot{m}$.

In view of the last remark, the best choice of orthogonal functions $\Phi_1, ..., \Phi_m$ is in general not an easy problem even in the relatively simple case when the alternative H_1 is fixed and thus the function a is completely determined.

2.2. We shall confine ourselves to the case when the class of alternatives H_1 may be determined by the functions a which can be represented as finite combinations of certain orthogonal functions $\Phi_1, ..., \Phi_p$ (satisfying the conditions (IV.1.11)):

$$(4) \qquad a(\lambda) = h_1\Phi_1(\lambda) + \cdots + h_p\Phi_p(\lambda).$$

If these orthogonal functions are used for the construction of the critical region (1) also, then clearly

(5)
$$\sum_{k=1}^{m} \mu_k^2 = \sum_{k=1}^{\min(m,\,p)} h_k^2 \leqslant \sum_{k=1}^{p} h_k^2 = \frac{1}{4\pi} \int_{-\pi}^{\pi} a^2(\lambda)d\lambda.$$

The equality sign in (5) holds only in the case when $m \geqslant p$. From here and the remark at the end of the preceding Subsection we deduce that from the aspect of the asymptotic value of the power it is not desirable to choose $m > p$ in (1) since even in the case when $m = p$ the noncentrality parameter $\mu'\mu$ attains its maximal value equal to $(1/4\pi)\int_{-\pi}^{\pi}a^2(\lambda)d\lambda$ (cf. (3)). This, however, in general, does not imply that for $m = p$, the quantity (2) is also maximized, since the unknown coefficients $h_1, \dots, h_p$ in (4) may take values such that for $m < p$, the quantity (2) may be larger than for $m = p$ (in spite of the fact that for $m < p$ the noncentrality parameter which is equal to $h_1^2 + \dots + h_m^2$ is less than $(1/4\pi)\int_{-\pi}^{\pi}a^2(\lambda)d\lambda$). Nevertheless, it is recommended in statistical literature[3] to choose, in similar cases, the value of degrees of freedom m to equal p (in the particular case when the hypothesis about the spectrum of an autoregressive process using Quenouille's method [72] is considered this problem is discussed in [127]). This suggestion is also stipulated by the fact that when X_t is a Gaussian process with the spectral density f_Θ satisfying the conditions of the Remark 6 in Section 1 of Chapter IV, the test-statistic Φ_n considered herein with the critical region (1) for $m = p$ coincides with Rao's statistic $\hat{\Phi}_n$ for testing the hypothesis $H_0: f_\Theta = f_0$ against the alternative $H_1: f_\Theta = f_{n-1/2}{}_h$ where $h = (h_1, \dots, h_p)'$.

2.3. Now we consider several specific examples of constructing goodness-of-fit tests Φ_n.

---

[3] It is easy to observe that the problem considered here is related to the problem of testing hypotheses about a distribution based on an independent sample using Neyman's ψ_k^2-test [89] in the case when the alternative is selected in the form given in [16]. In the papers [16, 89] as well as in the book [71], Sections 30.37-30.45, an interested reader may find a discussion of the problem of determining the best choice -- from the asymptotic power aspect -- of the number of degrees of freedom.

Example 1. We start with the most important case, from an applications point of view, of testing the hypothesis H_0 that X_t is an autoregressive process of order q satisfying the difference equation (II.4.1) where ε_t, $t = ...,-1,0,1,...$ is a sequence of independent identically distributed random variables with zero expectation $E(\varepsilon_t) = 0$ and positive variance $E(\varepsilon_t^2) = \sigma^2 > 0$. As it is known (cf. Subsection 4.1 in Chapter II) under the null hypothesis H_0 the spectral density f of the process X_t is of the form (II.4.3).

For the problem at hand, the most natural alternative hypothesis seems to be H_1 that X_t is an autoregressive process of order q' (where $q' \geqslant q$) with spectral density f_n of the form

(6) $$f_n(\lambda) = \frac{\sigma^2}{2\pi}|h_q(z) - n^{1/2}h_q^{(1)}(z)|^{-2},$$

where $h_q(z)$ is given by the formula (II.4.2) and $h_q^{(1)}(z) = \iota_1^{(1)}z + \cdots + \iota_{q'}^{(1)}z^{q'}$. Then

(7) $$\lim_{n\to\infty} n^{1/2} \frac{f_n(\lambda)-f(\lambda)}{f(\lambda)} = a(\lambda) = 2\mathrm{Re}\left[\frac{h_q^{(1)}(z)}{h_q(z)}\right].$$

We introduce the system of Quenouille orthogonal functions (cf. [138] and also [139] page 95) satisfying the conditions (IV.1.11)

(8) $$\Phi_j(\lambda) = \begin{cases} 2\mathrm{Re}\dfrac{w_{j1}z + \cdots + w_{jq}z^q}{h_q(z)}, & \text{for } j = 1, ..., q, \\[3mm] 2\mathrm{Re}\dfrac{h_q(\bar{z})}{h_q(z)}z^j, & \text{for } j = q+1,q+2,.. \end{cases}$$

where $w_{k\ell}$, $k,\ell = 1, ..., q$ are elements of a triangular matrix W uniquely determined by the conditions: $w_{k\ell} = 0$ for $\ell > k$ and $W\Gamma_{\perp}^{(q)}W' = I_q$ ($\Gamma^{(q)}$ is defined here by the formula (II.4.8)). It is easy to verify that in view of (7) and (8) the function a can be represented here in the form (4) where $p = q'$. Thus, in the case under consideration, the suggestion of the preceding subsection reduces, when constructing the critical region of form (1), to the utilization of the first q' orthogonal functions from the system (8). The asymptotic value of the power is presented here by the formula (2) where the noncentrality parameter equals

$$\mu'\mu = \frac{1}{4\pi}\int_{-\pi}^{\pi} a^2(\lambda)d\lambda = \iota'_{(1)}\Gamma_\iota^{(q')}\iota_{(1)};$$

$\iota_{(1)}$ is a q'-dimensional column-vector with the k-th component $\iota_k^{(1)}$ and the $(q' \times q')$-matrix $\Gamma_\iota^{(q')}$ is, as above, defined by the formula (II.4.8) with $q = q'$.

Following the paper [127] we could have used a more general alternative[4] which assumes that the spectral density f_n is of the form

(9) $$f_n(\lambda) = \frac{\sigma^2}{2\pi}|1-n^{-1/2}g_{r'}^{(1)}(z)|^2|h_q(z)-n^{-1/2}h_{q'}^{(1)}(z)|^{-2},$$

where $g_{r'}^{(1)}(z) = \alpha_1^{(1)}z + \cdots + \alpha_{r'}^{(1)}z^{r'}$. However, instead we shall consider in the next example an even more general problem when under the null hypothesis H_0 the spectral density f of a linear process X_t is a rational function of the form (II.4.28) and under the alternative H_1

(10) $$f_n(\lambda) = \frac{\sigma^2}{2\pi}[g_r(z)-n^{-1/2}g_{r'}^{(1)}(z)]^2|h_q(z)-n^{-1/2}h_{q'}^{(1)}(z)|^{-2}$$

where $r' \geqslant r$ and $g_r(z)$ is given by the formula (II.4.27).

In connection with the problem discussed in this example we also observe that for constructing the critical region (1), Bartlett and Diananda in [13] in addition to Quenouille's orthogonal functions (cf. (8)) suggested to also use the functions

(11) $$\Phi_j(\lambda) = 2\cos j\lambda, \quad j = 1,2, \ldots$$

(cf. also [138] or [139] page 94). Since the k-component of the vector Φ_n equals in this case

(12)
$$\frac{\sqrt{n}}{\sigma^2}\int_{-\pi}^{\pi}\cos k\lambda|h_q(z)|^2I_n(\lambda)d\lambda$$
$$= \frac{\sqrt{n}}{2\sigma^2}\sum_{j,\ell=0}^{q}\iota_j\iota_\ell[\beta_n^*(j-\ell+k) + \beta_n^*(j-\ell-k)],$$

where $\iota_0 = -1$, the critical region (1) becomes

---

[4]Such alternatives are admissible only in the case when all the roots of the polynomials in the denominator of f_n differ substantially from the roots of the polynomial in the numerator. Otherwise it should be assumed that either $r'-r = 0$ or $q'-q = 0$.

(13)
$$\left\{ x: \frac{n}{4\sigma^4} \sum_{k=1}^{m} \left[\sum_{j,\ell=0}^{q} \iota_j \iota_\ell (\beta_n^*(j-\ell+k) + \beta_n^*(j-\ell-k)) \right]^2 > d_\alpha \right\}$$

However, taking into accunt the arguments presented in the preceding subsection we conclude that under the alternative of the form (6) utilizing the functions (11) is less appropriate since in this case the function a of the form (7) cannot be represented in the form of a finite series (4) (evidently provided only if $h_q(z) \not\equiv 1$). One arrives at the same conclusion also under more general alternatives of the form (9) (where $h_q^{(1)}(z) \neq 0$) provided

(14)
$$\lim_{n \to \infty} n^{1/2} \frac{f_n(\lambda) - f(\lambda)}{f(\lambda)} = a(\lambda)$$
$$= 2\mathrm{Re} \left[\frac{h_q^{(1)}(z)}{h_q(z)} - g_r^{(1)}(z) \right].$$

In the case when in formula (9) $\iota_1^{(1)} = \cdots = \iota_q^{(1)} = 0$, it follows from (14) that $a(\lambda) = -2[\alpha_1^{(1)} \cos \lambda + \cdots + \alpha_r^{(1)} \cos r' \lambda]$. If now, following the general suggestions of the preceding subsection, we use the first r' functions given by (11) for the construction of the critical region (1) then the noncentrality parameter in (2) is maximized and takes the value $\alpha'_{(1)} \alpha_{(1)}$ (where $\alpha_{(1)} = (\alpha_1^{(1)}, ..., \alpha_r^{(1)})'$) and the number of degrees of freedom equals r'.

On the other hand, the function $a(\lambda) = -2\mathrm{Re}\, g_r^{(1)}(z)$ can be represented in the form (4), where Φ_j is a function of the form (8) and $p = q + r'$. Thus if for the construction of the test-statistic Φ_n one uses the first $q + r'$ Quenouille functions as given in (8), then its power for $n \to \infty$ will be smaller than the power of the preceding test-statistic (since the noncentrality parameters of the limiting values of powers of the two tests coincide here, while the number of degrees of freedom for the second one is larger and equal to $q + r'$).

Example 2. Assume now that the linear process X_t, $t = ...,-1, 0,1,...,$ is a mixed autoregressive-moving average process and that under H_0 its spectral density f is of the form (II.4.28) while under H_1 it is of the form (10).[5] Then, clearly, the function

---

[5]Cf. the footnote on page 281.

(15) $$a(\lambda) = \lim_{n\to\infty} \frac{f_n(\lambda)-f(\lambda)}{f(\lambda)} = 2\text{Re}\left[\frac{h_{q'}^{(1)}(z)}{h_q(z)} - \frac{g_{r'}^{(1)}(z)}{g_r(z)}\right]$$

can be represented in the form (4) where $p = q + r + \max\{(q'-q), (r'-r)\}$,

(16) $$\Phi_j(\lambda) = \begin{cases} 2\text{Re}\left[\dfrac{\overset{q}{\underset{k=1}{\Sigma}} w_{jk}z^k}{h_q(z)} - \dfrac{\overset{r}{\underset{\ell=1}{\Sigma}} w_{j,\,q+\ell}z^\ell}{g_r(z)}\right], & j=1,...,q+r, \\[6mm] 2\text{Re}\ \dfrac{h_q(\bar z)g_r(\bar z)}{h_q(z)g_r(z)}\ z^j, & j=q+r+1,\ q+r+2,\ ..., \end{cases}$$

and $w_{k\ell}$ is the (k,ℓ)-th component of the triangular $(q+r) \times (q+r)$-matrix W, satisfying the condition

(17) $$W\begin{bmatrix} \Gamma_\iota^{(q)} & -\Omega^* \\ -\Omega & \Gamma_\alpha^{(r)} \end{bmatrix} W' = I_{q+r}$$

(cf. (II.4.33)). It follows from here that the power of the goodness-of-fit test Φ_n determined by the critical region of form (1), where $m = q + r + \max\{(q'-q),(r'-r)\}$ and $\Phi_1, ..., \Phi_m$ are the first m-orthogonal functions appearing in (16), converges as $n \to \infty$ to the quantity (2) where the number of degres of freedom equals $q + r + \max\{(q'-q),(r'-r)\}$ and the noncentrality parameter equals

(18) $$\frac{1}{4\pi}\int_{-\pi}^{\pi} a^2(\lambda)d\lambda = [\iota'_{(1)},\alpha'_{(1)}]\begin{bmatrix} \Gamma_\iota^{(q')} & -\Omega^* \\ -\Omega & \Gamma_\alpha^{(r')} \end{bmatrix}\begin{bmatrix} \iota_{(1)} \\ \alpha_{(1)} \end{bmatrix}$$

(cf. with Example 2 in Section 1 of Chapter IV).

Example 3. It makes sense to utilize the first r' orthogonal functions of Bartlett and Diananda (11) when constructing the goodness-of-fit test Φ_n for testing the hypothesis H_0 that the spectral density is of the form (II.4.47) versus the alternative H_1 that the spectral density is of the form

$$f_n(\lambda) = \frac{\sigma^2}{2\pi}\exp\left\{2\sum_{j=1}^{r} \gamma_j\cos j\lambda + 2n^{-1/2}\sum_{j=1}^{r'} \gamma_j^{(1)}\cos j\lambda\right\};$$

indeed, in this case

$$a(\lambda) = 2 \sum_{j=1}^{r'} \gamma_j^{(1)} \cos j\lambda.$$

The asymptotic power of the test-statistic ϕ_n, as $n \to \infty$, takes on in this case the value (2), where the number of degrees of freedom m equals r' and the noncentrality parameter is

$$\frac{1}{4\pi} \int_{-\pi}^{\pi} a^2(\lambda) d\lambda = \sum_{j=1}^{r'} [\gamma_j^{(1)}]^2.$$

3. Goodness-of-Fit Test for Testing Composite Hypotheses about the Spectrum of a Linear Process

3.1. In this section we consider the problem -- important from the practical point of view -- of testing the composite hypothesis H_0, that a spectral density f of a linear process X_t belongs to the family of spectral densities $f = f_\Theta$, $\Theta \in \Theta$ where Θ is an open set of a p-dimensional Euclidean space R_p. A precise formulation of this problem will be presented in the second subsection of this section. Here we consider the more general case when the n-dimensional distribution $P^{(n)}$ of the observed random vector $X = (X_1, ..., X_n)'$ depends on an unknown value of a p-dimensional parameter Θ (which, however, does not completely determine the distribution), i.e., $P^{(n)} = P_\Theta^{(n)}$ and we shall prove several lemmas which will be useful in the sequel.

Let an m-dimensional random column-vector $\phi_{n,\Theta} = \phi_{n,\Theta}(X)$ (depending on the value of parameter Θ) satisfy the following assumptions:

Assumption 1. *If $\Theta_* = \Theta_*(X)$ is a p-dimenisonal random column-vector which is an estimator of the parameter Θ, such that the vector $\sqrt{n}(\Theta_* - \Theta)$ is bounded in $P_\Theta^{(n)}$ probability (cf. definition on pp. 199-200) then*

(1) $\phi_{n,\Theta_*} - \phi_{n,\Theta} + B\sqrt{n}(\Theta_* - \Theta) \to 0 \quad as \ n \to \infty$

in $P_\Theta^{(n)}$ probability, where B is an $(m \times p)$-dimensional nonrandom matrix of rank $p, m > p$.

Assumption 2. *If X possesses the distribution $P_\Theta^{(n)}$ then the* *distribution of the random vector $\bullet = \bullet_{n,\theta}$ converges as $n \to \infty$ to an m-dimensional normal distribution $N(\mu, I_m)$ with expectation μ and unit covariance matrix I_m.*

Then the following lemma is valid.

Lemma 1. *Let an estimator $\hat{\theta} = \hat{\theta}(X)$ of the parameter θ exist such that the vector $\sqrt{n}(\hat{\theta}-\theta)$ is bounded in $P_\theta^{(n)}$ probability and*

(2) $\sqrt{n}(\hat{\theta}-\theta) - (B'B)^{-1}B'\bullet \to 0, \quad as\ n \to \infty$

in $P_\theta^{(n)}$ probability, where $B'B$ is a nonsingular matrix. Then the distribution of the random variable $\hat{\bullet}'\hat{\bullet}$, where $\hat{\bullet} = \hat{\bullet}_{n,\hat{\theta}}$ as $n \to \infty$, approaches the noncentral χ^2-distribution with $m-p$ degrees of freedom and noncentrality parameter $\mu'A\mu$, where $A = I_m - B(B'B)^{-1}B$.

Proof. It follows from (1) and (2) that

$$\hat{\bullet} - [I_m - B(B'B)^{-1}B']\bullet = \hat{\bullet} - \bullet + B\sqrt{n}(\hat{\theta}-\theta)$$

$$- B[\sqrt{n}(\hat{\theta}-\theta) - (B'B)^{-1}B'\bullet] \to 0 \quad as\ n \to \infty$$

in $P_\theta^{(n)}$ probability. Consequently, the distributions of random vectors $\hat{\bullet}$ and $A\bullet$ -- in view of Assumption 2 and the obvious idempotency of matrix A (i.e., $A = A^2$) -- coincides as $n \to \infty$ with the normal distribution $N(A\mu, A)$. The proof of Lemma 1 now follows from Lemma 1 of the paper [147] (cf. also [144], Lemma 4.2) since $tr(A) = m-p$.

Remark 1. If the entries of matrix B are continuous functions in θ, i.e., $B = B_\theta$ and the vector $\sqrt{n}(\hat{\theta}-\theta)$ is bounded in $P_\theta^{(n)}$ probability where $\hat{\theta}$ is the root with respect to θ of the equation $B_\theta'\bullet_{n,\theta} = 0$, relation (2) is then satisfied since in view of (1)

$$\hat{B}'\hat{\bullet} - B'\bullet + B'B\sqrt{n}(\hat{\theta}-\theta) = \hat{B}'[\hat{\bullet} - \bullet + B\sqrt{n}(\hat{\theta}-\theta)]$$

$$+ (\hat{B}-B)'[\bullet - B\sqrt{n}(\hat{\theta}-\theta)] \to 0$$

in $P_\theta^{(n)}$ probability as $n \to \infty$ where $B = B_{\hat{\theta}}$ and $\hat{\bullet} = \bullet_{n,\hat{\theta}}$.

Remark 2. Let a random matrix B_* exist whose entries

converge to the corresponding entries of the matrix B in $P_{\Theta}^{(n)}$ probability. Then the estimator

(3) $\hat{\Theta} = \Theta_* + n^{-1/2}(B'_*B_*)^{-1}B'_*\Phi_*$

of the parameter Θ satisfies the conditions of Lemma 1. Here Θ_* is the vector as in Assumption 1, and $\Phi_* = \Phi_{n,\Theta^*}$.
    Indeed, in view of (1) we have

$$\sqrt{n}(\hat{\Theta}-\Theta) - (B'B)^{-1}B'\Phi = (B'B)^{-1}B'[\Phi_*-\Phi+B\sqrt{n}(\Theta_*-\Theta)]$$

$$+ [(B'_*B_*)^{-1}B'_* - (B'B)^{-1}B']\Phi_* \to 0 \quad \text{as } n \to \infty$$

in $P_{\Theta}^{(n)}$ probability.

Lemma 2. *Let* $A_* = I_m - B_*(B'_*B_*)^{-1}B'_*$ *be a random* $(m\times m)$-*matrix such that its entries converge to the corresponding entries of matrix A in $P_{\Theta}^{(n)}$ probability as $n \to \infty$. Then as $n \to \infty$ the random variable $\Phi'_*A_*\Phi_*$ possesses a noncentral χ^2-distribution with $m-p$ degrees of freedom and noncentrality parameter $\mu'A\mu$.*

Proof. In view of (1) and (3)

$$\hat{\Phi} - A_*\Phi_* = \hat{\Phi} - \Phi + B\sqrt{n}(\hat{\Theta}-\Theta) - [\Phi_*-\Phi+B\sqrt{n}(\Theta_*-\Theta)]$$
(4)
$$+ (B_*-B)(B'_*B_*)^{-1}B'_*\Phi_* \to 0 \quad \text{as } n \to \infty$$

in $P_{\Theta}^{(n)}$ probability. The proof of Lemma 2 now follows directly from (4), Lemma 1 and the idempotency of the matrix A_*.   □

    Now let the following assumptions be satisfied as well.

Assumption 3. *There exists an estimator* $\tilde{\Theta}$ *of the parameter* Θ *such that*

(5) $\sqrt{n}(\tilde{\Theta}-\Theta) - W^{-1}L_{n,\Theta} \to 0 \quad \text{as } n \to \infty$

in $P_{\Theta}^{(n)}$ probability where W is a nondegenerate $(p\times p)$-matrix and $L = L_{n,\Theta} = L_{n,\Theta}(X)$ is a p-dimensional column-vector such that as $n \to \infty$ the distribution of $(m+p)$-dimensional vector $(\Phi,L)'$ converges to the normal distribution with mathematical expectation and covariance matrix equal to

$$(6) \qquad \begin{bmatrix} I_m & B \\ B' & W \end{bmatrix} \begin{bmatrix} \mu \\ \kappa \end{bmatrix} \quad \text{and} \quad \begin{bmatrix} I_m & B \\ B' & W \end{bmatrix}$$

respectively, where μ and κ are m and p-dimensional column-vectors and B and W are (m×p) and (p×p)-matrices.

Then the following lemma is valid.

Lemma 3. *If in (6) μ = 0 and κ = 0, then the distribution of the random variable ◆'◆, where ◆ = ◆$_{n,\theta}$ as n → ∞, coincides with the distribution of the quantity*

$$(7) \qquad \xi_1^2 + \xi_2^2 + \cdots + \xi_{m-p}^2 + v_1 \xi_{m-p+1}^2 + \cdots + v_p \xi_m^2,$$

where ξ$_j$ is the j-th entry of the vector ξ possessing the normal distribution N(0,I$_m$), and v$_1$, ..., v$_p$ satisfying the inequality 0 ≤ v$_j$ < 1, j = 1, ..., p are roots relative to v of equation

$$(8) \qquad \det[(1 - v)W - B'B] = 0.$$

Proof. In view of (1) and (5)

$$(9) \qquad ◆ - ◆ + BW^{-1}L \to 0 \quad \text{as } n \to \infty$$

in $P_\theta^{(n)}$ probability. This relation can be written also in the form

$$(10) \qquad ◆ - \eta_0 - (\zeta_0 - \zeta) \to 0, \quad \zeta_0 = B(B'B)^{-1}B'◆,$$
$$\zeta = BW^{-1}L, \quad \eta_0 = A◆$$

Clearly, $\eta_0'\zeta_0 = \eta_0'\zeta = 0$ since AB = 0. From here and (10) it follows that the limits (as n → ∞) of distributions of random variables ◆'◆ and

$$(11) \qquad [\eta_0+(\zeta_0-\zeta)]'[\eta_0+(\zeta_0-\zeta)] = \eta_0'\eta_0 + (\zeta_0-\zeta)'(\zeta_0-\zeta)$$

coincide, as $\eta_0'\eta_0$ and $(\zeta_0-\zeta)'(\zeta_0-\zeta)$ in the limit are mutually independent since in view of Assumption 3 $E[\eta_0(\zeta_0-\zeta)'] \to 0$. As it was shown above (cf. proof of Lemma 1), $\eta_0'\eta_0$ possesses, as n → ∞, a χ²-distribution with m-p degrees of freedom. As far as the second summand on the r.h.s. of (11) is concerned its distribution, as $n \to \infty$, coincides with the distribution of the random variables

$$\nu_1 \xi^2_{m-p+1} + \nu_2 \xi^2_{m-p+2} + \cdots + \nu_p \xi^2_m,$$

where $\nu_1, \nu_2, ..., \nu_p$ are nonzero eigenvalues of the matrix in the r.h.s. of the following relation

$$E[(\zeta_0 - \zeta)(\zeta_0 - \zeta)'] \to B[(B'B)^{-1} - W^{-1}]B'.$$

(cf. [144], Lemma 1, or [147] Lemma 4.2). In view of Lemma 4.3 in [147] and the fact that the matrix $W - B'B$ is nonnegative definite where W and $B'B$ are nondegenerate (cf. Lemma 5 below) it follows that the eigenvalues $\nu_1, ..., \nu_p$ are roots with respect to ν of the equation (8) and satisfy the conditions $0 < \nu_j < 1$, $j = 1, ..., p$. Lemma 3 is thus proved.   □

The assertion of Lemma 3 is clearly a generalization to the general case considered here of the well-known results by Chernoff and Lehmann [144].

Assumption 4. *Let the vector* $\mathbf{L}_{n,\Theta}$ *satisfy in addition to Assumption 3 the relation*

(12) $\mathbf{L}_* - \mathbf{L} + W\sqrt{n}(\Theta_* - \Theta) \to 0$

as $n \to \infty$ *in* $P^{(n)}_\Theta$ *probability, where* $\mathbf{L}_* = \mathbf{L}_{n,\Theta^*}$, $\mathbf{L} = \mathbf{L}_{n,\Theta}$, *and* Θ_* *be as above.*

We shall now prove the following:

Lemma 4. *Let* W_* *be an estimator of matrix* W *such that all its entries as* $n \to \infty$ *converge to the corresponding entries of the matrix* W *in* $P^{(n)}_\Theta$ *probability. Then, under the conditions stated above, the distribution of the random variable*

$$[\Phi_* - B_* W_*^{-1} \mathbf{L}_*]' C_*^{-1} [\Phi_* - B_* W_*^{-1} \mathbf{L}_*],$$

where $C_* = I_m - B_* W_*^{-1} B_*'$ *as* $n \to \infty$, *tends to a noncentral* χ^2*-distribution with* m *degrees of freedom and noncentrality parameter* $\mu' C \mu$, *where* $C = I_m - BW^{-1}B'$.

Proof. Since

$$(\spadesuit_*-B_*W_*^{-1}L) - (\spadesuit BW^{-1}L) = [\spadesuit_*-\spadesuit+B\sqrt{n}(\theta_*-\theta)]$$

$$- BW^{-1}[L_*-L+W\sqrt{n}(\theta_*-\theta)] + (BW^{-1}-B_*W_*^{-1})L_* \to 0$$

in $P_\theta^{(n)}$ probability as $n \to \infty$ (cf. (1) and (12)) it is sufficient to prove that the random variable

$$[\spadesuit BW^{-1}L]'C^{-1}[\spadesuit BW^{-1}L]$$

as $n \to \infty$ possesses the distribution indicated in the statement of Lemma 4, i.e., the limiting distribution of the random vector $\spadesuit - BW^{-1}L$ is a normal distribution with expectation $C\mu$ and covariance matrix C. The latter fact easily follows from Assumption 3.   □

Lemma 5. *Under the Assumption 3, the matrix C-A (where as above $C = I_m-BW^{-1}B'$, $A = I_m-B(B'B)^{-1}B'$) is a nonnegative definite matrix.*

Proof. Since in view of (6)

$$E[L - B'\spadesuit][L - B'\spadesuit]' \to W - B'B$$

as $n \to \infty$, the matrices $W-B'B$ and $(B'B)^{-1}-W^{-1}$ are nonnegative definite. Consequently, $C-A = B[(B'B)^{-1}-W^{-1}]B'$ is a nonnegative definite matrix.   □

3.2. Assume now that X_t, $t = ...,-1,0,1,...$ is a linear process of the form (II.6.1) with the spectral density (II.6.2), where the function g (cf. formula (II.6.3)) depends on an unknown p-dimensional parameter $\theta \in \Theta$ (Θ is an open set in the Euclidean space R_p), i.e., $g = g_\theta = g_\theta(\lambda)$. Furthermore, we asume that the distribution of the random variables ε_t in (II.6.1) is unknown (it is only known that they possess mathematical expectation and finite moments up to the fourth order inclusively) and consider the problem of testing the composite hypothesis H_0 that the function g belongs to the family g_θ, $\theta \in \Theta$.

Assuming that the function g_θ is differentiable with respect to the components θ_k, $k = 1, ..., p$, of the vector θ, consider the $(m+p)$-dimensional random column-vector $\Psi(\theta,\sigma^2)$ whose k-th components is of the form

$$(13) \qquad \frac{\sqrt{n}}{2\sigma^2} \int_{-\pi}^{\pi} \psi_{k,\Theta}(\lambda) \frac{I_n(\lambda)}{g_\Theta(\lambda)} \, d\lambda,$$

where $\psi_{k,\Theta} = \Phi_{k,\Theta}$, $k = 1, ..., m$, satisfy the conditions (IV.1.11) for $\Theta \in \Theta$, while

$$\psi_{m+k,\Theta} = \frac{\partial}{\partial\Theta_k} \log g_\Theta, \qquad k = 1, ..., p.$$

Let the function $\psi_{k,\Theta}$ of the argument λ be such that in view of Corollary A1.3 at the end of Appendix 2 to Chapter II, the random vector Ψ_{Θ,σ^2} possesses (as $n \to \infty$) an $(m+p)$-dimensional normal distribution $N(0,J_\Theta)$ with zero expected value and covariance matrix J_Θ, whose $(k \times \ell)$-th entry is of the form

$$\frac{1}{4\pi} \int_{-\pi}^{\pi} \psi_{k,\Theta}(\lambda) \psi_{\ell,\Theta}(\lambda) d\lambda$$

(since it follows from (IV.1.11) and (II.1.3) that $\int_{-\pi}^{\pi} \psi_{k,\Theta}(\lambda) d\lambda = 0$ for $k = 1, ..., m+p$ and $\Theta \in \Theta$).

Remark 3. If σ_*^2 is a consistent estimator of the parameter σ^2 then the distribution of the random vector Ψ_{Θ,σ_*^2} approaches (as $n \to \infty$) the distribution $N(0,J_\Theta)$ as well.

Remark 4. Using standard arguments based on Taylor's expansion, provided the functions $\psi_{k,\Theta}$ and g_Θ are sufficiently smooth, from

$$(14) \qquad \sqrt{n}\left[g_{\Theta_*} - g_\Theta - \sum_{j=0}^{p} (\Theta_{j*} - \Theta_j) \frac{\partial}{\partial\Theta_j} g_\Theta \right] \to 0 \quad \text{as } n \to \infty$$

in $P_\Theta^{(n)}$ probability, where $\Theta_* = (\Theta_{1*}, ..., \Theta_{p*})'$ is an $\sqrt{n}$-consistent estimator of parameter $\Theta \in \Theta$, we can derive the following asymptotic relation:

$$(15) \qquad \Psi_{\Theta_*,\sigma^2} - \Psi_{\Theta,\sigma^2} + \begin{bmatrix} B_\Theta \\ W_\Theta \end{bmatrix} \sqrt{n}(\Theta_* - \Theta) \to 0 \quad \text{as } n \to \infty$$

in $P_\Theta^{(n)}$-probability, where B_Θ and W_Θ are matrices with entries

(16)
$$\frac{1}{4\pi}\int_{-\pi}^{\pi}\Phi_{k,\Theta}(\lambda)\frac{\partial}{\partial\Theta_{\ell}}\log g_{\Theta}(\lambda)d\lambda,$$

$$k = 1, ..., m, \quad \ell = 1, ..., p$$

and

(17)
$$\frac{1}{4\pi}\int_{-\pi}^{\pi}\frac{\partial}{\partial\Theta_{k}}\log g_{\Theta}(\lambda)\frac{\partial}{\partial\Theta_{\ell}}\log g_{\Theta}(\lambda)d\lambda,$$

$$k,\ell = 1, ..., p,$$

respectively.

It follows from the Remarks 3 and 4 that in particular, under the conditions of this subsection, the column-vector $\Phi_{n,\Theta}$ and $L_{n,\Theta}$ of dimensionality m and p whose k-th components coincide with the k-th and respectively the $(m+k)$-th component of the vector $\Psi_{\Theta,\sigma_{*}^{2}}$ satisfy the Assumptions 1-4 of the preceding subsection where $B = B_{\Theta}$ and $W = W_{\Theta}$. Moreover, in view of the results of Section 6 of Chapter II, $\tilde{\Theta}$ is a least square estimator of the parameter Θ determined from the condition (II.6.4) (or any asymptotically equivalent estimator).

Let the $(p \times p)$-matrix $B_{\Theta}^{!}B_{\Theta}$ whose (k,ℓ)-th entry equals

(18)
$$\frac{1}{4\pi}\iint_{-\pi}^{\pi}\frac{\partial}{\partial\Theta_{\ell}}\log g_{\Theta}(\lambda)\frac{\partial}{\partial\Theta_{k}}\log g_{\Theta}(\mu)\Psi_{\Theta}(\lambda,\mu)d\lambda d\mu,$$

where

$$\Psi_{\Theta}(\lambda,\mu) = \frac{1}{4\pi}\sum_{k=1}^{m}\Phi_{k,\Theta}(\lambda)\Phi_{k,\Theta}(\mu)$$

be nondegenerate for $\Theta \in \Theta$ and let an estimator $\tilde{\Theta}$ exist satisfying condition (2), where $B^{!}\Phi = B_{\Theta}^{!}\Phi_{n,\Theta}$ is a p-dimensional column-vector whose k-th component is of the form

(19)
$$\frac{1}{2\sigma_{*}^{2}}\iint_{-\pi}^{\pi}\frac{\partial}{\partial\Theta_{k}}\log g_{\Theta}(\mu)\Psi_{\Theta}(\lambda,\mu)\frac{I_{n}(\lambda)}{g_{\Theta}(\lambda)}d\lambda d\mu,$$

(cf. Remarks 1 and 2). In this case, in view of Lemmas 1 and 2 two different goodness-of-fit tests for testing the hypothesis H_{0} that the function g belongs to the family g_{Θ}, $\Theta \in \Theta$ can be determined by means of the critical regions

(20) $$\{x: \Phi_{n,\tilde{\Theta}}^{!}\Phi_{n,\tilde{\Theta}} > d_{\alpha}\}$$

and

(21) $\{x: \Phi_*'[I_m - B_*(B_*'B_*)^{-1}B_*']\Phi_* > d_\alpha\}$

respectively, where the k-th component of the m-dimensional random vector $\Phi_{n,\Theta}$ is of the form (13) for $\psi_{k,\Theta} = \Phi_{k,\Theta}$ and $\sigma^2 = \sigma_*^2$, while Θ_* is a $\sqrt{n}$-consistent estimator of Θ, B_* is a consistent estimator of matrix B_Θ with entries of the form (16) (for example $B_* = B_{\Theta_*}$), and d_α is defined by relation (IV.1.4) with m-p degrees of freedom.

Utilizing the assertion of Lemma 4 we can define yet another goodness-of-fit test with the critical region of the form

(22) $\{x: (\Phi_*-B_*W_*^{-1}L_*)'(I_m-B_*W_*^{-1}B_*')^{-1}(\Phi_*-B_*W_*^{-1}L_*) > d_\alpha\}$,

where the k-th component of the p-dimensional vector $L_{n,\Theta}$ is determined by the formula (13) with

$$\psi_{k,\Theta} = \frac{\partial}{\partial\Theta_k}\log g_\Theta$$

and $\sigma^2 = \sigma_*^2$, W_* is a consistent estimator of the matrix W with entries (17) (for example, $W_* = W_{\Theta_*}$) and d_α is defined by the relation (IV.1.4) with m degrees of freedom.

3.3. In order to determine the powers of tests introduced in the preceding subsection we shall assume that under the alternative H_1, X_t is a linear process as described in Subsection 1.2. Then it is easy to verify that under H_1 the random vector Ψ_{Θ,σ^2}, as $n \to \infty$, also possesses an $(m+p)$-dimensional normal distribution but with a nonzero expectation which equals an $(m+p)$-dimensional vector whose k-th component is of the form

$$\frac{1}{4\pi}\int_{-\pi}^{\pi} \psi_{k,\Theta}(\lambda)a(\lambda)d\lambda$$

(cf. Subsection 1.2) while the covariance matrix of this vector under H_1 will be the same as under H_0 (i.e., it equals J_Θ).

Furthermore, let us assume that the quantities Θ_* and σ_*^2 appearing in the preceding subsection are such that the random vector $\sqrt{n}(\Theta_*-\Theta)$ and the random variable $\sqrt{n}(\sigma_*^2-\sigma^2)$ are bounded in $P_1^{(n)}$ probability, where $P_1^{(n)}$ is the unknown distribution of the vector $X = (X_1, ..., X_n)'$ under the alternative H_1. Then, as it is also the case in the preceding subsection, the asymptotic distribution (as $n \to \infty$) of the random vector Φ_{Θ,σ^2} and Φ_{Θ,σ_*^2} will be, once again, the same

and relation (15) can be deduced from the relation (14) (where, however the convergence to 0 is now in the $P_1^{(n)}$ probability). Utilizing this fact and the results of Lemmas 1,2, and 4 we conclude that the power of the tests determined by the critical regions (20)-(22), as $n \to \infty$, converges to the quantity

$$(23) \qquad \int_{d_\alpha}^{\infty} \mathbf{1}_k(x, \, \mu'(I_m - BDB')\mu)dx$$

(here, as above, $\mathbf{1}_k(x,a)$ is the density of a noncentral χ^2-distribution with k degrees of freedom and noncentrality parameter a). In (23) $k = m-p$, $D = (B'B)^{-1}$ for the first two tests and $k = m$ and $D = W^{-1}$ for the last one, while the components of the vector μ are defined by the formula (1.13).

Remark 5. In the particular case where the functions $\Phi_{1,\Theta}, ...,$ $\Phi_{m,\Theta}$ are orthogonal to the functions $(\partial/\partial\Theta_\ell)\log g_\Theta$, $\ell = 1, ..., p$, and all the entries (16) of the matrix B_Θ are zero, only the last of the above defined tests remains valid. In this particular case the critical region (22) reduces to the simple form:

$$(24) \qquad \left\{ x: \Phi_*'\Phi_* = \sum_{k=1}^{m} \left[\frac{\sqrt{n}}{2\sigma_*^2} \int_{-\pi}^{\pi} \Phi_{k,\Theta_*}(\lambda) \frac{I_n(\lambda)}{g_{\Theta_*}(\lambda)}d\lambda \right] > d_\alpha \right\}.$$

The power of this test, as $n \to \infty$, evidently converges to the quantity (2.2), where μ is defined by the formula (1.13).

Example 1. Assume that the linear process X_t, $t = ...,-1,0,1, ...,$ is an autoregressive process of order q satisfying the difference equation (II.4.1), i.e., $g_\Theta(\lambda) = |h_q(z)|^{-2}$, where $\Theta = (\iota_1, ..., \iota_q)'$, $q = p$ (cf. formula (II.4.2)). Consider the problem of testing the hypothesis H_0 that the function g_Θ belongs to the family g_Θ, $\Theta \in \Theta$. We begin with the construction of goodness-of-fit tests based on the utilization of certain m simple orthogonal functions of the form (2.11) (which do not depend on the values of the unknown parameters $\iota_1, ..., \iota_q$ and σ^2), say for definiteness:

(25) $\qquad \Phi_k(\lambda) = 2 \cos(m_0 + k)\lambda, \qquad k = 1, ..., m,$

where $m_0 \geqslant 1$. The problem of the best choice of values of m_0 and m (from the aspect of maximizing the asymptotic value of the power) will be discussed below. Here we shall note that if $m_0 = 0$ then the k-th component of the m-dimensional random vector $\Phi_{n,\theta}$ will be of the form (2.12) depending on the values of the parameters $\sigma^2, \iota_1, ..., \iota_q$. However, if the least squares estimators $\tilde{\sigma}^2, \tilde{\iota}_1, ..., \tilde{\iota}_q$ of these values are known, determined by the relations (II.4.6) and (II.4.7), then one can, in principle, construct a goodness-of-fit test determined by the critical region of the form (2.13) where $\sigma^2 = \tilde{\sigma}^2$, $\iota_j = \tilde{\iota}_j$, and d_α is the quantile of the asymptotic distribution (as $n \to \infty$) of the random variable $\Phi'\Phi$ (the k-th component of the vector Φ is also of the form (2.12), here with $\sigma^2 = \tilde{\sigma}^2$ and $\iota_j = \tilde{\iota}_j$). As it was pointed out by many authors, including Bartlett and Diananda [13] who were the first to suggest the method of constructing the goodness-of-fit test discussed herein, an explicit form of its distributions is a very difficult problem. For example, Walker, in his paper [127] was able to determine the asymptotic value (as $n \to \infty$) of the characteristic function of the random variable $\Phi'\Phi$. However, it is easy to verify that in view of the general results stated in Lemma 3 in this section, the distribution of the random variable $\Phi'\Phi$ as $n \to \infty$ coincides with the distribution of the random variable (7), where $v_1, ..., v_p$ ($p = q$) are the roots of the characteristic equation (8). In view of (16) and (17) the (k,ℓ)-th entry of the $(m \times q)$-matrix B appearing in (8) equals $g_{k-\ell}$ here for $k-\ell > 0$, one for $k = \ell$, and zero otherwise (g_k is the k-th coefficient in the expansion of $h_q^{-1}(z)$ in the power of z). In this case also $W = \Gamma_\iota^{(q)}$ (cf. (II.4.8)). We shall not discuss the difficulties involved in determining the quantile d_α of the distribution of random variable (7) here, since these are of the same nature as the ones in the case of independent observations considered by Chernoff and Lehmann [144] (cf. also [71]).

Now let $\hat{\iota} = (\hat{\iota}_1, ..., \hat{\iota}_q)$ be an estimator of the parameter $\iota = (\iota_1, ..., \iota_q)'$ which under the null hypothesis H_0 satisfies the relation (cf. (2)):

(26) $\qquad \sqrt{n}(\hat{\iota} - \iota) - (B'B)^{-1}B'\Phi_n \to 0$

(in the sense of convergence in probability), where Φ_n is the

m-dimensional column-vector with entries of the form (2.12) for $k = m_0+1, ..., m_0+m$; the (k,ℓ)-th entry of the $(m\times q)$-matrix B equals $g_{k-\ell+m_0}$ for $k-\ell+m_0 > 0$, one for $k-\ell = m_0$, and zero

in other cases. Observe in this connection that the least squares estimator $\hat{\iota}$ does not satisfy relation (26) unlike the estimator of the form

(27) $\qquad \hat{\iota} = \tilde{\iota} + n^{-1/2}B^{-1}\Psi,$

where the entries of the matrix B and vector Ψ are given by the formulas (18) and (19) respectively with

$$\Phi_{k,\Theta}(\lambda) = 2\cos(m_0+k)\lambda,$$
$$\frac{\partial}{\partial\Theta_k}\log g_\Theta(\lambda) = 2\,\mathrm{Re}[z^k/h_q(z)],\quad \sigma_*^2 = \tilde{\sigma}^2 \text{ and } \iota = \tilde{\iota}.$$

Now it follows from Lemma 1, Remark 2, and formula (20) that a test for testing the hypothesis H_0 can be determined by the critical region of the form (2.13) where $\sigma^2 = \tilde{\sigma}^2$, $\iota_j = \tilde{\iota}_j$, and d_α is defined by the relation (IV.1.4) with $m-q$ degrees of freedom.

In view of Lemma 2 and formula (21) yet another test asymptotically equivalent to the preceding one can be determined by the critical region of the form

(28) $\qquad \{x: \Phi'\Phi - \Psi B^{-1}\Psi > d_\alpha\}.$

Assume that under the alternative H_1, X_t is an autoregressive process of order q' (where $q' > q$) with the spectral density of the form (2.6) where

(29) $\qquad h_{q'}^{(1)}(z) = \iota_1^{(1)}z^{q+1} + \cdots + \iota_{q'-q}^{(1)}z^{q'}$

and that relation (26) once again is valid. (It can be shown that the last assumption holds at least for estimators of the form (27) since the random vector $\sqrt{n}(\tilde{\iota}-\iota)$ is indeed bounded under H_1). Then, in view of Lemmas 1 and 5 and the relations (2.7), (2.9), and (1.13) the power of the tests defined above, evidently converges to expression (23) with $k = m-q$, $D = (B'B)^{-1}$ and $\mu = B_1\iota^{(1)}$ where $\iota^{(1)} = (\iota_1^{(1)}, ..., \iota_{q'-q}^{(1)})$ and the $m \times (q'-q)$-matrix B_1 is such that its (k,ℓ)-th entry equals

g_{k+m_0-q-1} for $k+m_0-q-j > 0$, one for $k+m_0 = q+j$, and zero

otherwise (we set $m+m_0 \geqslant q'-q$ so that the matrix B_1 will possess no zero columns).

Thus in this case the degrees of freedom in (23) equals $m-q$ and the noncentrality parameter is of the form

$$(30) \qquad \iota^{(1)}{}'[B_1'B_1 - B_1'B(B'B)^{-1}B'B_1]\iota^{(1)}.$$

Denote by B_0 the $(m \times q')$-matrix $[B \ B_1]$. Then utilizing the well-known formula for the inverse of block matrices (cf. (IV.2.17)) it is easy to verify that the matrix which is the inverse of the $(q'-q) \times (q'-q)$-submatrix appearing in the right-hand side corner of the expression for $(B_0'B_0)^{-1}$ coincides with the matrix appearing in the square brackets of (30). It is also easy to verify that in view of the relation

$$\beta(k) = \sigma^2 \sum_{j=0}^{\infty} g_j g_{j+k}, \qquad k = 0,1, ..., \qquad g_0 = 1,$$

for $m_0 = 0$ and $m \to \infty$, all the entries of the matrix $B_0'B_0$

converge to the entries $\beta(k-1)/\sigma^2$ of the $(q' \times q)$-matrix $\Gamma_\iota^{(q')}$ which coincides with the matrix $\Gamma_{(y,0)}$ (cf. (IV.3.6)) where $y = \iota$, $s = q$, and $s+k = q'$. From here it follows that as $m_0 = 0$ and $m \to \infty$ the noncentrality parameter (30) tends to the noncentrality parameter $\iota^{(1)}{}'C_y\iota^{(1)}$ in the asymptotic expression (as $n \to \infty$) for the power of the optimal test (in the sense defined in Section 2 of Chapter IV) which was utilized in Example 1 for testing the composite hypothesis H_0 that the order of the Gaussian autoregressive process X_t equals q against the alternative H_1: the order equals q'.

Thus, choosing m_0 equal to zero and m to be sufficiently large the value of the noncentrality parameter can be made arbitrarily close to the largest of its possible vaues $\iota^{(1)}{}'C_y\iota^{(1)}$. It should, however, be kept in mind that the increase in the number of degrees of freedom $(m-p)$ at the same time decreases the asymptotic value of the power of the tests.

Yet another goodness-of-fit test for testing the composite hypothesis H_0 discussed here can be determined by the critical region of the form (28). As it was mentioned above (cf. Remark 5) the critical region (28) is of an especially simple form (24) in the case when functions $\Phi_{1,\theta}, ..., \Phi_{m,\theta}$ are such that all the entries of the matrix B are zero. It is easy to verify that the last condition is in particular fulfilled if

(31) $$\Phi_{j,\,\Theta}(\lambda) = 2 \text{ Re } \frac{h_q(\overline{z})}{h_q(z)} z^{q+j}, \quad j = 1, \dots, m.$$

In this case (24) becomes

(32) $$\left\{ x: \sum_{k=1}^{m} \left[\frac{\sqrt{n}}{\tilde{\sigma}^2} \int_{-\pi}^{\pi} \text{Re}[\tilde{h}_q(z)]^2 \overline{z}^{q+k} I_n(\lambda) d\lambda \right]^2 > d_\alpha \right\},$$

where $\tilde{h}_q(z) = 1 - \tilde{\tau}_1 z - \dots - \tilde{\tau}_q z^q$. In view of (1.13), (2.7), (29), (31), and Remark 5 the power of the test determined by the critical region (32) converges (as $n \to \infty$) to the value (2.2) where μ is an m-dimensional vector, whose k-th component equals

$$\mu_k = \frac{1}{2\pi} \int_{-\pi}^{\pi} \frac{z^{k+q} h_q^{(1)}(\overline{z})}{h_q(z)} d\lambda$$

for $k = 1, \dots, q'-q$ and zero for $k > q'-q$. Consequently, for $m \geqslant q'-q$ the noncentrality parameter achieves its maximal value $\sum_{k-1}^{q'-q} \mu_k^2$. Since the limiting value (2.2) of the power of the test is directly proportional to the noncentrality parameter and is inverse proportional to the number of degrees of freedom, it follows that one should set $m = q'-q$ in (32).

Evidently in this case the function a which is of the form (2.7) where $h_q^{(1)}(z)$ is defined by (26) may be represented as a linear combination of the first q' orthogonal functions (2.8). It follows from (2.8) and (31) that the last $q'-q$ coefficients in this representation coincide with $\mu_1, \dots, \mu_{q'-q}$. Thus, in view of (2.7) and (2.9) the noncentrality parameter for $m = q'-q$ is equal to

$$\sum_{k=1}^{q'-q} \mu_k^2 = \frac{1}{2\pi} \int_{-\pi}^{\pi} a^2(\lambda) d\lambda$$

(33) $$- \sum_{j=1}^{q} \left\{ \frac{1}{2\pi} \int_{-\pi}^{\pi} \text{Re} \frac{w_{j1} z + \dots + w_{jq} z^q}{h_q(z)} a(\lambda) d\lambda \right\}^2$$

$$= \iota^{(1)'} C_\gamma \iota^{(1)},$$

where $\iota^{(1)} = (\iota_1^{(1)}, \dots, \iota_{q'-q}^{(1)})'$, $\gamma = (\iota_1, \dots, \iota_q)'$, $C_\gamma = G_\gamma - H_\gamma' J_\gamma^{-1} H_\gamma$, and G_γ, H_γ, and J_γ are defined by the formula

(IV.3.6) with $s = q$ and $h = q'-q$. Thus, it follows from the results of Example 1 in Section 3 of Chapter IV that the value of the power of the test determined by the critical region (30) as $n \to \infty$ coincides with the values of powers of the tests introduced in Chapter IV for testing the hypothesis H_0: $\iota_{(1)} = 0$ versus the alternative H_1: $\iota_{(1)} \neq 0$. Since this value is not less than the value (as $n \to \infty$) of the powers of the preceding test-statistics we shall confine ourselves in the next example to a generalization of the test with the critical region (32) to the general case of a mixed autoregressive-moving average process.

Example 2. Assume that X_t, $t = ...,-1,0,1, ...,$ is a linear process with spectral density of the form (II.4.28). Consider the problem of testing the hypothesis H_0 that $g_\Theta(\lambda) = |g_r(z)|^2|h_q(z)|^{-2}$, $\Theta = (\iota_1, ..., \iota_q, \alpha_1, ..., \alpha_r)'$, $p = q+r$, belongs to the family of functions g_Θ, $\Theta \in \Theta$.
    Let

(34) $$\Phi_{j,\Theta}(\lambda) = 2 \operatorname{Re} \frac{h_q(\bar{z})g_r(\bar{z})}{h_q(z)g_r(z)} z^{q+r+j}, \quad j = 1, ..., m.$$

Then it is easy to verify that in view of (16) and (34) all the entries of the matrix B_Θ equal zero. Let $\iota_{1*}, ..., \iota_{q*}, \alpha_{1*}, ..., \alpha_{r*}$, and σ_*^2 be consistent estimators of the parameters $\iota_1, ..., \iota_q, \alpha_1, ..., \alpha_r$, and σ^2 (for example, those which were defined in Subsection 2.3 of Chapter III) and let $h_q^*(z) = 1-\iota_{1*}z - \cdots - \iota_{q*}z^q$ and $g_r(z) = 1-\alpha_{1*}z - \cdots - \alpha_{r*}z^r$. Then the critical region (24) becomes

(35) $$\left\{ x: \sum_{k=1}^m \left[\frac{\sqrt{n}}{\sigma_*^2} \int_{-\pi}^{\pi} \operatorname{Re}\left[\frac{h_q^*(\bar{z})}{g_r(z)} \right]^2 z^{q+r+k} I_n(\lambda)d\lambda \right]^2 > d_\alpha \right\}.$$

If the alternative hypothesis H_1 is the same here as in Example 2 of the preceding Section with the only difference that

(36) $$g_r^{(1)}(z) = \alpha_1^{(1)} z^{r+1} + \cdots + \alpha_{r'-r}^{(1)} z^{r'},$$
$$h_q^{(1)}(z) = \iota_1^{(1)} z^{q+1} + \cdots + \iota_{q'-q}^{(1)} z^{q'}$$

in (2.10) and (2.15) then in view of Remark 5 and the formulas (1.13), (2.15), (34), and (36) the power of the test determined by the critical region (35) as $n \to \infty$ converges to the value (2.2), where μ is an m-dimensional vector whose k-th component equals

(37)
$$\mu_k = \sum_{j=1}^{q'-q} \iota_j^{(1)} \frac{1}{2\pi} \int_{-\pi}^{\pi} \frac{g_r(\bar{z})z^{r+k-j}}{h_q(z)g_r(z)} \, d\lambda$$
$$- \sum_{j=1}^{r'-r} \iota_j^{(1)} \frac{1}{2\pi} \int_{-\pi}^{\pi} \frac{h_q(\bar{z})z^{r+k-j}}{h_q(z)g_r(z)} \, d\lambda.$$

The first summand on the r.h.s. of (37) equals 0 for $k > q'-q$ and the second vanishes for $k > r'-r$ so that for $m \leqslant \max\{(q'-q), (r'-r)\}$ the noncentrality parameter takes on the maximal value

$$\sum_{1 \leqslant k \leqslant \max\{(q'-q), (r'-r)\}} \mu_k^2.$$

From here it follows that the asymptotic value (2.2) of the power of the test will be maximal provided $m = \max\{(q'-q), (r'-r)\}$ in (35).

In this case the function a which is defined here by the formulas (2.15) and (36) can evidently be represented as a linear combination of the first $r+q+\max\{(q'-q),(r'-r)\}$ orthogonal functions presented in (2.16). A comparison of (2.16) and (34) yields that the last $\max\{(q'-q), (r'-r)\}$ coefficients in this representation coincide with μ_j, $j = 1,2,$ $\ldots\max\{(q'-q), (r'-r)\}$. From here and the formulas (2.16) and (2.17) it follows that (using the notation of Example 2 in Section 3 of Chapter IV) the value of the noncentrality parameter in (2.2) when the degrees of freedom is $m = \max\{(q'-q), (r'-r)\}$, is equal to

$$\sum_{1 \leqslant k < \max\{(q'-q), (r'-r)\}} \mu_k^2$$

$$= \frac{1}{4\pi} \int_{-\pi}^{\pi} a^2(\lambda)d\lambda - \sum_{j=1}^{j+q} \left\{ \frac{1}{2\pi} \int_{-\pi}^{\pi} a(\lambda) \mathrm{Re} \left[\frac{\sum_{k=1}^{q} w_{jk}z^k}{h_q(z)} \right. \right.$$

$$\left. \left. - \frac{\sum_{i=1}^{r} w_{j,q+r}z^{\ell}}{g_r(z)} \right] d\lambda \right\}^2 = [\iota'_{(1)}, \alpha'_{(1)}](G_\gamma - H'_\gamma J_\gamma^{-1} H_\gamma) \begin{bmatrix} \iota_{(1)} \\ \alpha_{(1)} \end{bmatrix}.$$

where $\iota'_{(1)} = (\iota_1^{(1)}, \ldots, \iota_{q'-q}^{(1)})$ and $\alpha'_{(1)} = (\alpha_1^{(1)}, \ldots, \alpha_{r'-r}^{(1)})$.

Example 3. Finally, consider the case when the hypothesis H_0 is that the spectral density f of a linear process X_t, $t = \ldots-1,0,1, \ldots$, is of the form (II.4.47) where $y = (y_1, \ldots, y_r)' \in \Gamma$ (here Γ is an open subset in the space R_r).

Let orthogonal functions $\Phi_1, \ldots, \Phi_m$ be of the form (25). Clearly then all the entries of the matrix B equal zero, provided only $m_0 \geqslant r$ and the critical region becomes of the form (IV.3.7) where, however, the unknown value of the parameter σ^2 is replaced by its consistent estimator σ_*^2 (cf. Subsection 2.4 in Chapter III), $s = r$ and j takes on values $m_0+1, m_0+2, \ldots, m_0+m$.

If under the alternative hypothesis H_1 the spectral density is again of the form (II.4.47) where, however, $r = r'$ ($r' > r$) and $y_{r+1} = n^{-1/2} y_1^{(1)}, \ldots, y_{r'} = n^{-1/2} y_{r'-r}^{(1)}$ then it is easy to verify that the power of the test determined by such a critical region, as $n \to \infty$, converges to the value (2.2), where the noncentrality parameter equals

$$[y_{m_0-r+1}^{(1)}]^2 + [y_{m_0-r+2}^{(1)}]^2 + \cdots + [y_{\min\{(r'-r,m\}}^{(1)}]^2.$$

Thus the asymptotic value of the power is maximal at $m_0 = r$ and $m = r'-r$.

Appendix 1. Remarks and Bibliography

Section 1

In the special case of a Gaussian process X_t, $t = \ldots,-1,0,1,\ldots$, with a square integrable spectral density, the measure P_n on $C[0,\pi]$ generated by the process

$$\zeta_n(\lambda) = n^{1/2} \int_0^\lambda [I_n(\lambda) - f(\lambda)]d\lambda, \quad 0 \leqslant \lambda \leqslant \pi,$$

converges weakly to the measure P_0 generated by the process $\{\zeta_0(\lambda), 0 \leqslant \lambda \leqslant \pi\}$, where $\zeta_0(\lambda)$ is a Gaussian random process with zero mean and covariance function

$$E\{\zeta_0(\lambda)\zeta_0(\mu)\} = 2\pi \int_0^{\min(\lambda,\mu)} f^2(\lambda)d\lambda$$

(cf. [85]). In the case of a linear process it was shown by Grenander and Rosenblatt [36] that if the spectral density f of a linear process X_t is absolutely continuous, the coefficients g_s in (II.6.1) are such that $g_s = O(s^\beta)$ for $\beta < -3/2$ and, moreover $E\varepsilon_0^8 < \infty$, then the above stated result remains valid where, however, now

$$E[\zeta_0(\lambda)\zeta_0(\mu)] = 4\pi \int_{-\pi}^{\min(\lambda, \mu)} f^2(\lambda)d\lambda$$

$$+ \kappa_4 \int_{-\pi}^{\lambda} f(\lambda)d\lambda \int_{-\pi}^{\mu} f(\mu)d\mu$$

Later, Ibragimov and Tovstik [69] proved that actually it is sufficient to require the existence of $\delta > 0$ such that

$$\int_{-\pi}^{\pi} f^{2+\delta}(\lambda)d\lambda > \infty, \quad E(|\varepsilon_t|^\alpha) < \infty, \quad \alpha > 4 + \frac{8}{\delta}.$$

Below we shall show that it is possible to avoid the introduction of even these broader conditions of the paper [69]. For this purpose we shall utilize the assertions of the theorems of the papers [132,36] which can be conveniently stated in the form of the following lemmas.

Lemma 1. *Let X_t, $t = ...-1,0,1,...$, be a linear process satisfying the conditions presented in the beginning of Subsection 1.1. Then*

$$n^{1/2} \max_{0 \leqslant \lambda \leqslant \pi} |T_n(\lambda)| \to 0 \quad as\ n \to \infty$$

in probability where

$$T_n(\lambda) = I_n(\lambda) - 2\pi f(\lambda)I_{n,\varepsilon}(\lambda)/\sigma^2$$

and

$$I_{n,\varepsilon}(\lambda) = \frac{1}{2\pi n}|\sum_{t=1}^{n} \varepsilon_t e^{i\lambda t}|^2$$

is the periodogram of the process ε_t.

Lemma 2. *Let P_0 be a measure on $C[0,\pi]$ generated by a normal process $\xi_0(\lambda)$ with zero mean and covariance function*

$$E(\xi_0(\lambda)\xi_0(\mu)) = \frac{1}{2\pi}\min(\lambda,\mu) + \frac{\kappa_4}{4\pi^2}\lambda\mu.$$

Then the measure P_n on $C[0,\pi]$ generated by the process

$$\xi_{n,\varepsilon}(\lambda) = n^{1/2}\left[\frac{1}{\sigma^2}\int_0^{\lambda} I_{n,\varepsilon}(\lambda)d\lambda - \frac{\lambda}{2\pi}\right], \quad 0 \leqslant \lambda \leqslant \pi,$$

converges weakly to the measure P_0 as $n \to \infty$.

We note that Lemma 1 is also proved in the paper [36] but under the much more restrictive conditions on g_s stated above.

In place of the process ζ_0 we shall now consider the process

$$\xi_n(\lambda) = n^{1/2} \int_0^\lambda \frac{I_n(\lambda) - f(\lambda)}{f(\lambda)} d\lambda, \qquad 0 \leqslant \lambda \leqslant \pi.$$

A direct application of these lemmas leads to the following result.

Lemma 3. *Under the conditions of Lemma 1, the measure P_n on $C[0,\pi]$ generated by the process $\xi_n(\lambda)$ converges weakly to the measure P_0 appearing in the condition of Lemma 2.*

Indeed since

$$\xi_n(\lambda) = n^{1/2} \int_0^\lambda \frac{T_n(\lambda)}{2\pi f(\lambda)} d\lambda + \xi_{n,\,\varepsilon}(\lambda),$$

it is sufficient to verify that

$$\frac{n^{1/2}}{2\pi} \int_0^\lambda \frac{T_n(\lambda)}{f(\lambda)} d\lambda \to 0$$

as $n \to \infty$ in probability, in view of the condition $f > 0$ and Lemma 1 since

$$\frac{n^{1/2}}{2\pi} |\int_0^\lambda \frac{T_n(\lambda)}{f(\lambda)} d\lambda| \leqslant \max_\lambda f^{-1}(\lambda) n^{1/2} \max_\lambda |T_n(\lambda)|.$$

2. When introducing the alternatives H_1 in Subsection 1.2 we aim at the simplest possible form of assuring conditions for the validity of Proposition 2. An attempt at a more natural formulation of "close" alternatives in the spirit of the general asymptotic theory would lead us to arguments similar to those which were presented in Appendix 3 to Chapter II, Remark 3 to Section 6.

Proposition 2 is a corollary of the fact that, first, the assertion of the Proposition 1 holds also when the alternative hypothesis H_1 is valid, i.e., when in their statement f is replaced by f_n so that

$$\xi_n(\lambda) = \frac{n^{1/2}}{2\pi} \int_{-\pi}^\lambda \frac{I_n(\lambda)}{f_n(\lambda)} d\lambda$$

and second, under these conditions

$$\frac{1}{2\pi}\int_0^\lambda \frac{I_n(\lambda)}{f_n(\lambda)} a_n(\lambda)d\lambda \to \frac{1}{2\pi}\int_0^\lambda a(\lambda)d\lambda,$$

$$a_n(\lambda) = n^{1/2} \frac{f_n(\lambda)-f(\lambda)}{f(\lambda)}$$

as $n \to \infty$ in probability.

3. For computational convenience the integrals in formula (1) can often be replaced by the corresponding Riemann sums. Indeed, consider instead of the random function $\zeta_n(\tau)$, $0 \leqslant \tau \leqslant 1$, a sequence of random variables

$$\zeta_n\!\left(\frac{\ell}{n}\right) = (1/2n)^{1/2}\left[\sum_{k=1}^{\ell} \frac{I_n(\pi k/n)}{f(\pi k/n)} - \frac{\ell}{n}\sum_{k=1}^{n} \frac{I_n(\pi k/n)}{f(\pi k/n)}\right],$$

$$\ell = 1, ..., n.$$

Then

$$P\{\max_{1\leqslant \ell\leqslant n} |\zeta_n(\ell/n)| < \varepsilon\} \to \sum_{k=-\infty}^{\infty} (-1)^k e^{-2k^2 \varepsilon^2}.$$

Since $\sum_{k=1}^{n} I_n(\pi k/n)/nf(\pi/kn)$ is the Riemann sum of the

integral $\int_0^\pi I_n(\lambda)/\pi f(\lambda)d\lambda$, as $n \to \infty$ this sum approaches 1 in probability. Consequently, in the last relation the random variable $\zeta_n(\ell/n)$ can be replaced by the quantity

$$(n/2)^{1/2}\left\{ \frac{\sum_{k=1}^{\ell} I_n(\pi k/n)/f(\pi k/n) - (\ell/n)}{\sum_{k=1}^{n} I_n(\pi k/n)/f(\pi k/n)} \right\}$$

(cf. [12] or [139], p. 121).

Section 2

1. A comparison of Quenouille's and Bartlett-Diananda's tests from the aspect of asymptotic (as $n \to \infty$) values of their powers in applications to the problem considered in Example 1 is presented in the paper [127] (under a more restrictive

class of alternative hypotheses).

2. The general results presented in Section 2 can be easily carried over to the case of Gaussian random processes X_t, $-\infty <$ $t < \infty$, with continuous time t. The problem of hypothesis testing relative to multidimensional autoregressive processes is treated, for example, in [14,73] as well as in [140], Chapter VI, Section 7.

Section 3

The results of Subsection 1 are generalizations to the case considered herein of the well-known results dealing with the particular case when X_1, ..., X_n are independent, identically distributed (with distribution function $F(x,\Theta)$, $-\infty < x < \infty$ depending on the unknown parameter Θ) random variables, and $\Phi_{n,\Theta}$ is a random vector whose k-th component is of the form $[\nu_k - np_{k,\Theta}](np_{k,\Theta})^{-1/2}$, where ν_k is the number of values of X_i located in the interval (a_{k-1},a_k) (here $-\infty = a_0 < a_1 < \cdots$ $< a_m = \infty$) and

$$p_{k,\Theta} = \int_{a_{k-1}}^{a_k} dF(x,\Theta)$$

(cf. for example, the books [71,76,106] as well as [52,91,144,147]). Actually in this last particular case, the limiting distribution of the vector $\Phi_{n,\Theta}$ (when the null hypothesis that the distribution function belongs to the family of functions $F(x,\Theta)$, $\Theta \in \Theta$ is valid) is an m-dimensional normal distribution $N(0,I_m - \Phi_\Theta \Phi'_\Theta)$ where Φ_Θ is a vector whose

k-th component is equal to $\sqrt{p_{k,\Theta}}$. Due to the presence of the term $\Phi_\Theta \Phi'_\Theta$ in the expression of the covariance matrix of this limiting distribution the degrees of freedom of statistics considered herein is decreased by one. Observe also that in the particular case considered herein the (k,ℓ)-th entry of the $(m \times p)$-matrix B is of the form $p_{k,\Theta}^{-1/2}(\partial/\partial\Theta_\ell)p_{k,\Theta}$, W is the Fisher's information matrix, $\bar{\Theta}$ is the usual maximum likelihood estimator (or some estimator asymptotically equivalent to it), and $\hat{\Theta}$ is a multinomial maximum likelihood estimator (or some estimator asymptotically equivalent to it) of the parameter Θ.

2. In applications to Gaussian random processes (with discrete as well as with continuous time t) the tests determined by the critical regions (20) and (21) were proposed in papers by Osidze [95] and [96] respectively, while the test with the critical region of the form (22) is presented in the author's paper [44].

BIBLIOGRAPHY

Translator's Remark

Whenever possible, the English version of the paper or book is presented. However, the order of the entries follows the original Russian edition. This has been done to minimize the possibility of printing errors. Since all the references to the literature in the text are indicated by the ordinal numbers, this ordering should cause no confusion.

1. Adenstedt, R. K. (1974). On large-sample estimation for the mean of a stationary random sequence, *Ann. Statist.*, **2**, No. 6, 1095-1107.

2. Akaike H. (1973). Maximum likelihood identification of Gaussian autoregressive moving average models, *Biometrika*, **60**, 255-266.

3. Anderson, O. D. (1976). On the inverse of the autocovariance matrix for a general moving average process, *Biometrika*, **63**, 391-394.

4. Anderson, T. W. (1971). *The Statistical Analysis of Time Series*, J. Wiley, New York.

5. Anderson, T. W. (1973). Asymptotically efficient estimation of covariance matrices with linear structure, *Ann. Statist.*, **1**, 1, 135-141.

6. Anderson, T. W. (1975). Estimation of maximum likelihood in autoregressive moving average models in the time and frequency domains, Department of Statistics, Stanford University, Stanford, California, Tech. Rpt. No. 20.

7.   Anderson, T. W. (1977). Estimation for autoregressive moving average models in time and frequency domains, *Ann. Statist.*, **5**, 5, 842-865.

8.   Arato, M. (1970). Exact formulas for densities of measures of elementary Gaussian processes, *Studia Scient. Math. Hungarica*, **5**, 17-27.

9.   Astrom, K. J. and Bohlin, T. (1965). Numerical identification of linear dynamic system from normal operating records, *Proceedings of the Second IFAC Symposium on the Theory of Self-Adaptive Control Systems*, Sept. 14-17, 1965, New York Plenum Press, New York, 1966.

10.   Bartlett, M. S. (1946). On the theoretical specification and sampling properties of autocorrelated time-series, *J. R. Stat. Soc.*, **1**, 27-41.

11.   Bartlett, M. S. (1953). Approximate confidence intervals, II. More than one unknown parameter, *Biometrika*, **40**, 3-4, 306-317.

12.   Bartlett, M. S. (1954). Problèmes de l'analyse spectrale des series temporelles stationnaires, *Publ. Inst. Statist.* (Univ. de Paris) 3, fasc. 3, 119-134.

13.   Bartlett, M. S. and Diananda, P. H. (1953). Extensions of Quenouille's test for autoregressive schemes, *J. Roy. Statist. Soc.*, Ser. B, **15**, 1, 107-124.

14.   Bartlett, M. S. and Rajalakshman, D. S. (1953). Goodness of fit test for simultaneous autoregressive series, *J. Roy. Statist. Soc.*, Ser. B, **15**, 107-124.

15.   Bartoo, T. B. and Puri, P. S. (1967). On optimal asymptotic tests of composite hypotheses, *Ann. Math. Statist.*, **38**, 6, 1845-1852.

16.   Barton, D. E. (1953). On Neyman's smooth test of goodness of fit and its power with respect to a particular system of alternatives, *Scand. Aktuartidskr.*, **36**, 24-36.

17.   Bentkus, R. (1972). On the error in estimates of spectral functions of stationary processes, *Lit. Mat. Sbornik*, **12**, No. 1, 55-71.

18.   Bentkus, R. (1972). On the asymptotic normality of estimators of a spectral function, *ibid*, **12**, No. 3, 5-18.

19.   Bentkus, R. (1977). Cumulants of polylinear forms of stationary sequences, *Lithuanian Math. Journal*, **17**, 1, 16-31 (Russian version 27-46).

20.   Bentkus, R. Ju. and Žurbenko, I. G. (1976). Asymptotic normality of spectral estimates. *Soviet Mathematics Dokl.*, **17**, No. 4, 943-946 (Russian original 229, No. 1, 11-14).

21. Blackman, R. B. and Tukey, J. W. (1959). *The Measurement of Power Spectra from the Point of View of Communications Engineering*, Dover, New York.

22. Bloomfield, P. (1976). *Fourier Analysis of Time Series: An Introduction*, Wiley, New York.

23. Bloomfield, P. (1973). An exponential model for the spectrum of a scalar time series, *Biometrika*, 60, 2, 217-226.

24. Box, G. E. P. and Jenkins, G. M. (1976). *Time Series Analysis. Forecasting and Control* (Revised edition), Holden-Day, San Francisco.

25. Bol'shev, L. N. and Smirnov, N. V. (1965). *Tables of Mathematical Statistics*, Nauka, Moscow.

26. Brillinger, D. R. (1975). *Time Series: Data Analysis and Theory*, Holt, Rinehart and Winston, New York.

27. Brillinger, D. R. (1969). Asymptotic properties of spectral estimates of second order, *Biometrika*, 56, 2, 375-390.

28. Buhler, W. J. and Puri, P. S. (1966). On optimal asymptotic tests of composite hypotheses with several constraints, Z. *Wahrsheinlichkeitstheorie und verw. Gebiete*, 5, 1, 71-88.

29. Wald, A. (1943). Tests of statistical hypotheses concerning several parameters when the number of observations is large, *Trans. Amer. Math. Soc.*, 54, 3, 426-482.

30. Hajek, J. (1962). On linear statistical problems in stochastic processes, *Czechoslov. Math. J.*, 12, 404-444.

31. Hajek, J. and Sidak, Z. (1967). *Theory of Rank Tests*, Academic Press, New York.

32. Guyon, X. and Prum, B. (1977). Estimations et tests relatifs aux processus spatiaux reguliers du second ordre, Publ. Univ. Orsay, No. 201, France.

33. Guyon, X. and Prum, B. (1977). Statistique de processus a parametre multidimensionnel, *C. R. Acad. Sci.*, Paris, 284, Ser. A, 327-330.

34. Giersch, W. and Sharpe, D. (1973). Estimation of power spectra with finite order autoregressive models, *IEEE Trans. Automat. Contr.*, AC-18, 367-369.

35. Grenander, U. and Rosenblatt, M. (1957). *Statistical Analysis of Stationary Time Series*, John Wiley, New York.

36. Grenander, U. and Rosenblatt, M. (1953). Statistical spectral analysis of time series arising from stationary stochastic processes, *Ann. Math. Statist.*, 24, 537-558.

37. Grenander, U. and Szegö, G. (1958). *Toeplitz forms and their applications*, Univ. of Calif. Press, Berkeley, California.

38. Dacunha-Castelle, D. (1979). Remarque sur l'etude asymptotique du rapport de vraisemblance de deux processus gaussiens stationnaires, *C. R. Acad. Sci.* Paris, **288**, Ser. A, 225-228.

39. Dunsmuir, W. (1979). A central limit theorem for parameter estimation in stationary vector time series and its application to models for a signal observed with noise, *Ann. Statist.*, **3**, 7, 490-506.

40. Dunsmuir, W. and Hannan, E. J. (1976). Vector linear time series models, *Adv. Appl. Prob.*, **8**, 339-364.

41. Dzhaparidze, K. O. (1973). Methods of estimating parameters of stationary stochastic signals with a rational spectrum *Prob. Inf. Trans.*, **9**, 4, May 1975, 295-301 (Russian original Oct.-Dec. 1973, 33-42).

42. Dzhaparidze, K. O. (1974). On simplified estimators of unknown parameters with good asymptotic properties. *Theory Prob. and Applic.*, **19**, 347-358.

43. Dzhaparidze, K. O. (1974). A new method for estimating spectral parameters of a stationary regular time series, *Theory Prob. and Applic.*, **19**, 1, 120-130.

44. Dzhaparidze, K. O. (1977). Tests of composite hypotheses for random variables and stochastic processes, *Theory Prob. and Applic.*, **22**, 104-118.

45. Dzhaparidze, K. O. (1977). Estimation of parameters of spectral density with fixed zeros, *Theory Prob. and Applic.*, **22**, 708-729.

46. Dzhaparidze, K. O. (1974). *Lectures on Statistics of Random Processes*, Jena University (in Russian).

47. Dzhaparidze, K. O. (1977). *Asymptotically efficient estimation of parameters of a spectrum of Gaussian time series*, Tbilisi University Press, Tbilisi, GSSR (in Russian).

48. Dzhaparidze, K. O. (1970). On the estimation of the spectral parameters of a Gaussian stationary process with rational spectral density, *Theory Prob. and Applic.*, **15**, 531-538.

49. Dzhaparidze, K. O. (1971). On methods for obtaining asymptotically efficient spectral parameter estimates for a stationary Gaussian process with rational spectral density, *Theory Prob. and Applic.*, **16**, 550-554.

50. Dzhaparidze, K. O. and Marr, G. I. (1974). On the evaluation of the likelihood ratio for a generalized Gaussian process with rational spectral density, *Theory Prob. and Applic.*, **19**, 407-409.

51. Dzhaparidze, K. O. and Marr, G. I. (1978). Estimation of spectrum parameters of random processes on the basis of observations in noise, *Probl. of Inform. Transmission*, July 1978, 26-34 (Russian original **14**, No. 1, Jan.-March 1978, 37-49).

52. Dzhaparidze, K. O. and Nikulin, M. S. (1974). On a modification of a standard statistics of Pearson, *Theor. Prob. and Applic.*, **19**, 851-852.

53. Dzhaparidze, K. O. and Yaglom, A. M. (1973). Asymptotically efficient estimation of spectrum parameters of stationary stochastic processes, *Proceedings of the Prague Symposium on Asymptotic Statistics*, 55-105, Charles University, Prague.

54. Dzhaparidze, K. O. and Yaglom, A. M. (1975). Application of a modified "scoring method" of Fisher to the estimation of spectral parameters of random processes, *Soviet Mathematics Dokl.*, **15**, No. 4, 1077-1082 (Russian original **217**, 512-515 (1974)).

55. Dzhaparidze, K. O. and Yaglom, A. M. (1977). Estimation of parameters of the spectral density of random processes with stationary increments and stationary processes with vanishing spectral density. Abstracts of papers presented at the *Second International Vilnius' Conference on the Theory of Probability and Mathematical Statistics*, Vol. I, 121-122.

56. Jenkins, G. N. and Watts, D. G. (1968). *Spectral Analysis and its Applications*, Holden-Day, San Francisco.

57. Deistler, M., Dunsmuir, W., and Hannan, E. J. (1978). Vector linear time series models corrections and extensions, *Adv. Appl. Probab.*, **10**, 360-372.

58. Doob, J. L. (1953). *Stochastic Processes*, J. Wiley, New York.

59. Durbin, J. (1960). The fitting of time-series models, *Rev. Inst. Intern. Statist.*, **28**, 233-234.

60. Durbin, J. (1959). Efficient estimation of parameters in moving-average models, *Biometrika*, **46**, 306-316.

61. Davies, R. B. (1973). Asymptotic inference in stationary Gaussian time-series, *Adv. Appl. Probab.*, **5**, 469-497.

62. Davis, H. T. and Jones, R. H. (1968). Estimation of the innovation variance of a stationary time series, *JASA*, **63**, 321, 141-149.

63. Zhurbenko, I. G. and Zuev, N. H. (1975). Higher order spectral densities of stationary processes with mixing, *Ukrainian Math. Journal*, 27, No. 4, July-August 1976, 364-373

(Trans. of the Russian original, July-August 1975, 442-464).

64. Zacks, S. (1970). *The Theory of Statistical Inference*, J. Wiley, New York.

65. Zygmund, A. (1959). *Trigonometric Series I*. (Second edition), Cambridge University Press, Cambridge, England.

66. Ibragimov, I. A. (1963). On estimation of the spectral function of a stationary Gaussian process, *Theory Prob. and Applic.*, **8**, 366-401.

67. Ibragimov, I. A. (1968). On a theorem of G. Szego, *Mathematical Notes*, **3**, 442-448.

68. Ibragimov, I. A. and Rosanov, Yu. A. (1970). *Gaussian Random Processes*, Nauka, Moscow.

69. Ibragimov, I. A. and Tovstik, T. M. (1964). On an estimator for spectral densities, *Vestnik LGU* (Herald of Leningrad State University), 1, 42-57.

70. Ibragimov, I. A. and Has'minskii, R. Z. (1981). *Statistical Estimation Asymptotic Theory*, Springer-Verlag, New York.

71. Kendall, M. G. and Stuart, A. (1967). *The Advanced Theory of Statistics, Vol. 2 Inference and Relationship*, (Fourth edition (1979)), Griffin (Hafner), London.

72. Quenouille, M. H. (1947). A large-sample test for goodness of fit of autoregressive schemes, *J. Roy. Stat. Soc.*, Ser. A, **110**, 123-129.

73. Quenouille, M. H. (1957). *The Analysis of Multiple Time Series*, Griffin, London.

74. Clevenson, M. L. (1970). Asymptotically efficient estimates of the parameters of a moving average time series, Stanford University, Statistics Dept. Techn. Rep. No. 15, Stanford, California.

75. Kohn, R. (1978). Asymptotic properties of time domain Gaussian estimators, *Adv. Appl. Probab.*, 2, 10, 339-359.

76. Crámer, H. (1946). *Mathematical Methods of statistics*, Princeton University Press, Princeton, New Jersey.

77. Koopmans, L. H. (1974). *The Spectral Analysis in Time Series*, Academic Press, New York.

78. Levin, M. J. (1965). Power spectrum parameter estimation, *IEEE Trans. Inform. Theory* IT-11, 1, 100-107.

79. LeCam, L. (1956). On the asymptotic theory of estimation and testing hypothesis, *Proc. Third Berkeley Sympos. Math. Statist. Probab.*, 1, 129-156.

80. LeCam, L. (1960). Locally asymptotically normal families of distributions, *Univ. of California Publ. Statist.*, 3,

No. 2, 37-98.

81.   LeCam, L. (1969). *Théorie Asymptotique de la Décision Statistique*, Les Presses de l'Universite de Montreal, Montréal.

82.   LeCam, L. (1974). Notes on Asymptotic Methods in Statistical Decision Theory, Centre de Recherches Mathématiques, Université de Montréal, Montréal.

83.   Lehmann, E. (1958). *Testing Statistical Hypotheses*, J. Wiley, New York.

84.   Ljung, G. M. and Box, G. E. P. (1979). The likelihood function of stationary autoregression-moving average models, *Biometrika*, **2**, 66, 265-270.

85.   Malevich, T. L. (1964). The asymptotic behavior of an estimate for the spectral function of a stationary Gaussian process, *Theor. Prob. and Applic.*, **9**, No. 2, 350-353.

86.   Mann, H. and Wald, A. (1943). On the statistical treatment of linear stochastic difference equations, *Econometrics*, **11**, 173-220.

87.   Mentz, R. P. (1976). On the inverse of some covariance matrices of Toeplitz type, *SIAM J. Appl. Math.*, **3**, 31, 426-437.

88.   Moran, P. A. (1970). On asymptotically optimal tests of composite hypotheses, *Biometrika* 57, 1, 47-55.

89.   Neyman, J. (1937). "Smooth Test" for goodness of fit, *Skand. Aktuartidsk.*, **20**, 149-199.

90.   Neyman, J. (1959). Optimal asymptotic tests of composite statistical hypotheses, *The H. Cramér Volume*, 213-234, Almquist and Wiksell, Uppsala.

91.   Nikulin, M. S. (1973). Chi-square test for continuous distributions with shift and scale parameters, *Theor. Prob. and Applic.*, **18**, 559-568.

92.   Nicholls, D. F. and Hall, A. D. (1979). The likelihood function of stationary autoregression-moving average models, *Biometrika*, **66**, 259-264.

93.   Newbold, P. (1974). The exact likelihood function for a mixed autoregressive-moving average process, *Biometrika*, **61**, 3, 423-426.

94.   Ortega, J. M. and Rheinboldt, W. C. (1970). *Iterative Solutions of Nonlinear Equations in Several Variables* (Computer Science and Applied Math. Series), Academic Press, New York.

95.   Osidze, A. G. (1979). On χ^2 criterion for testing hypothesis about the spectral density of a Gaussian random process with unknown parameter, *Reports of Akad. Nauk Georgian SSR*, **75**, No. 2, 273-275.

96. Osidze, A. G. (1974). On a goodness of fit test in the case of dependence of spectral density of Gaussian processes on unknown parameters, *ibid.*, **74**, No. 2, 273-275.

97. Osidze, A. G. (1975). On a statistic for testing the composite hypothesis regarding the form of a spectral density of a stationary Gaussian random process, *ibid.*, **77**, No. 2, 313-315.

98. Pagano, M. (1972). Estimation of models of autoregressive signal plus white noise, State University of New York at Buffalo, Research Report No. 57.

99. Pagano, M. (1974). Estimation of models of autoregressive signal plus white noise, *Ann. Statist.*, **2**, 1, 99-108.

100. Parzen, F. (1971). Efficient estimation of stationary time series mixed schemes, *Bull. Intern. Stat. Inst.*, **44**, Book 2, 315-319.

101. Parzen, E. (1961). An approach to time series analysis, *Ann. Math. Statist.*, **32**, 951-989.

102. Pisarenko, V. F. (1962). On the estimator of parameters of Gaussian stationary process with spectral density $|P(i\lambda)|^{-2}$, *Lit. Matem. Sbornik*, **2**, No. 2, 159-167 (in Russian).

103. Pisarenko, V. F. (1965). On the computation of the relation of likelihood for Gaussian processes with rational spectrum, *Theory Prob. and Applic.*, **10**, 299-303.

104. Murthy, Prabhakar D. N. (1973). Method of maximum likelihood for stationary time series models, *IEEE Trans. Autom. Contr.*, **AC-18**, 4, 397-398.

105. Rao, C. R. (1948). Large sample tests of statistical hypotheses concerning several parameters with applications to problems of estimation, *Proc. Cambr. Phil. Soc.*, **44**, 1, 50-57.

106. Rao, C. R. (1973). *Linear Statistical Inference and its Applications*, 2nd ed., J. Wiley, New York.

107. Rasulov, N. P. (1976). On asymptotically efficient estimates of regression coefficients under spectral density of noise degenerates, *Theory Probab. and Applic.*, **21**, 316-324.

108. Rozanov, I. A. (1971). *Infinite-Dimensional Gaussian Distributions*, Amer. Mth. Society, Providence, R.I.

109. Rosenblatt, M. (1956). A central limit theorem and a strong mixing condition, *Proc. Nat. Acad. Sci. U.S.A.*, **42**, 43-47.

110. Roussas, G. G. (1972). *Contiguity of Probability Measures*, Cambridge University Press, Cambridge.

111. Siddiqui, M. M. (1958). On the inversion of the sample covariance matrix of a stationary autoregressive process, *Ann. Math. Statist.*, **29**, 585-588.

112. Striebel, Ch. T. (1959). Densities for stochastic processes, *Ann. Math. Statist.*, **30**, 559-567.

113. Tiao, G. C. and Ali, M. M. (1971). Analysis of correlated random effects: linear model with two random components, *Biometrika*, **58**, 37-51.

114. Thomas, J. B. and Zadeh, L. A. (1961). Note on an integral equation occurring in the prediction, detection and analysis of multiple time series, *IRE Trans.*, **II-7**, 2, 118-120.

115. Tretter, A. and Steiglitz, K. (1967). Power-spectrum identification in terms of rational models, *IEEE Trans. Automat. Contr.* **AC-12**, 185-188.

116. Tuan Pham Dinh, M. (1974). Sur le calcul de la fonction de vraisemblance liée a l'estimation des paramètres d'un processus gaussien stationnaire centre de densité spectrale rationnelle, *C. R. Acad. Sc. Paris*, **273**, No. 22, 1441-1444.

117. Tuan Pham Dinh, M. (1978). L'adéquation du processus multivariable Gaussian continu stationnaire centre de densité spectrale rationnelle, *Lecture Notes in Mathematics*, **636**, *Journees de Statistique des Processus Stochastiques*, *Proceedings*, Grenoble, Juin 1977, Springer-Verlag, Berlin-Heidelberg-New York, 1978.

118. Wise, J. (1955). The autocorrelation function and the spectral density function, *Biometrika*, **42**, 151-159.

119. Wilks, S. S. (1938). The large sample distribution of the likelihood ratio for testing composite hypotheses, *Ann. Math. Statist.*, **9**, 1, 60-62.

120. Wilson, G. (1969). Factorization of the covariance generation function of a pure moving averages process, *SIAM J. Numer. Anal.*, **6**, 1-7.

121. Whittle, P. (1952). Estimation and information in time series analysis, *Skand. Aktuar.*, **35**, 48-60.

122. Whittle, P. (1952). Tests of fit in time series, *Biometrika*, **39**, 3-4, 309-318.

123. Whittle, P. (1953). The analysis of multiple time series, *J. Roy. Statist. Soc.*, Ser. B, **15**, 125-139.

124. Whittle, P. (1954). In H. Wold's *A Study in the Analysis of Stationary Time Series*, Appendix 2, Almquist and Wiksell, Uppsala.

125. Whittle, P. (1962). Gaussian estimations in stationary time series, *Bull. Inst. Internat. Statist.*, **39**, 105-129.

126. Ulrich, T. J., and Bishop, T. N. (1975). Maximum entropy spectral analysis and autoregressive decomposition, *Rev. Geophys. and Space Phys.* Vol. 13, No. 1, 183-200.

127. Walker, A. M. (1952). Some properties of the asymptotic power functions of goodness-of-fit tests for linear autoregressive schemes, *J. Roy. Statist. Soc.*, Ser. B, 14, 117-134.

128. Walker, A. M. (1960). Some consequences of superimposed error in time series analysis, *Biometrika*, 47, 1 and 2, 33-43.

129. Walker, A. M. (1961). Large sample estimation of parameters for moving-average models, *Biometrika*, 48, 343-357.

130. Walker, A. M. (1962). Large-sample estimation of parameters for autoregressive proesses with moving-average residuals, *Biometrika*, 49, 117-132.

131. Walker, A. M. (1964). Asymptotic properties of least-squares estimates of parameters of the spectrum of a stationary nondeterministic time series, *J. Austral. Math. Soc.*, 4, 363-384.

132. Walker, A. M. (1965). Some asymptotic results for the periodogram of a stationary time series, *J. Austral. Math. Soc.*, 5, 107-108.

133. Uppuluri, V. R. R. and Carpenter, J. A. (1969). The inverse of a matrix occurring in first-order moving average models, *Sankhyā*, A, 31, 79-82.

134. Fikhtengol'tz, G. M. (1969). *Course in differential and integral calculus*, Vol. IV, (in Russian) Nauka, Moscow.

135. Fisher, R. A. (1935). The detection of linkage with dominant abnormalities, *Ann. Eugen.*, 6, 187-201.

136. Brodlie, K. W. (1977). Unconstrained minimization, in *The State of the Art in Numerical Analysis* (D. Jacobs, ed.), 229-269, Academic Press, London.

137. Harris, B. (ed.) (1967). *Spectral Analysis of Time Series*, John Wiley, New York

138. Hannan, E. J. (1958). The asymptotic powers of certain tests on goodness-of-fit for time series, *J. Roy. Statist. Soc.*, Ser. B, 20, 1, 143-151.

139. Hannan, E. J. (1960). *Time Series Analysis*, Methuen, London.

140. Hannan, E. J. (1970). *Multiple Time Series*, J. Wiley, New York.

141. Hannan, E. J. (1969). The estimation of mixed autoregressive moving-average systems, *Biometrika*, 56,

579-592.

142. Hannan, E. J. (1976). The asymptotic distribution of serial covariances, *Ann. Statist.*, 4, 396-399.

143. Hosoya, Y. (1979). High-order efficiency in the estimation of linear processes, *Ann. Statist.*, 7, 516-530.

144. Chernoff, H. and Lehman, E. L. (1954). The use of maximum likelihood estimates in χ^2-tests for goodness of fit, *Ann. Math. Statist.*, 25, 3, 579-586.

145. Chibisov, D. M. (1962). Application of Neyman's criteria to the verification of composite hypotheses, *Theor. Prob. and Applic.*, 7, No. 3, 345-346.

146. Chibisov, D. M. (1967). A theorem on admissible tests and its applications to an asymptotic problem of testing hypotheses, *ibid.*, 12, No. 1, 90-103.

147. Chibisov, D. M. (1971). Certain chi-square type tests for continuous distributions, *Theory Prob. and Applic.*, 16, No. 1, 1-22.

148. Shaman, P. (1969). On the inverse of the covariance matrix of a first-order moving average, *Biometrika*, 56, 595-600.

149. Shaman, P. (1973). On the inverse of the covariance matrix for an autoregressive-moving average process, *Biometrika*, 60, 1, 193-196.

150. Yaglom, A. M. (1962). *An Introduction to the Theory of Stationary Random Functions*, Prentice-Hall, Englewood Cliffs, New Jersey.

151. Yaglom, A. M. (1955). Extrapolation, interpolation and filtering of stationary random processes with a rational spectral density, *Proc. Moscow Math. Soc.*, 4, 333-374.

152. Dzhaparidze, K. O., Yaglom, A. M. (1983). Spectrum Parameter Estimation in Time Series Analysis, in *Developments in Statistics* (P. R. Krishnaiah, ed.), Vol. 4, Chapter 1, pp. 1-96, Academic Press, New York.

153. Ginovyan, M. S. (1984). Asymptotic behavior of the Toeplitz determinant, *Journal of Soviet Mathematics*, 24, No. 5, 494-500.

154. Ginovyan, M. S. (1983). $\sqrt{n}$-approximation of the likelihood function, *Journal of Soviet Mathematics*, 21, No. 1, 20-30.

155. Ginovyan, M. S. (1984). Asymptotic behavior of the logarithm of the likelihood function when the spectral density has polynomial zeros, *Journal of Soviet Mathematics*, 25, No. 3, 1113-1125.

156. Ingster, Yu. I. (1981). Asymptotic regularity of a family of measures corresponding to a Gaussian random process which contains a white noise component for a finite-parameter family of spectral densities, *Journal of Soviet Mathematics*, 25, No. 3, 1165-1181.

157. Hirshman, I. I. (1971). Recent Developments in the Theory of Finite Toeplitz Operators, *Advances in Probability* (P. Ney, ed.), 1, 105-167.

158. Solev, V. N. (1983). Approximation of the likelihood function, *Theory of Probab. and Applic.*, 28, No. 1, 201-203.

159. Guyon, X. (1982). Parameter estimation for a stationary process on a d-dimensional lattice, *Biometrika*, 69, 1, 95-105.

160. Coursol, J. and Dacunha-Castelle, D. (1982). Remarks on the approximation of the likelihood function of a stationary Gaussian process, *Theory Probab. and Applic.*, 27, No. 1, 162-167.

161. Grenander, U. (1981). *Abstract Inference*, John Wiley, New York.

162. Hall, P. and Heyde, C. C. (1980). *Martingale Limit Theory and Its Application*, Academic Press, New York.

163. Kac, M. (1954). Toeplitz matrices, translation kernels and a related problem in probability theory, *Duke Mathematical Journal*, 21, 501-509.

164. Kac, M. (1959). Probability and Related Topics in Physical Sciences, *Lectures in Appl. Mathematics*, Vol. I, Interscience, London.

165. Hosoya, Y. and Taniguchi, M. (1982). A central limit theorem for stationary processes and the parameter estimation of linear processes, *Ann. Statist.*, 10, No. 1, 132-153.

166. · Turin, G. L. (1960). The characteristic function of Hermitian quadratic forms in complex normal variables, *Biometrika*, 47, 199-201.

167. Ibragimov, I. A. (1962). Some limit theorems for stationary processes, *Theory Probab. Appl.*, 7, 349-382.

168. Ibragimov, I. A. (1975). A note on the central limit theorem for dependent random variables, *Theory Probab. Appl.*, 20, 135-141.

169. Ibragimov, I. A. and Linnik, Yu. V. (1971). *Independent and Stationary Sequences of Random Variables*, Wolters-Noordhoff, Groningen.

318   Bibliography

170.   Billingsley, P. (1968). *Convergence of Probability Measures*, John Wiley, New York.
171.   Dahlhaus, R. (1983). Parameter estimation of stationary processes with spectra containing strong peaks, Universität Essen. Fachbereich Mathematik.
172.   Beran, R. (1976). Adaptive estimates for autoregressive processes, *Ann. Inst. Statist. Math.*, **28**, No. 1, 77-89.
173.   Billingsley, P. (1961). The Lindeberg-Levy theorem for martingales, *Proc. Amer. Math. Soc.*, **12**, 788-792.
174.   Brown, B. M. (1971). Martingale central limit theorems, *Ann. Math. Statist.*, **42**, 59-66.
175.   Hájek, J. (1970). A characterization of limiting distributions of regular estimates, *Z. Wahrscheinlichkeitstheorie und Verw. Gebiete*, **14**, 323-330.
176.   Huber, P. J. (1981). *Robust Statistics*, John Wiley, New York.
177.   Martin, R. D. (1982). The Cramér-Rao bound and robust M-estimates for autoregressions. *Biometrika*, **69**, 437-442.
178.   Kreiss, J.-P. (1984). On adaptive estimation in stationary ARMA-processes, Submitted to *Ann. Statist.*
179.   Begun, J. M., Hall, W. J., Huang, Wei-Min, and Wellner, J. A. (1983). Information and asymptotic efficiency in parametric-nonparametric models, *Ann. Statist.*, **11**, 432-452.
180.   Dacunha-Castelle, D. (1981). Inversion des operateurs de Toeplitz et statistiques des champs aléatoires gaussiens, in *Statistical and Physical Aspects of Gaussian Processes, Colloq. Internat.* CNRS, 307, 231-241, CNRS, Paris.
181.   Shaman, P. (1976). Approximations for stationary covariance matrices and their inverses with application to ARMA models, *Ann. Statist.*, **4**, 292-301.
182.   Pagano, M. (1973). When is an autoregressive scheme stationary? *Comm. in Statist.*, **1**, 533-544.
183.   Basawa, I. V. and Brockwell, P. J. (1984). Asymptotic conditional inference for regular nonergodic models with an application to autoregressive processes, *Ann. Statist.*, **12**, 161-171.
184.   Kawashima, H. (1980). Parameter estimation of autoregressive integrated processes by least squares, *Ann. Statist.*, **8**, 423-435.
185.   Dobrushin, R. L. (1980). Gaussian random fields -- Gibbsian point of view, in *Multicomponent Random Systems* (R. L.

Dobrushin and Ya. G. Sinai, eds.), 119-151, Dekker, New York.

186. Brillinger, D. R. (1974). Fourier analysis of stationary processes, *Proc. IEEE*, **62**, 1628-1643.

187. Rice, J. (1979). On the estimation of the parameters of a power spectrum, *J. Multivariate Anal.*, **9**, 378-392.

188. Olshen, R. A. (1967). Asymptotic properties of the periodogram of a discrete stationary process, *J. Appl. Probab.*, **4**, 508-528.

189. Kabaila, P. (1983). On the asymptotic efficiency of estimators of the parameters of an ARMA process, *Journal of Time Series Analysis*, **4**, No. 1, 37-48.

190. Lai, T. L. and Wei, C. Z. (1982). Least squares estimates in stochastic regression models with applications to identification and control of dynamic systems, *Ann. Statist.*, **10**, 154-166.

191. Lai, T. L. and Wei, C. Z. (1983). Asymptotic properties of general autoregressive models and strong consistency of least-squares estimates of their parameters, *J. Mult. Analysis*, **13**, 1-23.

192. Shiryayev, A. N. (1984). *Probability*, Springer-Verlag, New York.

193. Hannan, E. J. (1983). Limit theorems for autocovariances and Fourier coefficients, in *Recent Trends in Statistics, Proceedings of the Anglo-German Statistical Meeting* (S. Heiler, ed.), pp. 132-142, Dortmund, 24-26 May 1982, Vandenhoeck & Ruprecht. Göttingen.

194. Solo, V. (1984). Consistency for the least squares estimator in a transfer function model, *J. Appl. Prob.*, **21**, 88-97.

195. Wu, C. F. (1981). Asymptotic theory of nonlinear least squares estimation, *Ann. Statist.*, **9**, 501-513.

196. Beinicke, G. and Dzhaparidze, K. O. (1982). On parameter estimation by the Davidon-Fletcher-Powell method, *Theor. Prob. and Applic.*, **27**, 396-402.

197. Dzhaparidze, K. O. (1983). On iterative procedures of asymptotic inference. *Statistica Neerlandica*, **37**, 181-189.

198. Millar, P. W. (1983). The minimax principle in asymptotic statistical theory, in *Lecture Notes in Mathematics*, Vol. 976, (P. L. Hennequin, ed.), Springer-Verlag,, Berlin-Heidelberg- New York, 76-267.

199. Kallenberg, W. C. M., ed., (1984). Testing Statistical Hypotheses: Worked Solutions, CWI Syllabus, **3**, Centre for Mathematics and Computer Science, Amsterdam.

200. Bernshtein, A. V. (1984). An asymptotically complete subclass in a class of all tests in the problem of distinguishing between composite hypotheses, *Theory Prob. and Applic.*, 29, No. 1, 179-180.

201. Ingster, Yu. I. (1983). Asymptotically opitmal Bayes tests for composite hypotheses, *Theory Prob. and Applic.*, 28, 775-794.

202. Pham-Dinh, T. (1978). Estimation of parameters in the ARMA model when the characteristic polynomial of the MA operator has a unit zero, *Ann. Statist.*, 6, 1369-1389.

203. Fabian, V. and Hannan, J. (1982). On estimation and adaptive estimation for locally asymptotically normal families, *Z. Wahrscheinlichkeitstheorie veiw Gebiete*, 59, 459-478.

INDEX